Mary Jones and Geoff Jones

IGCSE
Biology
Coursebook
Second edition

Completely Cambridge – Cambridge resources for Cambridge qualifications

Cambridge University Press works closely with University of Cambridge International Examinations
(CIE) as parts of the University of Cambridge. We enable thousands of students to pass their CIE exams by
providing comprehensive, high-quality, endorsed resources.

To find out more about University of Cambridge International Examinations
visit www.cie.org.uk

To find out more about Cambridge University Press
visit www.cambridge.org/cie

CAMBRIDGE
UNIVERSITY PRESS

CAMBRIDGE UNIVERSITY PRESS
Cambridge, New York, Melbourne, Madrid, Cape Town,
Singapore, São Paulo, Delhi, Tokyo, Mexico City

Cambridge University Press
The Edinburgh Building, Cambridge CB2 8RU, UK

www.cambridge.org
Information on this title: www.cambridge.org/9780521147798

First published 2010
6th printing 2011

Printed in the United Kingdom at the University Press, Cambridge

A catalogue record for this publication is available from the British Library

ISBN 978-0-521-14779-8 Paperback with CD-ROM

Cover image: The blue-cheeked butterflyfish, *Chaetodon semilarvatus*, found in
the Red Sea. © blickwinkel/Alamy

Contents

Introduction

This book has been written to help you to do well in your Cambridge International Examinations IGCSE Biology examination (0610). We hope that you enjoy using it.

The book also covers the Biology content of the new CIE IGCSE Combined Science syllabus (0653) and Coordinated Science syllabus (0654), and can be used with the CIE 'O' level Biology syllabus (5090).

Core and Supplement

Your teacher will tell you whether you are studying just the Core part of the Biology syllabus, or whether you are studying the Supplement as well. If you study the Core only, you will be entered for Papers 1, 2 and one of the practical exams, and can get a maximum of Grade C. If you also study the Supplement, your teacher may enter you for Paper 3 instead of Paper 2, and you will be able to get a maximum of Grade A*. The Supplement material in this book is marked by a letter 'S' and green bars in the margin.

Definitions

There are quite a lot of definitions in the IGCSE syllabus that you need to learn by heart. These are all in this book, at appropriate points in each chapter, inside boxes with a heading 'Key definition'. Make sure you learn these carefully.

Questions

Each chapter has several sets of Questions within it. Most of these require quite short answers, and simply test if you have understood what you have just read (or what you have just been taught).

At the end of each chapter, there are some longer Revision questions testing a range of material from the chapter.

Activities

Each chapter contains Activities. These will help you to develop the practical skills that will be tested in your IGCSE Biology examination. It's really important that you should try to do as many of these as possible. Most of them need to be done in a laboratory.

There are three possible exams to test your practical skills, called Paper 4, Paper 5 and Paper 6. Your teacher will tell you which of these you will be entered for. They are equally difficult, and you can get up to Grade A* on any of them. You should try to do the Activities no matter which of these papers you are entered for.

Key ideas

At the end of each chapter, there is a short list of the main points covered in the chapter. Remember, though, that these are only very short summaries, and you'll need to know more detail than this to do really well in the exam.

The CD-ROM

There is a CD-ROM in the back of the book. You'll also find the Key ideas on the CD-ROM. You can use the revision check lists on the CD-ROM to check off how far you have got with learning and understanding each idea.

The CD-ROM also contains a set of interactive multiple-choice questions testing whether you know and understand the material from each chapter.

You'll find some self-assessment check lists on the CD-ROM too, which you can print off and use to assess yourself each time you observe and draw a specimen, construct a results chart, draw a graph from a set of results or plan an experiment. These are all very important skills, and by using these check lists you should be able to improve your performance until you can do them almost perfectly every time.

There are some suggestions on the CD-ROM about how you can give yourself the very best chance of doing well in your exams, by studying and revising carefully.

Workbook

There is a workbook to go with this textbook. If you have one, you will find it really helpful in developing your skills, such as handling information and solving problems, as well as some of the practical skills.

Acknowledgements

Thanks to the following for permission to reproduce photographs:

Cover, p. 240 blickwinkel/Alamy; pp. 2, 27, 66*t*, 107, 183, 214*t*, 226, 238*tr*, 238*br* Geoff Jones; pp. 6, 129, 205*l* Wendy Lee; pp. 14, 15*t* Eleanor Jones; pp. 15*b*, 51, 80, 102, 163*t*, 163*b* Biophoto Associates/SPL; p. 21 Jack Sullivan/Alamy; p. 35 Visual&Written SL/Alamy; pp. 52, 100 Andrew Syred/SPL; pp. 56, 143, 242 Nigel Cattlin/Alamy; p. 66*b* Viel/photocuisine/Corbis; p. 67*l* Alex Segre/Alamy; p. 67*r* Images of Africa Photobank/Alamy; p. 68*l* Gideon Mendel for Action Aid/Corbis; p. 68*tr* David South/Alamy; p. 68*br* David R. Frazier Photolibrary, Inc./Alamy; p. 70 mediablitzimages (UK) Limited/Alamy; p. 89 Janine Wiedel Photolibrary/Alamy; p. 90 Prof. P. Motta/Dept. of Anatomy/University "La Sapienza", Rome/SPL; pp. 92, 96, 97, 184*t* Phototake Inc./Alamy; p. 101 J.C. Revy/SPL; p. 108 Jayanta Dey/epa/ Corbis; p. 123 Rick Rickman/NewSport/Corbis; p. 130 Visual Ideas/Nora/Corbis; pp. 160, 166*l*, 195*t* CNRI/SPL; p. 161*l* Zuma Press/Zuma/Corbis; p. 161*r* St Bartholomew's Hospital/SPL; p. 166*r* Microfield Scientific Ltd/SPL; p. 171 Steve Gschmeissner/SPL; p. 180 Juergen Berger/SPL; p. 184*b* Susumu Nishinaga/SPL; p. 185 Pictox/Alamy; p. 194 Cheryl Power/SPL; p. 195*b* Leonard Lessin/FBPA/SPL; p. 205*tr* imagebroker/Alamy; p. 205*br* Sam Sangster/Alamy; p. 207 Mary Evans Picture Library/Alamy; p. 208 Pat & Tom Leeson/SPL; pp. 209*l*, 209*r* Stephen Dalton/NHPA; p.212 Agence Nature/NHPA; p. 214*b* Terry Matthews/Alamy; p. 235 Lou Linwei/Alamy; p. 236 Jim West/Alamy; p. 238*l* Sylvia Cordaiy Photo Library Ltd/Alamy

Abbreviations
SPL = Science Photo Library
t = top, *b* = bottom, *l* = left, *r* = right

1 Classification

In this chapter, you will find out:

- the seven characteristics that distinguish living things from non-living objects
- about the binomial system of naming species
- the features of the five main classes of vertebrates
- the features of flowering plants, arthropods, annelids, nematodes and molluscs
- the features of viruses, bacteria and fungi
- how to use a dichotomous key to identify an unknown organism.

Characteristics of living things

1.1 All living things have seven characteristics.

Biology is the study of living things, which are often called **organisms**. Living organisms have seven features or **characteristics** which make them different from objects that are not alive (Figure **1.1**, overleaf). The list below Figure **1.1** provides a definition of each characteristic, which you should learn. You will find out much more about each of them later in this book.

Classification

1.2 Classification involves grouping things.

Classification means putting things into groups. There are many possible ways we could group living organisms. For example, we could put all the organisms with legs into one group, and all those without legs into another. Or we could put all red organisms into one group, and all blue ones into another. The first of these ideas would be much more useful to biologists than the second.

The main reason for classifying living things is to make it easier to study them. For example, we put humans and dogs and horses and mice into one group (the mammals) because they share certain features (for example, having

hair) that are not found in other groups. We think that all mammals share these features because they have all descended from the same ancestors, long ago. We think that all mammals are related to one another. We would therefore expect all mammals to have bodies that have similar structures and that work in similar ways. If we find a new animal that has hair, then we know that it belongs in the mammal group. We will already know a lot about it, even before we have studied it at all.

1.3 Biologists classify living things.

The first person to try to classify living things in a scientific way was a Swedish naturalist called Linnaeus. He introduced his system of classification in 1735. He divided all the different kinds of living things into groups called **species**. He recognised 12 000 species. Linnaeus' species were groups of organisms which had a lot of features in common. We still use this system today.

Species are grouped into genera (singular: **genus**). Each genus contains several species with similar characteristics (Figure **1.2**, page **3**). Several genera are then grouped into a **family**, families into **orders**, orders into **classes**, classes into **phyla** and finally phyla into **kingdoms**. Some of the more important groups are described in this chapter.

Growth All organisms begin small and get larger, by the growth of their cells and by adding new cells to their bodies.

Movement All organisms are able to move to some extent. Most animals can move their whole body from place to place, and plants can slowly move parts of themselves.

Nutrition Organisms take substances from their environment and use them to provide energy or materials to make new cells.

Excretion All organisms produce unwanted or toxic waste products as a result of their metabolic reactions, and these must be removed from the body.

Reproduction Organisms are able to make new organisms of the same species as themselves.

Sensitivity All organisms pick up information about changes in their environment, and react to the changes.

Respiration All organisms break down glucose and other substances inside their cells, to release energy that they can use.

Figure 1.1 Characteristics of living organisms.

Key definitions

excretion removal from organisms of toxic materials, the waste products of metabolism (chemical reactions in cells including respiration) and substances in excess of requirements

growth a permanent increase in size and dry mass by an increase in cell number or cell size or both

movement an action by an organism or part of an organism causing a change of position or place

nutrition the taking in of nutrients which are organic substances and mineral ions, containing raw materials or energy for growth and tissue repair, absorbing and assimilating them

reproduction the processes that make more of the same kind of organism

respiration the chemical reactions that break down nutrient molecules in living cells to release energy

sensitivity the ability to detect or sense changes in the environment (stimuli) and to make responses

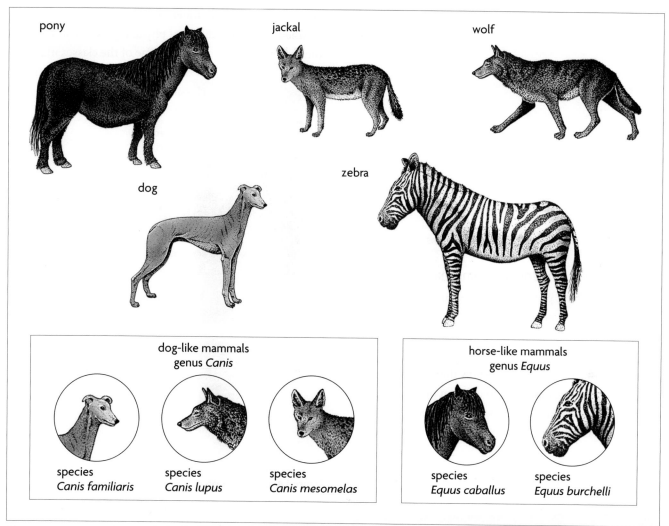

Figure 1.2 Sorting and naming five mammals.

Figure **1.2** shows five animals that all belong to the mammal order. You can see that they all have hair, which is a characteristic feature of all mammals. The animals have been classified into two groups – horse-like mammals and dog-like mammals. (What features do you think differ between these two groups?) The horse-like ones all belong to the genus *Equus*. The dog-like ones all belong to the genus *Canis*.

1.4 Each species has a binomial.

Linnaeus gave every living organism two names, written in Latin. The first name is the name of the genus it belongs to, and always has a capital letter. The second is the name of its species, and always has a small letter. This two-word name is called a **binomial**.

For example, a wolf belongs to the genus *Canis* and the species *lupus*. Its binomial is *Canis lupus*. These names are

printed in italics. When you write one, you cannot really write in italics, so you should underline any Latin names. The genus name can be abbreviated like this: *C. lupus*.

1.5 Organisms are divided into five kingdoms.

All living things are placed in one of five kingdoms (Figure **1.3**, overleaf). Each of these kingdoms has its own set of characteristics.

There is also another group that is of interest to biologists. They are the **viruses**. Biologists disagree about whether viruses should be classed as living things. Most believe that they should not, because the only characteristic that they share with living organisms is being able to reproduce – and they can't even do that on their own. A virus has to get into another living cell and hijack it before it can produce new viruses.

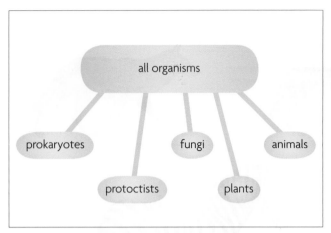

Figure 1.3 The five kingdoms of living organisms.

1.6 The animal kingdom contains many phyla.

You probably think you can recognise an animal when you see one. However, it is not always so easy. For a very long time, people thought that sea anemones were plants, because they tend to stay in one place and their tentacles look rather like petals. Now we know that they are animals. One of the best ways to tell if an organism is an animal is to look at its cells under the microscope. Animal cells never have cell walls (section **2.4**).

There is a very large number of different kinds of animals, classified into many different phyla. Figure **1.4** shows just some of these phyla, and also some of the classes in two of these phyla. Figures **1.5** to **1.9** show how some of these groups of animals are classified.

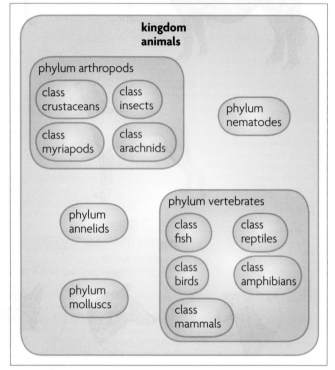

Figure 1.4 Classification of the animal kingdom.

Questions

1.1 **a** Without looking back at page 1, decide which **five** of these characteristics are found in **all** living things.

movement	blood system	sight
growth	photosynthesis	nutrition
sensitivity	speech	excretion

   b List the other **two** characteristics of all living organisms.

1.2 The table shows how two organisms – a monarch butterfly and a giant pangolin – are classified.

kingdom	animal	animal
phylum	arthropods	vertebrates
class	insects	mammals
order	Lepidoptera (butterflies and moths)	Pholidota
family	Danaidae	Manidae
genus	*Danaus*	*Manis*
species	*Danaus plexippus*	*Manis gigantea*

   a Suggest whether these two organisms are not related at all, distantly related or closely related. Explain how you made your decision.

   b Write down the genus of the giant pangolin.

   c Use the Internet or a textbook to find out how a human is classified. Write it down in a table like the one above.

Phylum Vertebrates

These are animals with a supporting rod running along the length of the body. The most familiar ones have a backbone and are called vertebrates.

Class Fish

The fish all live in water, except for one or two like the mudskipper, which can spend short periods of time breathing air.

Characteristics:
- vertebrates with scaly skin
- have gills
- have fins.

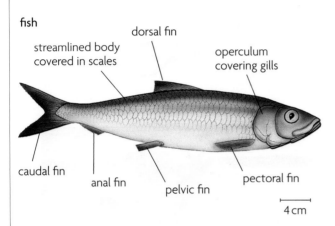

fish

streamlined body covered in scales

dorsal fin

operculum covering gills

caudal fin

anal fin

pelvic fin

pectoral fin

4 cm

Class Amphibians

Although most adult amphibians live on land, they always go back to the water to breed. Frogs and toads are amphibians.

Characteristics:
- vertebrates with moist, scale-less skin
- eggs laid in water, larva (tadpole) lives in water
- adult often lives on land
- larva has gills, adult has lungs.

frog

thin, moist skin

ear drum

1 cm

Class Reptiles

These are the crocodiles, lizards, snakes, turtles and tortoises. Reptiles do not need to go back to the water to breed because their eggs have a waterproof shell which stops them from drying out.

Characteristics:
- vertebrates with scaly skin
- lay eggs with rubbery shells.

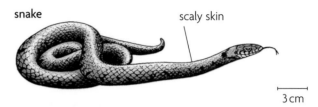

snake

scaly skin

3 cm

Class Birds

The birds, like reptiles, lay eggs with waterproof shells.

Characteristics:
- vertebrates with feathers
- forelimbs have become wings
- lay eggs with hard shells
- homeothermic
- have a beak.

bird

beak

body covered with feathers

5 cm

Class Mammals

This is the group that humans belong to.

Characteristics:
- vertebrates with hair
- have a placenta
- young feed on milk from mammary glands
- homeothermic
- have a diaphragm
- heart has four chambers
- have different types of teeth (incisors, canines premolars and molars)
- cerebral hemispheres are very well developed.

ocelot

20 cm

Figure 1.5 Classification of vertebrates.

Phylum Arthropods

Arthropods are animals with jointed legs, but no backbone. They are a very successful group, because they have a waterproof exoskeleton that has allowed them to live on dry land. There are more kinds of arthropod in the world than all the other kinds of animal put together.

Characteristics:
- several pairs of jointed legs
- exoskeleton.

Crustaceans

These are the crabs, lobsters and woodlice. They breathe through gills, so most of them live in wet places and many are aquatic.

Characteristics:
- arthropods with more than four pairs of jointed legs
- not millipedes or centipedes
- breathe through gills.

edible crab, *Cancer pagurus*

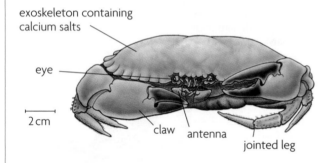

Arachnids

These are the spiders, ticks and scorpions. They are land-dwelling organisms.

Characteristics:
- arthropods with four pairs of jointed legs
- breathe through gills called book lungs.

spider, *Araneus diadematus*

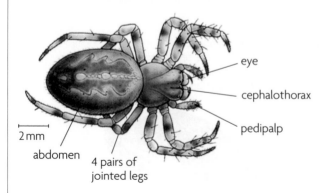

Insects

Insects are a very successful group of animals. Their success is mostly due to their exoskeleton and tracheae, which are very good at stopping water from evaporating from the insects' bodies, so they can live in very dry places. They are mainly terrestrial.

Characteristics:
- arthropods with three pairs of jointed legs
- two pairs of wings (one or both may be vestigial)
- breathe through tracheae.

locust

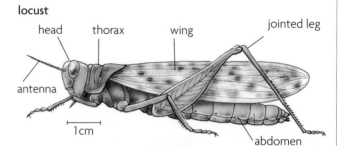

moth

Myriapods

These are the centipedes and millipedes.

Characteristics:
- body consists of many segments
- each segment has jointed legs.

centipede

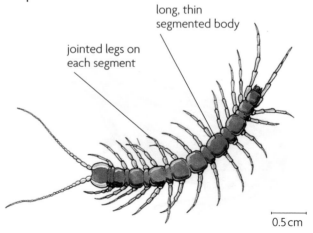

Figure 1.6 Classification of arthropods.

Phylum Annelids

Annelids are worms, with bodies made up of ring-like segments. Most of them live in water, though some, like the earthworm, live in moist soil.

Characteristics:
- animals with bodies made up of ring-like segments.

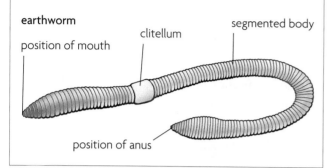

Figure 1.7 Classification of annelids.

Phylum Nematodes

Nematodes are worms, but unlike annelids their bodies are not divided into segments. They are usually white, long and thin. They live in many different habitats. Many nematodes live in the soil.

Characteristics:
- animals with long, thin, unsegmented bodies.

Figure 1.9 Classification of nematodes.

Phylum Molluscs

Molluscs are soft-bodied animals, sometimes with a shell – for example, snails; or without – for example, slugs.

Characteristics:
- animals with soft, unsegmented bodies
- may have a shell.

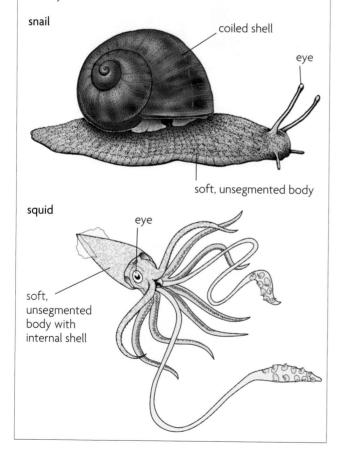

Figure 1.8 Classification of molluscs.

Activity 1.1
Making biological drawings

skills

C2 Observing, measuring and recording

Biologists need to be able to look closely at specimens – which might be whole organisms, or just part of an organism – and note significant features of them. It is also important to be able to make simple drawings to record these features.

You don't have be good at art to be good at biological drawings. A biological drawing needs to be simple but clear.

You will be provided with a specimen of an animal to draw.

(... continued on page 8)

(... continued from page 7)

1   Look carefully at the specimen, and decide what group of animals it belongs to, using Figures **1.5–1.9** to help you. Jot down the features of the organism that helped you to classify it.

2   Make a large, clear drawing of your organism.

Here are some points to bear in mind when you draw.

- Make good use of the space on your sheet of paper – your drawing should be large. However, do leave space around it so that you have room for labels.

- Always use a sharp HB pencil and have a good eraser with you.

- Keep all lines single and clear.

- Don't use shading unless it is **absolutely** necessary.

- Don't use colours.

- Take time to get the outline of your drawing correct first, showing the right proportions.

- Now label your drawing to show the features of the organism that are characteristic of its classification group. You could also label any features that help the organism to survive in its environment. These are called **adaptations**. For example, if your organism is a fish, you could label 'scales overlapping backwards, to provide a smooth, streamlined surface for sliding through the water'.

Here are some points to bear in mind when you label a diagram.

- Use a ruler to draw each label line.

- Make sure the end of the label line actually touches the structure being labelled.

- Write the labels horizontally.

- Keep the labels well away from the edges of your drawing.

Activity 1.2
Calculating magnification

skills
C2 Observing, measuring and recording

Drawings of biological specimens are usually made at a different size from the real thing. It is important to show this on the diagram.

The **magnification** of a diagram is how much larger it is than the real thing.

$$\text{magnification} = \text{size of drawing} \div \text{size of real object}$$

For example, measure the length of the spider's body in Figure **1.6**. You should find that it's 40 mm long.

The real spider was 8 mm long. So we can calculate the magnification like this:

$$\text{magnification} = \text{length in drawing} \div \text{length of real spider}$$
$$= 40 \div 8$$
$$= \times 5$$

The following are two very important things to notice.

- You must use the same units for all the measurements. Usually, millimetres are the best units to use.

- You should not include any units with the final answer. Magnification does not have a unit. However, you **must** include the 'times' sign. If you read it out loud, you would say 'times five'.

Now try a calculation yourself.

1   Measure the length of the lowest 'tail' (it is really called an appendage) on the centipede in Figure **1.6**.

2   The real length of the appendage was 10 mm. Use this, and your answer to question **1**, to calculate the magnification of the drawing of the centipede.

Questions

1.3 What is a binomial, and what does it tell you about an organism?

1.4 Explain how the skin covering of a reptile differs from that of an amphibian.

1.5 State **two** characteristics that all arthropods have in common.

1.6 State **two** characteristics that differ between insects and arachnids.

1.7 Flowering plants may be monocots or dicots.

Plants are organisms that have cells with cell walls made of cellulose (section **2.4**). At least some parts of a plant are green. The green colour is caused by a pigment called chlorophyll, which absorbs energy from sunlight. The plant uses this energy to make glucose, using carbon dioxide and water from its environment. This is called photosynthesis.

Plants include small organisms such as mosses, as well as ferns and flowering plants. Figure **1.10** shows the classification of flowering plants. They are split into two groups, the **monocotyledonous** plants – usually known as monocots – and the **dicotyledonous** plants, known as dicots.

1.8 Viruses are not true living things.

You have almost certainly had an illness caused by a virus. Viruses cause common diseases such as colds and influenza, and also more serious ones such as AIDS.

Viruses are not normally considered to be alive, because they cannot do anything other than just exist, until they get inside a living cell. They then take over the cell's machinery to make multiple copies of themselves. These new viruses burst out of the cell and invade others, where the process is repeated. The host cell is usually killed when this happens.

Figure **1.11** shows one kind of virus. It is not made of a cell – it is simply a piece of RNA surrounded by some protein molecules. It is hugely magnified in this diagram. The scale bar represents a length of 10 nanometres. One nanometre is 1×10^{-6} mm. In other words, you could line up more than 15 000 of these viruses between two of the millimetre marks on your ruler.

Phylum Flowering plants

Characteristics:
- have roots, stems and leaves
- have xylem and phloem
- reproduce by producing seeds
- seeds produced inside ovary, inside flower.

Monocotyledonous plants:
- have strap-shaped leaves with parallel veins
- have one cotyledon inside each seed.

Dicotyledonous plants:
- have leaves which can be broad, and which have a network of branching veins
- have two cotyledons inside each seed.

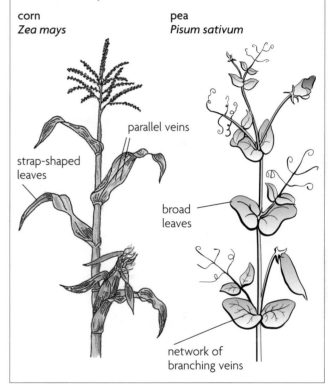

Figure **1.10** Classification of flowering plants.

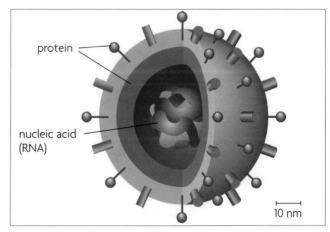

Figure **1.11** An influenza virus.

1.9 Bacteria have cells without nuclei.

Figure **1.12** shows some bacteria. Bacteria have cells that are very different from the cells of all other kinds of organism. The most important difference is that they do not have a nucleus.

You will meet bacteria at various stages in your biology course. Some of them are harmful to us and cause diseases such as tuberculosis (TB) and cholera. Many more, however, are helpful. You will find out about their useful roles in the carbon cycle and the nitrogen cycle, in making foods such as yoghurt and cheese, in the treatment of sewage to make it safe to release into the environment and in making insulin for the treatment of people with diabetes.

Kingdom Bacteria
These are the bacteria. Some bacteria can carry out photosynthesis. The oldest fossils belong to this kingdom, so we think that they were the first kinds of organism to evolve.

Characteristics:
● often unicellular (single-celled)
● have no nucleus
● have cell walls.

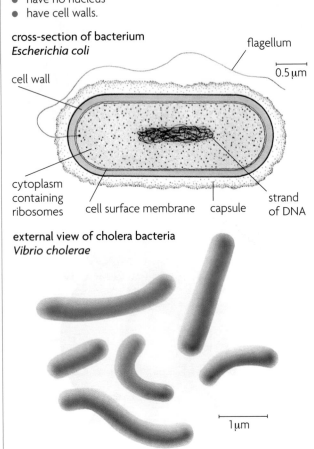

cross-section of bacterium
Escherichia coli

flagellum

0.5 μm

cell wall

cytoplasm containing ribosomes

cell surface membrane

capsule

strand of DNA

external view of cholera bacteria
Vibrio cholerae

1 μm

Figure 1.12 Bacteria.

1.10 Fungi do not have chlorophyll.

For a very long time, fungi were classified as plants. However, we now know that they are really very different, and belong in their own kingdom. Figure **1.13** shows the characteristic features of fungi.

Like bacteria, we have found many different uses to make of fungi. We eat them as mushrooms, and also as so-called single cell protein (section **7.12**). We use the unusual fungus yeast to make alcohol and bread. We obtain antibiotics such as penicillin from various different fungi. Some fungi, however, are harmful. Some of these cause food decay, while a few cause diseases, including ringworm and athlete's foot.

Kingdom Fungi
Fungi do not have chlorophyll and do not photosynthesise. Instead they feed saprophytically, or parasitically, on organic material like faeces, human foods and dead plants or animals.

Characteristics:
● multicellular (many-celled)
● have nuclei
● have cell walls
● do not have chlorophyll
● feed by saprophytic or parasitic nutrition.

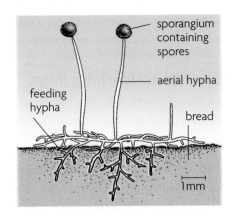

Mucor hiemalis

sporangium containing spores

aerial hypha

feeding hypha

bread

1 mm

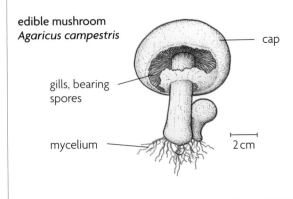

edible mushroom
Agaricus campestris

cap

gills, bearing spores

mycelium

2 cm

Figure 1.13 Fungi.

Questions

1.7 Why are viruses not generally considered to be living things?

1.8 How do the cells of bacteria differ from the cells of plants and animals?

1.9 State **one** similarity and **one** difference between the cells of a fungus and the cells of a plant.

1.11 Keys are used to identify unknown organisms.

If you want to identify an organism whose name you do not know, you may be able to find a picture of it in a book. However, not every organism may be pictured, or your organism may not look exactly like any of the pictures. If this happens, you can often find a **key** that you can use to work out what your organism is.

A key is a way of leading you through to the name of your organism by giving you two descriptions at a time, and asking you to choose between them. Each choice you make then leads you on to another pair of descriptions, until you end up with the name of your organism. This kind of key is called a **dichotomous** key. 'Dichotomous' means 'branching into two', and refers to the fact that you have **two** descriptions to choose from at each step.

Here is a key that you could use to identify the organisms shown in Figure **1.14**.

1	jointed limbs	2
	no jointed limbs	5
2	more than 5 pairs of jointed limbs	centipede
	5 or fewer pairs of jointed limbs	3
3	first pair of limbs form large claws	crab
	no large claws	4
4	3 pairs of limbs	locust
	4 pairs of limbs	spider
5	body of many segments of similar size	earthworm
	unsegmented body with a shell	snail

To use the key, pick **one** of the animals that you are going to identify. Let's say you choose organism **A**. Decide which description in step **1** matches your organism. It obviously has no jointed limbs, so the key tells us to go to step **5**. Decide which description in step **5** matches organism **A**. It has a body of many segments of similar size, so **A** is an earthworm.

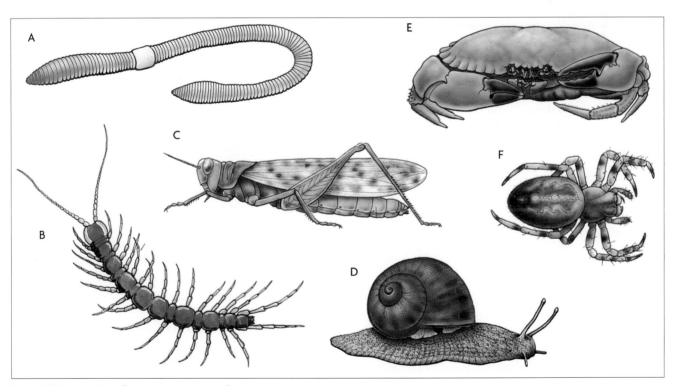

Figure 1.14 Organisms for practising using a key.

Key ideas

- All living things have seven characteristics: reproduction, nutrition, respiration, growth, excretion, movement and sensitivity.

- Living organisms are classified into groups according to how closely related they are. Each species of organism is given a unique two-word Latin name called a binomial. The first word of the binomial is the genus and the second word is the species.

- Vertebrates are classified into five classes: fish, amphibians, reptiles, birds and mammals. They each have their own distinctive set of features. For example, amphibians have a smooth skin, fish and reptiles have scales, birds have feathers and scales, and mammals have hair.

- Arthropods are invertebrates with jointed legs and segmented bodies. They can be further classified into arachnids, myriapods, insects and crustaceans.

- Annelids are worms with segmented bodies but no legs.

- Nematodes are worms with unsegmented bodies.

- Molluscs have unsegmented bodies, and often have a shell.

- Bacteria are single-celled organisms whose cells do not have nuclei.

- Fungi include moulds, mushrooms and toadstools. They have cells with cell walls but do not photosynthesise.

- Viruses are not generally considered to be alive at all. They are not made of cells and cannot carry out any of the characteristics of living things on their own.

- Flowering plants can be classified into monocotyledonous plants and dicotyledonous plants. Monocots have seeds with one cotyledon, and their leaves often have parallel veins. Dicots have seeds with two cotyledons, and their leaves generally have branching veins.

- A dichotomous key is a set of paired contrasting descriptions which lead you through to the identification of an unknown organism.

Revision questions

1 A student investigating an ecosystem found six different species of animals living there. The table shows some of their characteristics.

species	backbone present	body covering	jointed legs present
A	no	hard shell	no
B	yes	hair	yes
C	no	tough skeleton	yes
D	yes	scales	no
E	yes	smooth skin	yes
F	yes	scales	yes

  a To which kingdom do all these organisms belong?
  b Which organism could be a mollusc?
  c Which two organisms are reptiles?
  d Name the class to which organism **E** belongs.
  e State **one** other characteristic you would expect to see in organism **B**.

2 Three species of tree have the following binomials:

  Carpodiptera africana

  Commiphora africana

  Commiphora angolensis

  Which **two** of these species do biologists consider to be most closely related? Explain your answer.

3 Construct a dichotomous key to help someone to identify five of your teachers. Try to meet these criteria:
  - each pair of characteristics describes one contrasting feature – for example, 'has dark hair' versus 'has brown hair' (not 'has dark hair' versus 'has a long nose')
  - each person could be identified without having to compare them with another person – for example 'is more than 1.8 m tall' versus 'is less than 1.8 m tall' (not 'is tall' or 'is short')
  - contains no more than four pairs of points.

  When you have finished, swap your key with someone else to check if it works. If it does not, then make adjustments to it before handing it in to be marked.

2 Cells

Cell structure

2.1 All living things are made of cells.

Cells are very small, so large organisms contain millions of cells. Some organisms are **unicellular**, which means that they are made of just a single cell. Bacteria and yeast are examples of single-celled organisms.

2.2 Microscopes are used to study cells.

To see cells clearly, you need to use a microscope (Figure **2.1**). The kind of microscope used in a school laboratory is called a **light microscope** because it shines light through the piece of animal or plant you are looking at. It uses glass lenses to magnify and focus the image. A very good light microscope can magnify about 1500 times, so that all the structures in Figures **2.2** and **2.3** can be seen.

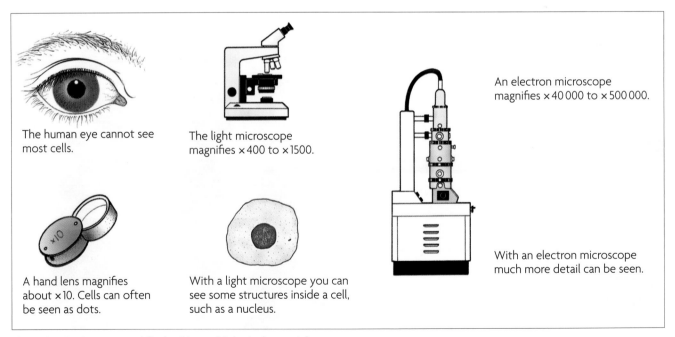

The human eye cannot see most cells.

The light microscope magnifies × 400 to × 1500.

An electron microscope magnifies × 40 000 to × 500 000.

A hand lens magnifies about ×10. Cells can often be seen as dots.

With a light microscope you can see some structures inside a cell, such as a nucleus.

With an electron microscope much more detail can be seen.

Figure 2.1 Equipment used for looking at biological material.

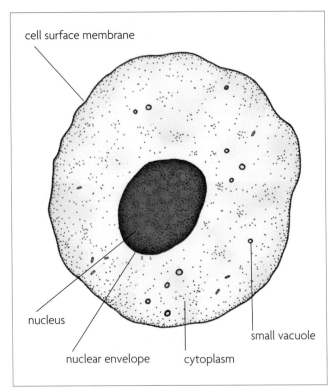

Figure 2.2 A typical animal cell – a liver cell – as seen with a light microscope.

Photomicrographs of plant and animal cells are shown in Figure 2.4 and Figure 2.5. A photomicrograph is a picture made using a light microscope.

To see even smaller things inside a cell, an electron microscope is used. This uses a beam of electrons instead of light, and can magnify up to 500 000 times. This means that a lot more detail can be seen inside a cell. We can see many structures more clearly, and also some structures that could not be seen at all with a light microscope.

Questions

2.1 How many times can a good light microscope magnify?

2.2 If an object was 1 mm across, how big would it look if it was magnified 10 times?

 FACT!

The largest cells in the human body are nerve cells (neurones) which can be as much as one metre long. The smallest cells in the human body are sperm cells, which are about 60 micrometres long. A micrometre is one thousandth of a millimetre.

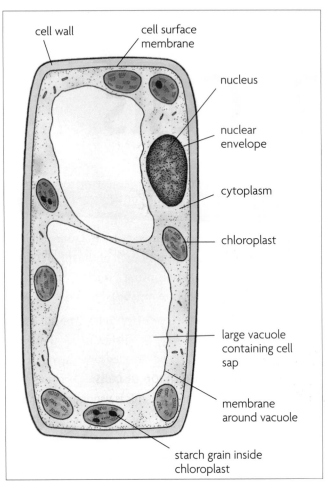

Figure 2.3 A typical plant cell – a palisade cell – as seen with a light microscope.

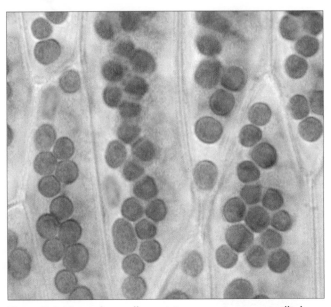

Figure 2.4 Many plant cells contain green structures, called chloroplasts. Even if it does not have any chloroplasts, you can still identify a plant cell because it has a cell wall around it (×2000).

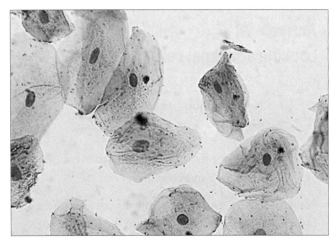
Figure 2.5 Cells from the trachea (windpipe) of a mammal, seen through a light microscope (×300).

2.3 All cells have a cell membrane.

Whatever sort of animal or plant they come from, all cells have a **cell membrane** around the outside. Inside the cell membrane is a jelly-like substance called **cytoplasm**, in which are found many small structures called **organelles**. The most obvious of these organelles is usually the **nucleus**.

2.4 Plant cells have a cell wall.

All plant cells are surrounded by a cell wall made mainly of **cellulose**. Paper, which is made from cell walls, is also made of cellulose. Animal cells never have cell walls made of cellulose. Cellulose belongs to a group of substances called polysaccharides, which are described in section **4.5**. Cellulose forms fibres which criss-cross over one another to form a very strong covering to the cell (Figure **2.6**). This helps to protect and support the cell. If the cell absorbs a lot of water and swells, the cell wall stops it bursting.

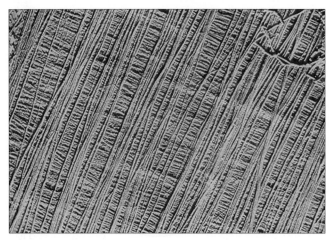
Figure 2.6 Cellulose fibres from a plant cell wall. This picture was taken using an electron microscope (×50 000).

Because of the spaces between fibres, even very large molecules are able to go through the cellulose cell wall. It is therefore said to be **fully permeable**.

2.5 Cell membranes are partially permeable.

All cells have a membrane surrounding the cell. It is called the **cell surface membrane** or **plasma membrane**. In a plant cell, it is very difficult to see, because it is right against the cell wall.

The cell surface membrane is a very thin layer of protein and fat. It is very important to the cell because it controls what goes in and out of it. It is said to be **partially permeable**, which means that it will let some substances through but not others.

2.6 Cytoplasm is a complex solution.

Cytoplasm is a clear jelly. It is nearly all water; about 70% is water in many cells. It contains many substances dissolved in it, especially proteins. Many different **metabolic reactions** (the chemical reactions of life) take place in the cytoplasm.

2.7 Most cells contain vacuoles.

A vacuole is a space in a cell, surrounded by a membrane, and containing a solution. Plant cells have very large vacuoles, which contain a solution of sugars and other substances, called **cell sap**. Animal cells have much smaller vacuoles, which may contain food or water.

2.8 Chloroplasts trap the energy of sunlight.

Chloroplasts are never found in animal cells, but most of the cells in the green parts of plants have them. They contain the green colouring or pigment called **chlorophyll**. Chlorophyll absorbs sunlight, and the energy of sunlight is then used for making food for the plant by **photosynthesis** (Chapter 4).

Chloroplasts often contain starch grains, which have been made by photosynthesis. Animal cells never contain starch grains.

2.9 The nucleus stores inherited information.

The nucleus is where the genetic information is stored which helps the cell to make the right sorts of proteins. The information is kept on the **chromosomes**, which are inherited from the organism's parents. The chromosomes are made of **DNA**.

Chromosomes are very long, but so thin that they cannot easily be seen even using the electron microscope. However, when the cell is dividing, they become short and thick, and can be seen with a good light microscope.

Table 2.1 A comparison between animal and plant cells

Similarities

1 Both have a cell surface membrane surrounding the cell.
2 Both have cytoplasm.
3 Both contain a nucleus.

Differences

	plant cells	animal cells
1	have a cellulose cell wall outside the cell membrane	have no cell wall
2	often have chloroplasts containing chlorophyll	have no chloroplasts
3	often have large vacuoles containing cell sap	have only small vacuoles
4	often have starch grains	never have starch grains; sometimes have glycogen granules
5	often regular in shape	often irregular in shape

Animal and plant cells obtain their food in different ways. Plants make their own food, so their cells contain chloroplasts. Starch granules store some of the food they make. Animals often have to move to find their food. This is made easier if their cells do not have a rigid wall.

Activity 2.1
Looking at animal cells

skills

C1 *Using techniques, apparatus and materials*

C2 *Observing, measuring and recording*

Safety Wash your hands thoroughly after handling the trachea and cells.

Some simple animal cells line the mouth and trachea (or windpipe). If you colour or stain the cells, they are quite easy to see using a light microscope (see diagram below and Figure 2.5).

1 Using a section lifter, gently rub off a little of the lining from the inside of the trachea provided.

2 Put your cells onto the middle of a clean microscope slide, and gently spread them out. You will probably not be able to see anything at all at this stage.

3 Put on a few drops of methylene blue.

4 Gently lower a coverslip over the stained cells, trying not to trap any air bubbles.

5 Use filter paper or blotting paper to clean up the slide, and then look at it under the low power of a microscope.

6 Make a labelled drawing of a few cells.

Questions

1 Which part of the cell stained the darkest blue?

2 Is the cell membrane permeable or impermeable to methylene blue? Explain your answer.

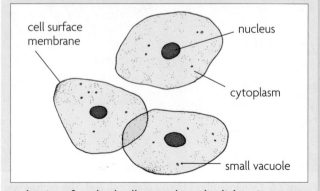

A drawing of tracheal cells seen through a light microscope.

Activity 2.2
Looking at plant cells

skills
C1 *Using techniques, apparatus and materials*
C2 *Observing, measuring and recording*

Safety Take care with the sharp blade when cutting the onion.

To be able to see cells clearly under a microscope, you need a very thin layer. It is best if it is only one cell thick (see diagram on the right). An easy place to find such a layer is inside an onion bulb.

1 Cut a small piece from an onion bulb, and use forceps to peel a small piece of thin skin, called epidermis, from the inside of it. Do not let it get dry.

2 Put a drop or two of water onto the centre of a clean microscope slide. Put the piece of epidermis into it, and spread it flat.

3 Gently lower a coverslip onto it.

4 Use filter paper or blotting paper to clean up the slide, and then look at it under the low power of a microscope.

5 Make a labelled drawing of a few cells.

6 Using a pipette, take up a small amount of iodine solution. Very carefully place some iodine solution next to the edge of the coverslip. The iodine solution will seep under the edge of the coverslip. To help it do this, you can place a small piece of filter paper next to the opposite side of the coverslip, which will soak up some of the liquid and draw it through.

7 Look at the slide under the low power of the microscope. Note any differences between what you can see now and what it looked like before adding the iodine solution.

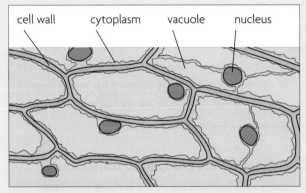
cell wall   cytoplasm   vacuole   nucleus

A drawing of onion epidermis cells seen through a light microscope after staining with iodine.

Questions

1 Name two structures which you can see in these cells, but which you could not see in the tracheal cells (Activity 2.1).

2 Most plant cells have chloroplasts, but these onion cells do not. Suggest a reason for this.

3 Iodine solution turns blue-black in the presence of starch. Did any of the onion cells contain starch?

Questions

2.3 What sort of cells are surrounded by a cell membrane?

2.4 What are plant cell walls made of?

2.5 What does fully permeable mean?

2.6 What does partially permeable mean?

2.7 What is the main constituent of cytoplasm?

2.8 What is a vacuole?

2.9 What is cell sap?

2.10 Chloroplasts contain chlorophyll. What does chlorophyll do?

2.11 What is stored in the nucleus?

2.12 Why can chromosomes be seen only when a cell is dividing?

Cells and organisms

2.10 Structures within a cell are called organelles.

A nucleus and a chloroplast are **organelles**. Organelles are structures in the cell with special functions. You will find out more about their functions later in this book. The nucleus and chloroplasts are each surrounded by their own membranes.

2.11 There is division of labour between cells.

A large organism such as yourself may contain many millions of cells, but not all the cells are alike. Almost all of them can carry out the activities which are characteristic of living things, but many of them specialise in doing some of these better than other cells do. Muscle cells, for example, are specially adapted for movement. Most cells in the leaf of a plant are specially adapted for making food by photosynthesis.

2.12 Similar cells are grouped to form tissues.

Often, cells which specialise in the same activity are found together. A group of cells like this is called a **tissue**. An example of a tissue is a layer of cells lining your stomach. These cells make enzymes to help to digest your food (Figure 2.7).

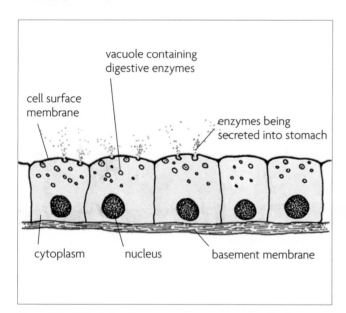

Figure 2.7 Cells lining the stomach – an example of a tissue.

The stomach also contains other tissues. For example, there is a layer of muscle in the stomach wall, made of cells which can move. This muscle tissue makes the wall of the stomach move in and out, churning the food and mixing it up with the enzymes.

Plants also have tissues. You may already have looked at some epidermis tissue from an onion bulb. Inside a leaf, a layer of cells makes up the palisade tissue, in which the cells are specialised to carry out photosynthesis.

2.13 An organ contains tissues working together.

All tissues in the stomach work together, although they have their own job to do. A group of tissues like this makes up an **organ**. The stomach is an organ. Other organs include the heart, the kidneys and the lungs.

In a plant, an onion bulb is an organ. A leaf is another example of a plant organ.

2.14 An organ system contains organs working together.

The stomach is only one of the organs which help in the digestion of food. The mouth, the intestines and the stomach are all part of the digestive system. The heart is part of the circulatory system, while each kidney is part of the excretory system.

The way in which organisms are built up can be summarised like this: **organelles** make up **cells**, which make up **tissues**, which make up **organs**, which make up **organ systems**, which make up **organisms**.

Key definitions

organelle a structure within a cell

tissue a group of cells with similar structures, working together to perform a shared function

organ a structure made up of a group of tissues, working together to perform specific functions

organ system a group of organs with related functions, working together to perform body functions

- Cells are the smallest units of living things. They are too small to be seen with the naked eye, so we need to use microscopes to see their structures.

- Cells have a cell membrane, cytoplasm and a nucleus. Plant cells also have a cell wall, and often have chloroplasts and a large vacuole containing cell sap.

- The cell membrane is partially permeable, and it controls what enters and leaves the cell.

- The cytoplasm is a jelly-like solution of many different substances in water. It is the site of many different metabolic reactions.

- The nucleus contains the chromosomes, which are made of DNA. This is the genetic information and it controls the activities of the cell.

- The cell wall of a plant cell is made of criss-crossing fibres of cellulose. It is fully permeable. It helps to support the cell, and prevents the cell bursting if it absorbs a lot of water.

- The vacuole of a plant cell contains cell sap, which is a solution of sugars and other substances in water.

- Chloroplasts contain the green pigment chlorophyll, which absorbs sunlight for photosynthesis. There may be starch grains inside the chloroplasts, which are the form in which plants store the food that they make in photosynthesis.

- A tissue is a group of similar cells which work together to carry out a particular function. Tissues are grouped into organs, and organs are grouped into organ systems.

1 Arrange these structures in order of size, beginning with the smallest:

   stomach   mitochondrion   starch grain
   tracheal cell   nucleus

2 For each of the following, state whether it is an organelle, a cell, a tissue, an organ, an organ system, or an organism.
   a heart
   b chloroplast
   c nucleus
   d trachea
   e onion epidermis
   f onion bulb
   g onion plant
   h mitochondrion
   i human being
   j lung

3 State which part of a plant cell:
   a makes food by photosynthesis
   b releases energy from food
   c controls what goes in and out of the cell
   d stores information about making proteins
   e contains cell sap
   f protects the outside of the cell.

4 Distinguish between each pair of terms.
   a chloroplast, chlorophyll
   b cell wall, cell membrane
   c organelle, organ

3 Movement in and out of cells

In this chapter, you will find out:

◆   how the random movement of molecules causes diffusion
◆   that osmosis is a special kind of diffusion
◆   why diffusion and osmosis are important to cells
◆   how cells can make substances move against their concentration gradients.

Diffusion and osmosis

3.1 Diffusion results from random movement.

Atoms, molecules and ions are always moving. The higher the temperature, the faster they move. In a solid substance the particles cannot move very far, because they are held together by attractive forces between them. In a liquid they can move more freely, knocking into one another and rebounding. In a gas they are freer still, with no attractive forces between the molecules or atoms. Molecules and ions can also move freely when they are in solution.

When they can move freely, particles tend to spread themselves out as evenly as they can. This happens with gases, solutions, and mixtures of liquids. Imagine, for example, a rotten egg in one corner of a room, giving off hydrogen sulfide gas. To begin with, there will be a very high concentration of the gas near the egg, but none in the rest of the room. However, before long the hydrogen sulfide molecules have spread throughout the air in the room. Soon, you will not be able to tell where the smell first came from – the whole room will smell of hydrogen sulfide.

The hydrogen sulfide molecules have spread out, or **diffused**, through the air.

Key definition

diffusion   the net movement of molecules from a region of their higher concentration to a region of their lower concentration down a concentration gradient, as a result of their random movement

3.2 Diffusion is important to living organisms.

Living organisms obtain many of their requirements by diffusion. They also get rid of many of their waste products in this way. For example, plants need carbon dioxide for photosynthesis. This diffuses from the air into the leaves, through the stomata. It does this because there is a lower concentration of carbon dioxide inside the leaf, as the cells are using it up. Outside the leaf in the air, there is a higher concentration. Carbon dioxide molecules therefore diffuse into the leaf, down this concentration gradient.

Oxygen, which is a waste product of photosynthesis, diffuses out in the same way. There is a higher concentration of oxygen inside the leaf, because it is being made there. Oxygen therefore diffuses out through the stomata into the air.

Activity 3.1
Demonstrating diffusion in a solution

skills
  C2 Observing, measuring and recording

1   Fill a gas jar with water. Leave it for several hours to let the water become very still.

2   Carefully place a small crystal of potassium permanganate into the water.

3   Make a labelled drawing of the gas jar to show how the colour is distributed at the start of your experiment.

4   Leave the gas jar completely undisturbed for several days.

5   Make a second drawing to show how the colour is distributed.

You can try this with other coloured salts as well, such as copper sulfate or potassium dichromate.

Questions

1   Why was it important to leave the water to become completely still before the crystal was put in?

2   Why had the colour spread through the water at the end of your experiment?

3   Suggest **three** things that you could have done to make the colour spread more quickly.

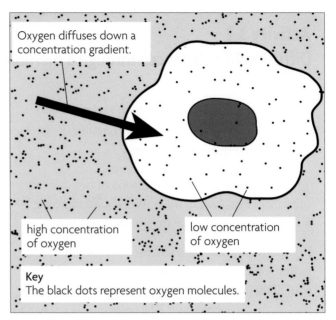

Oxygen diffuses down a concentration gradient.

high concentration of oxygen

low concentration of oxygen

Key
The black dots represent oxygen molecules.

Figure 3.1 Diffusion of oxygen into a cell.

Figure 3.2 The scent from flowers diffuses out into the air around them, even if there is no wind. The scent is made of tiny molecules, which stimulate receptors in the bee's body.

Diffusion is also important in gas exchange for respiration in animals and plants (Figure **3.1**). Some of the products of digestion are absorbed from the ileum of mammals by diffusion (section **7.31**), and flowering plants use diffusion to attract pollinators like bees (Figure **3.2**).

Questions

3.1   What effect does an increase in temperature have on the movement of molecules?

3.2   Define diffusion.

3.3   List **three** examples of diffusion in living organisms.

Activity 3.2
Diffusion of substances through a membrane

skills

C1 *Using techniques, apparatus and materials*

C2 *Observing, measuring and recording*

C3 *Interpreting and evaluating*

You are going to investigate diffusion of two different substances dissolved in water. When a substance is dissolved, its particles are free to move around.

In this investigation, you will use starch solution and iodine solution. The solutions will be separated by a membrane made out of Visking tubing. Visking tubing has microscopic holes in it. The holes are big enough to let water molecules and iodine molecules through, but not starch molecules, which are bigger than the holes.

1   Collect a piece of Visking tubing. Moisten it and rub it until it opens.

2   Tie a knot in one end of the tubing.

3   Using a pipette, carefully fill the tubing with some starch solution.

4   Tie the top of the tubing very tightly, using thread.

5   Rinse the tubing in water, just in case you got any starch on the outside of it.

6   Put some iodine solution into a beaker.

7   Gently put the Visking tubing into the iodine solution, so that it is completely covered, as shown in the diagram.

8   Leave the apparatus for about 10 minutes.

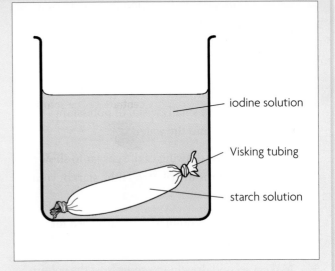

iodine solution

Visking tubing

starch solution

Questions

1   What colour were the liquids inside and outside the tubing at the start of the experiment?

2   What colour were the liquids inside and outside the tubing at the end of the investigation?

3   When starch and iodine mix, a blue-black colour is produced. Where did the starch and iodine mix in your experiment?

4   Did either the starch particles or the iodine particles diffuse through the Visking tubing? How can you tell?

5   Copy and complete these sentences.

At the start of the experiment, there were starch molecules inside the tubing but none outside the tubing. Starch particles are too .... to go through Visking tubing.

At the start of the experiment, there were iodine molecules .... the tubing but none .... the tubing. The iodine molecules diffused into the tubing, down their .... gradient.

When the starch and iodine molecules mixed, a .... colour was produced.

3.3 In osmosis, water diffuses through a partially permeable membrane.

Figure **3.3** illustrates a concentrated sugar solution, separated from a dilute sugar solution by a membrane. The membrane has holes or pores in it which are very small. An example of a membrane like this is Visking tubing.

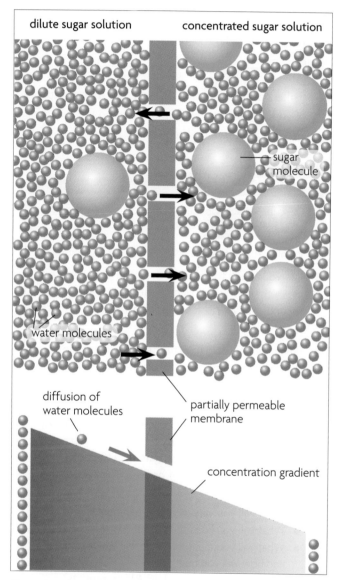

Figure 3.3 Osmosis.

Water molecules are very small. Each one is made of two hydrogen atoms and one oxygen atom. Sugar molecules are many times larger than this. In Visking tubing, the holes are big enough to let the water molecules through, but not the sugar molecules. It is called a **partially permeable membrane** because it will let some molecules through but not others.

There is a higher concentration of sugar molecules on the right-hand side of the membrane in Figure **3.3**, and a lower concentration on the left-hand side. If the membrane was not there, the sugar molecules would diffuse from the concentrated solution into the dilute one until they were evenly spread out. However, they cannot do this because the pores in the membrane are too small for them to get through.

There is also a concentration gradient for the water molecules. On the left-hand side of the membrane, there is a high concentration of water molecules. On the right-hand side, the concentration of water molecules is lower because a lot of space is taken up by sugar molecules.

It is actually rather confusing to talk about the 'concentration' of water molecules, because the term 'concentration' is normally used to mean the concentration of the solute dissolved in the water. It is much better to use a different term instead. We say that a dilute solution (where there is a lot of water) has a **high water potential**. A concentrated solution (where there is less water) has a **low water potential**.

In Figure **3.3**, there is a high water potential on the left-hand side and a low water potential on the right-hand side. There is a **water potential gradient** between the two sides. The water molecules diffuse down this gradient, from a high water potential to a low water potential.

Questions

3.4   Which is larger – a water molecule or a sugar molecule?

3.5   What is meant by a partially permeable membrane?

3.6   Give **two** examples of partially permeable membranes.

3.7   How would you describe a solution that has a high concentration of water molecules?

3.8   Define osmosis.

Activity 3.3
Measuring the rate of osmosis

1 Collect a piece of Visking tubing. Moisten it and rub it between your fingers to open it. Tie one end tightly.

2 Use a dropper pipette to put some concentrated sugar solution into the tubing.

3 Place a long, narrow glass tube into the tubing, as shown in the diagram. Tie it very, very tightly, using thread.

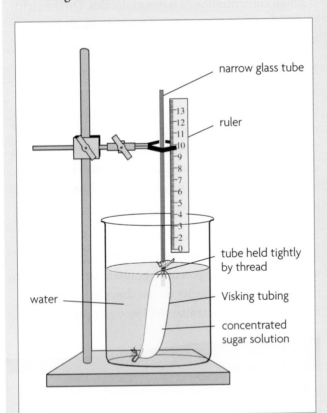

4 Place the tubing inside a beaker of water, as shown in the diagram.

5 Mark the level of liquid inside the glass tube.

6 Make a copy of this results chart. Every 2 minutes, record the level of the liquid in the glass tube.

time in minutes	0	2	4	6	8	10	12	14	16
height of liquid in mm									

7 Collect a sheet of graph paper. Draw a line graph of your results. Put *time in minutes* on the *x*-axis, and *height in mm* on the *y*-axis.

Questions

1 Describe what happened to the liquid level inside the glass tube.

2 Explain why this happened.

3 Use your graph to work out the mean (average) rate at which the liquid moved up the tube, in mm per second. (Ask your teacher for help if you are not sure how to do this.)

4 Predict what would have happened to the rate of osmosis in this experiment if you had used a kind of Visking tubing with ridges and grooves in it, giving it a larger surface area. Explain your answer.

5 When temperature rises, particles move more quickly. Describe how you could use this apparatus to carry out an experiment to investigate the effect of temperature on the rate of osmosis. Think about the following things.

• What will you vary in your experiment?
• What will you keep the same?
• What will you measure, when will you measure it and how will you measure it?
• How will you record and display your results?

Predict the results that you would expect.

3.4 Cell membranes are partially permeable.

Cell membranes behave very much like Visking tubing. They let some substances pass through them, but not others. They are partially permeable membranes.

There is always cytoplasm on one side of any cell membrane. Cytoplasm is a solution of proteins and other substances in water. There is usually a solution on the other side of the membrane, too. Inside large animals, cells are surrounded by tissue fluid (section **8.19**). In the soil, the roots of plants are often surrounded by a film of water.

So, cell membranes often separate two different solutions – the cytoplasm, and the solution around the cell. If the solutions are of different concentrations, then osmosis will occur.

3.5 Osmosis and animal cells.

Figure **3.4** illustrates an animal cell in pure water. The cytoplasm inside the cell is a fairly concentrated solution. The proteins and many other substances dissolved in it are too large to get through the cell membrane. Water molecules, though, can get through.

If you compare this situation with Figure **3.3** (page **23**), you will see that they are similar. The dilute solution in Figure **3.3** and the pure water in Figure **3.4** are each separated from a concentrated solution by a partially permeable membrane. In Figure **3.4**, the concentrated solution is the cytoplasm and the partially permeable membrane is the cell surface membrane. Therefore, osmosis will occur.

Water molecules will diffuse from the dilute solution into the concentrated solution. What happens to the cell? As more and more water enters the cell, it swells. The cell membrane has to stretch as the cell gets bigger, until eventually the strain is too much, and the cell bursts.

Figure **3.5** illustrates an animal cell in a concentrated solution. If this solution is more concentrated than the cytoplasm, then water molecules will diffuse out of the cell. Look at Figure **3.3** (page **23**) to see why.

As the water molecules go out through the cell membrane, the cytoplasm shrinks. The cell shrivels up.

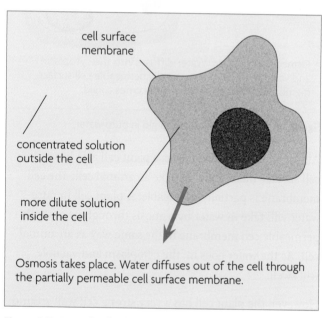

Osmosis takes place. Water diffuses out of the cell through the partially permeable cell surface membrane.

Figure 3.5 Animal cells shrink in a concentrated solution.

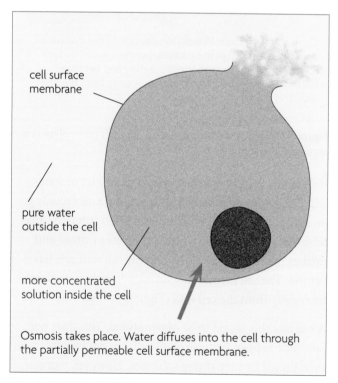

Osmosis takes place. Water diffuses into the cell through the partially permeable cell surface membrane.

Figure 3.4 Animal cells burst in pure water.

3.6 Osmosis and plant cells.

Plant cells do not burst in pure water. Figure 3.6 illustrates a plant cell in pure water. Plant cells are surrounded by a cell wall. This is fully permeable, which means that it will let any molecules go through it.

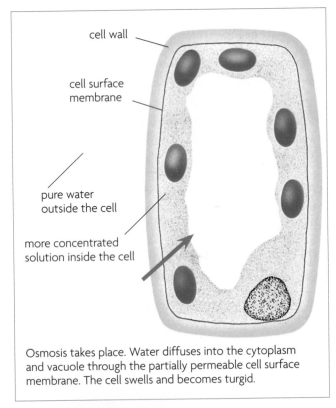

Osmosis takes place. Water diffuses into the cytoplasm and vacuole through the partially permeable cell surface membrane. The cell swells and becomes turgid.

Figure 3.6 Plant cells become turgid in pure water.

Although it is not easy to see, a plant cell also has a cell surface membrane just like an animal cell. The cell membrane is partially permeable. A plant cell in pure water will take in water by osmosis through its partially permeable cell membrane in the same way as an animal cell. As the water goes in, the cytoplasm and vacuole will swell.

However, the plant cell has a very strong cell wall around it. The cell wall is much stronger than the cell membrane and it stops the plant cell from bursting. The cytoplasm presses out against the cell wall, but the wall resists and presses back on the contents.

A plant cell in this state is rather like a blown-up tyre – tight and firm. It is said to be **turgid**. The turgidity of its cells helps a plant that has no wood in it to stay upright, and keeps the leaves firm. Plant cells are usually turgid.

Figure 3.7 illustrates a plant cell in a concentrated solution. Like the animal cell in Figure 3.5, it will lose water by osmosis. The cytoplasm shrinks, and stops pushing outwards on the cell wall. Like a tyre when some of the air has leaked out, the cell becomes floppy. It is said to be **flaccid**. If the cells in a plant become flaccid, the plant loses its firmness and begins to **wilt**.

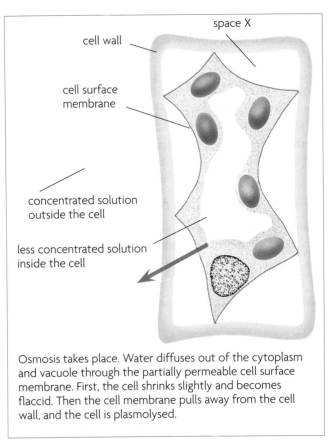

Osmosis takes place. Water diffuses out of the cytoplasm and vacuole through the partially permeable cell surface membrane. First, the cell shrinks slightly and becomes flaccid. Then the cell membrane pulls away from the cell wall, and the cell is plasmolysed.

Figure 3.7 Plant cells become flaccid and may plasmolyse in a concentrated solution.

If the solution is very concentrated, then a lot of water will diffuse out of the cell. The cytoplasm and vacuole go on shrinking. The cell wall, though, is too stiff to be able to shrink much. As the cytoplasm shrinks further and further into the centre of the cell, the cell wall gets left behind. The cell membrane, surrounding the cytoplasm, tears away from the cell wall (Figures 3.7 and 3.8).

A cell like this is said to be **plasmolysed**. This does not normally happen because plant cells are not usually surrounded by very strong solutions. However, you can make cells become plasmolysed if you do Activity 3.4. Plasmolysis usually kills a plant cell because the cell membrane is damaged as it tears away from the cell wall.

Activity 3.4
Experiment to investigate the effects of different solutions on plant cells

Safety Take care when using a sharp blade to cut the plant tissue.

1 Set up a microscope.

2 Take three clean microscope slides. Label them A, B and C.

3 Put a drop of distilled water onto the centre of slide **A**.

4 Put a drop of medium concentration sugar solution onto slide **B**.

5 Put a drop of concentrated sugar solution onto slide **C**.

6 Peel off a very thin layer of coloured epidermis from a *Rhoeo* leaf, or other leaf, or from a rhubarb petiole. To get good results, it should be as thin as possible (only one cell thick).

7 Cut three squares of this epidermis, each with sides about 5 mm long.

8 Put one square into the drop of solution on each of your three slides.

9 Carefully cover each one with a coverslip. Clean excess liquid from your slides with filter paper.

10 Look at each of your slides under the microscope. Make a labelled drawing of a few cells from each one.

Questions

1 Which part of the cell is coloured?

2 What has happened to the cells in pure water? Explain your answer.

3 What has happened to the cells in medium concentration sugar solution? Explain your answer.

4 What has happened to the cells in concentrated sugar solution? Explain your answer.

Questions

3.9 What happens to an animal cell in pure water?

3.10 Explain why this does not happen to a plant cell in pure water.

3.11 Which part of a plant cell is:
  a fully permeable
  b partially permeable?

3.12 What is meant by a turgid cell?

3.13 What is plasmolysis?

3.14 How can plasmolysis be brought about?

3.15 In Figure 3.7, what fills space X? Explain your answer.

3.16 Describe the events shown in Figures 3.6 and 3.7 in terms of water potential.

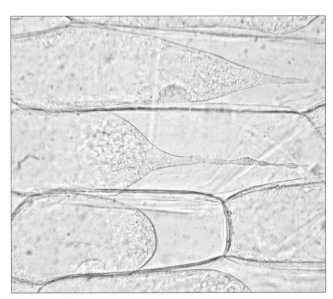

Figure 3.8 These onion cells have been placed in a concentrated solution. The cytoplasm has shrunk inwards, leaving big gaps between itself and the cell walls (× 2000).

Activity 3.5
Osmosis and potato strips

C1 Using techniques, apparatus and materials

C2 Observing, measuring and recording

C3 Interpreting and evaluating

Safety Take care when using a sharp blade to cut the potato.

1   Peel a potato or other plant tuber or root. Very carefully cut five strips from it, each exactly 40 mm long, 10 mm wide and 10 mm deep.

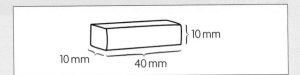

2   Make a copy of the results table below.

container	A	B	C	D	E
concentration of solution					
initial length of strip in mm					
final length of strip in mm					
change in length of strip in mm					

3   Take five containers and label them **A, B, C, D** and **E**. Pour a different solution into each one, as provided by your teacher. Write down the concentration of each solution in the results table.

4   Place one potato strip into each container, so that it is completely covered by the liquid, as in the diagram. Leave all the strips for at least half an hour.

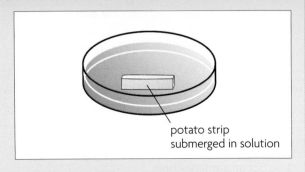

potato strip
submerged in solution

5   Remove the strip from container **A** and measure it. Write the results in the table.

6   Repeat for all the other strips.

7   Now calculate the change in length of each strip. If it got smaller, show this with a minus sign.

Questions

1   Which strips, if any, got shorter?

2   Copy and complete these sentences, to explain why these strips got shorter.

   Potato strips are made of plant cells. Each cell is surrounded by a partially permeable cell .... When the strip is in a solution that is more concentrated than the cytoplasm in the cells, water moves .... the potato cells by osmosis. This makes the cells get .... , so the whole strip becomes smaller.

3   Which strips, if any, got longer?

4   Write some sentences, like the ones in question 2, to explain why these strips got longer.

5   Describe how you could use this technique to find out the concentration of the cell contents in a potato strip.

Active transport

3.7 Cells take in substances by active transport.

There are many occasions when cells need to take in substances which are only present in small quantities around them. Root hair cells in plants, for example, take in nitrate ions from the soil. Very often, the concentration of nitrate ions inside the root hair cell is higher than the concentration in the soil. The diffusion gradient for the nitrate ions is out of the root hair, and into the soil.

Despite this, the root hair cells are still able to take nitrate ions in. They do it by a process called **active transport**. Active transport is an energy-consuming process by which substances are transported against their concentration gradient.

Key definition

active transport   the movement of ions in or out of a cell through the cell membrane, from a region of their lower concentration to a region of their higher concentration against a concentration gradient, using energy released during respiration

In the cell membrane of the root hair cells are special **carrier proteins**. Carrier proteins pick up nitrate ions from outside the cell, and then change shape in such a way that they push the nitrate ions through the cell membrane and into the cytoplasm of the cell.

As its name suggests, active transport uses energy. The energy is provided by respiration inside the root hair cells. (You can find out about respiration in Chapter **9**.) Energy is needed to produce the shape change in the carrier protein.

Most other cells can carry out active transport. In the human small intestine, for example, glucose can be actively transported from the lumen of the intestine into the cells of the villi.

Figure **3.9** shows how active transport of glucose takes place.

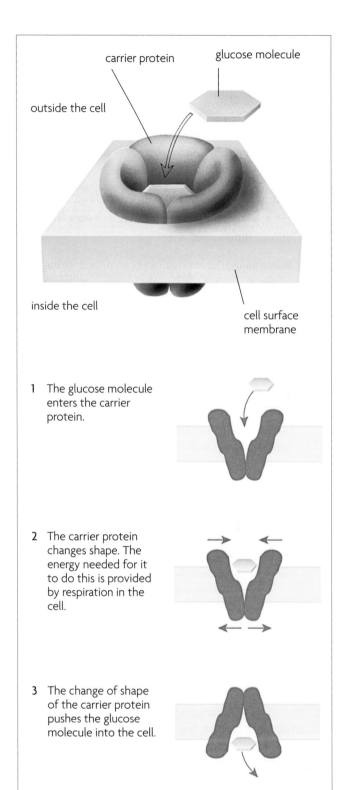

1   The glucose molecule enters the carrier protein.

2   The carrier protein changes shape. The energy needed for it to do this is provided by respiration in the cell.

3   The change of shape of the carrier protein pushes the glucose molecule into the cell.

Figure 3.9 Active transport.

◆ Particles in gases, liquids and solutions are in constant random motion. As a result of this, there is a net movement from where they are in a high concentration to where they are in a low concentration. This is diffusion.

◆ Diffusion is important to cells. For example, oxygen enters a respiring cell by diffusion, and carbon dioxide diffuses out of it.

◆ Water molecules are small and can diffuse through a partially permeable membrane. Larger molecules dissolved in the water cannot do this. The diffusion of water through a partially permeable membrane is called osmosis.

◆ Osmosis is important to cells. In a dilute solution, water passes into a cell through its partially permeable membrane. The cell gets bigger. Animal cells may burst, but plant cells do not burst because of their strong cell wall.

◆ In a concentrated solution, water passes out of a cell by osmosis through its partially permeable membrane. The cell shrinks. Plant cells may become plasmolysed – that is, the cell membrane pulls away from the cell wall.

◆ A solution containing a lot of water is said to have a high water potential. A solution containing only a little water has a low water potential. Water moves by osmosis down a water potential gradient, from a high water potential to a low water potential.

◆ Cells can use energy to move substances *up* their concentration gradient, from a low concentration to a high concentration. This is called active transport. It uses energy that the cells release by respiration.

Revision questions

1 Which of **a–d** below is an example of **(i)** diffusion, **(ii)** osmosis, or **(iii)** neither? Explain your answer in each case.
   a Water moves from a dilute solution in the soil into the cells in a plant's roots.
   b Saliva flows out of the salivary glands into your mouth.
   c A spot of blue ink dropped into a glass of still water quickly colours all the water blue.
   d Carbon dioxide goes into a plant's leaves when it is photosynthesising.

2 Each of these statements was made by a candidate in an examination. Each one contains at least one error. Decide what is wrong with each statement, and rewrite it correctly.
   a If Visking tubing containing a sugar solution is put into a beaker of water, the sugar solution moves out of the tubing by osmosis.

   b Plant cells do not burst in pure water because the cell wall stops water getting into the cell.
   c Animal cells plasmolyse in a concentrated sugar solution.
   d When a plant cell is placed in a concentrated sugar solution, water moves out of the cell by osmosis, through the partially permeable cell wall.

3 Explain each of the following.
   a Diffusion happens faster when the temperature rises.
   b Oxygen diffuses out of a plant leaf during daylight hours.
   c Water molecules can pass through Visking tubing, but starch molecules cannot.
   d An animal cell bursts if placed in pure water.
   e If a plant is short of water, its leaves lose their firmness and the plant wilts.

4 The chemicals of life

In this chapter, you will find out:

- why water is so important to living organisms
- what carbohydrates, lipids and proteins are made of, and their properties
- the roles of carbohydrates, lipids and proteins in living organisms
- how to test for the presence of carbohydrates, lipids and proteins
- some good sources of foods containing carbohydrates, lipids and proteins.

What are you made of?

The bodies of all living things are made of many different kinds of chemicals. Most of our bodies are made up of water. We also contain carbohydrates, proteins and fats. These substances are what our cells are made of. Each of them is vital for life.

In this chapter, we will look at each of these kinds of substances in turn. As you work through your biology course, you will keep meeting them over and over again.

It will help if you have a basic understanding of the meanings of the terms atom, element and molecule. If you are not sure about these, ask your biology or chemistry teacher to explain them to you.

4.1 Water dissolves substances in cells.

In most organisms, almost 80% of the body is made up of water. We have seen that cytoplasm is a solution of many different substances in water. The spaces between our cells are also filled with a watery liquid.

Inside every living organism, chemical reactions are going on all the time. These reactions are called **metabolism**. Metabolic reactions can only take place if the chemicals which are reacting are dissolved in water. This is one reason why water is so important to living organisms. If their cells dry out, the reactions stop, and the organism dies.

Water is also needed for other reasons. For example, plasma, the liquid part of blood, must contain a lot of water, so that substances like glucose can dissolve in it. These dissolved substances are transported around the body.

We also need water to help us to get rid of waste products. As you will see in Chapter 11, the kidneys remove the waste product urea from the body. The urea is dissolved in water, forming urine.

Water also helps us to keep cool. When we are too hot, the sweat glands in the skin release sweat, which is mostly water. The water in the sweat evaporates, and this cools us down.

Carbohydrates

4.2 Starch and sugars are carbohydrates.

Carbohydrates include starches and sugars. Their molecules contain three kinds of atom – carbon (C), hydrogen (H), and oxygen (O). A carbohydrate molecule has about twice as many hydrogen atoms as carbon or oxygen atoms.

4.3 Glucose is a simple sugar.

The simplest kinds of carbohydrates are the **simple sugars** or **monosaccharides**. **Glucose** is a simple sugar.

A glucose molecule is made of six carbon atoms joined in a ring, with the hydrogen and oxygen atoms pointing out from and into the ring (Figure **4.1**).

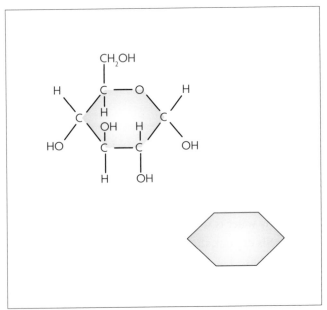

Figure 4.1 The structure of a glucose molecule. You do not need to remember this, but you should remember that glucose is made of carbon, hydrogen and oxygen atoms.

A glucose molecule contains six carbon atoms, twelve hydrogen atoms, and six oxygen atoms. To show this, its molecular formula can be written $C_6H_{12}O_6$. This formula stands for one molecule of this simple sugar, and tells you which atoms it contains, and how many of each kind.

Although they contain many atoms, simple sugar molecules are very small (Figure **4.2a**). They are soluble in water, and they taste sweet.

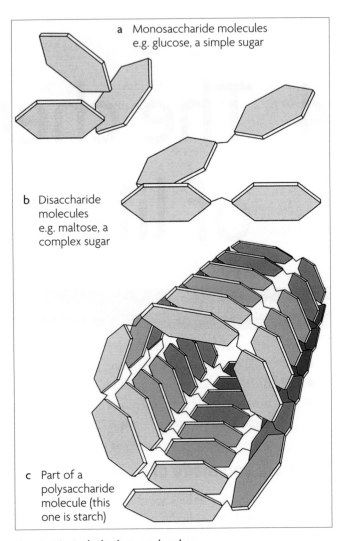

a Monosaccharide molecules e.g. glucose, a simple sugar

b Disaccharide molecules e.g. maltose, a complex sugar

c Part of a polysaccharide molecule (this one is starch)

Figure 4.2 Carbohydrate molecules.

4.4 Sucrose is a complex sugar.

If two simple sugar molecules join together, a larger molecule called a **complex sugar** or **disaccharide** is made (Figure **4.2b**). Two examples of complex sugars are **sucrose** (the sugar we use in hot drinks, or on breakfast cereal, for example) and **maltose** (malt sugar). Like simple sugars, they are soluble in water and taste sweet.

4.5 Starch is a polysaccharide.

If many simple sugars join together, a very large molecule called a **polysaccharide** is made. Some polysaccharide molecules contain thousands of sugar molecules joined together in a long chain. The **cellulose** of plant cell walls is a polysaccharide and so is **starch**, which is often found inside plant cells (Figure **4.2c**). Most polysaccharides are insoluble, and they do not taste sweet.

4.6 Living organisms get energy from carbohydrates.

Carbohydrates are needed for energy. One gram of carbohydrate releases 17 kJ (kilojoules) of energy. The energy is released by respiration (Chapter **9**).

The carbohydrate that is normally used in respiration is glucose. This is also the form in which carbohydrate is transported around an animal's body. Human blood plasma contains dissolved glucose, being transported to all the cells. The cells then use the glucose to release the energy that they need to carry out the processes of life.

Plants also use glucose in respiration, to provide them with energy. However, they do not transport glucose around their bodies. Instead, they transport sucrose. The cells change the sucrose to glucose when they need to use it.

Plants store carbohydrates as starch. It is quick and easy to change glucose into starch, or starch into glucose.

Some plants store large quantities of starch in their seeds or tubers, and we use these as food.

Animals do not store starch. Instead, they store carbohydrates in the form of a polysaccharide called **glycogen**. However, only small quantities of glycogen can be stored. It is mostly stored in the cells in the liver and the muscles.

4.7 Carbohydrates come mostly from plant-based foods.

Plants make their own carbohydrates, by photosynthesis. Animals and fungi get their carbohydrates from the food that they eat.

Figure **4.3** (overleaf) shows some foods that contain carbohydrates. Almost all of the carbohydrate that we get in our diet comes from foods derived from plants. This is because animals store only a very little carbohydrate in their bodies. So meat and fish contain hardly any carbohydrate at all.

Activity 4.1
Testing foods for sugars

skills

C1 *Using techniques, apparatus and materials*

C2 *Observing, measuring and recording*

Safety Wear eye protection if available.
If possible, heat the tubes using a water bath. If you have to heat directly over a Bunsen flame, use a test-tube holder and point the opening of the tube away from people.
Take care if using a sharp blade to cut the food.

All simple sugars are reducing sugars. This means that they will react with a blue liquid called Benedict's solution. We can use this reaction to find out if a food or other substance contains a reducing sugar.

1   Draw a results chart.

food	colour with Benedict's solution	simple sugar present

2   Cut or grind a little of the food into very small pieces. Put these into a test tube. Add some water, and shake it up to try to dissolve it.

3   Add some Benedict's solution. Benedict's solution is blue, because it contains copper salts.

4   Heat the tube strongly. If there is reducing sugar in the food, the solution will turn orange-red.

5   Record your result in your results chart. If the Benedict's solution does not change colour, do not write 'no change'. Write down the actual colour that you see – for example, blue. Then write down your conclusion from the result of the test.

This test works because the reducing sugar reduces the blue copper salts to a red compound. All simple sugars are reducing sugars, and also some complex sugars. For example, the complex sugar maltose is a reducing sugar.

Carbohydrates are needed for energy.

Figure 4.3 Carbohydrate foods.

Fats

4.8 Fats are made of glycerol and fatty acids.

Fats are also known as **lipids**. Like carbohydrates, fats contain only three kinds of atom – carbon, hydrogen and oxygen. A fat molecule is made of four molecules joined together. One of these is **glycerol**. Attached to the glycerol are three long molecules called **fatty acids** (Figure 4.4).

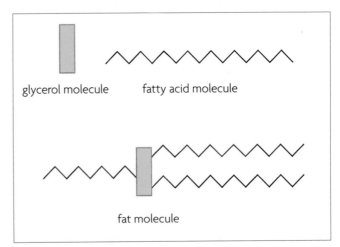

glycerol molecule   fatty acid molecule

fat molecule

Figure 4.4 A fat molecule.

Fats are insoluble in water. Fats that are liquid at room temperature are called oils.

4.9 Fats are good storage products.

Like carbohydrates, fats and oils can be used in a cell to release energy. A gram of fat gives about 39 kJ of energy. This is more than twice as much energy as that released by a gram of carbohydrate. However, most cells use carbohydrates first when they need energy, and only use fats when all the available carbohydrates have been used.

The extra energy which fats contain makes them very useful for storing energy. In mammals, some cells, particularly ones underneath the skin, become filled with large drops of fats or oils. These stores can be used to release energy when needed. This layer of cells is called **adipose tissue**. Adipose tissue also helps to keep heat inside the body – that is, it insulates the body. Animals such as seals and some whales, which live in very cold places, often have especially thick layers of adipose tissue, called blubber (Figure 4.5). Many plants store oils in their seeds – for example, peanut, coconut and castor oil. The oils provide a good store of energy for germination.

Figure 4.5 Animals that live in very cold places, such as this bowhead whale, may have thick blubber to keep in body heat.

4.10 Fats are in foods from plants and animals.

We have seen that plants make carbohydrates by photosynthesis. They use some of these carbohydrates to make fats. We get fats from eating plants or animals. Figure 4.6 shows some foods that contain fats.

Questions

4.7 Which three elements are found in all fats and oils?

4.8 State two uses of fats to living organisms.

4.9 We get cooking oil mostly from the seeds of plants. Why do plant seeds contain oil?

Fats are needed for energy and for keeping warm.

Figure 4.6 Foods containing fats.

Activity 4.3
Testing foods for fats

skills

C1 *Using techniques, apparatus and materials*

C2 *Observing, measuring and recording*

Safety Take care if using a sharp blade to cut the food.

The emulsion test

Fats will not dissolve in water, but they will dissolve in ethanol. If a solution of fat in ethanol is added to water, the fat forms tiny globules which float in the water. This is called an **emulsion**. The globules of fat make the water look milky.

1 Draw a results chart.

2 Chop or grind a small amount of food, and put some into a very clean, dry test tube. Add some absolute (pure) ethanol. Shake it thoroughly.

3 Put some distilled water in another tube.

4 Pour some of the liquid part, but not any solid, from the first tube into the water. A milky appearance shows that there is fat in the food.

Proteins

4.11 Proteins are long chains of amino acids.

Protein molecules contain some kinds of atoms which carbohydrates and fats do not (Figure 4.7). As well as carbon, hydrogen and oxygen, they also contain nitrogen (N) and small amounts of sulfur (S).

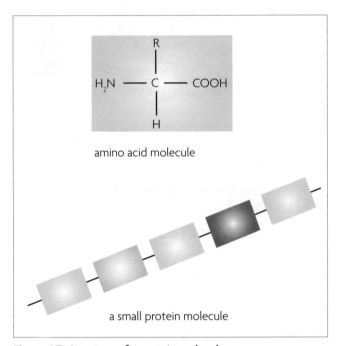

amino acid molecule

a small protein molecule

Figure 4.7 Structure of a protein molecule.

Like polysaccharides, protein molecules are made of long chains of smaller molecules joined end to end. These smaller molecules are called **amino acids**. There are about 20 different kinds of amino acid. Any of these 20 can be joined together in any order to make a protein molecule (Figure 4.7). Each protein is made of molecules with amino acids in a precise order. Even a small difference in the order of amino acids makes a different protein, so there are millions of different proteins which could be made.

Some proteins are soluble in water; an example is haemoglobin, the red pigment in blood. Others are insoluble in water; for example, keratin. Hair and fingernails are made of keratin.

4.12 Proteins are used for growth and repair.

Unlike carbohydrates, proteins are not normally used to provide energy. Many of the proteins in the food you eat are used for making new cells. New cells are needed for growing, and for repairing damaged parts of the body. In particular, cell membranes and cytoplasm contain a lot of protein.

Proteins are also needed to make antibodies. These fight bacteria and viruses inside the body. Enzymes are also proteins.

4.13 Plant foods and animal foods are good sources of protein.

Plants use some of their carbohydrates to make proteins. To do this, they need ammonium ions or nitrate ions. These ions contain nitrogen, which the plant has to combine with the carbohydrates to make amino acids. The amino acids are then linked into a long chain to make a protein.

We get proteins when we eat plants or animals. Some foods that are especially good sources of protein are shown in Figure 4.8.

Table 4.1 compares carbohydrates, lipids and proteins.

Proteins are needed for growth, repair and fighting disease.

Figure 4.8 Foods containing protein.

Activity 4.4
Testing foods for protein

Safety Wear eye protection if available.
Potassium hydroxide is a strong alkali.
If you get it on your skin, wash with
plenty of cold water.
Take care if using a sharp blade to cut
the food.

The biuret test

The biuret test uses potassium hydroxide solution
and copper sulfate solution. You can also use a
ready-mixed reagent called biuret reagent, which
contains these two substances already mixed
together.

1 Draw a results chart.

2 Put the food into a test tube, and add a
little water.

3 Add some potassium hydroxide solution.

4 Add two drops of copper sulfate solution.

5 Shake the tube gently. If a purple colour
appears, then protein is present.

Table 4.1 Carbohydrates, fats and proteins

	carbohydrates	fats	proteins
elements they contain	C, H, O	C, H, O	C, H, O, N
smaller molecules of which they are made	simple sugars (monosaccharides)	fatty acids and glycerol	amino acids
solubility in water	sugars are soluble	insoluble	some are soluble and some are insoluble
why organisms need them	easily available energy (17 kJ/g)	storage of energy (39 kJ/g); insulation; making cell membranes	making cells, antibodies, enzymes, haemoglobin and many other substances; also used for energy
some foods that contain them	bread, cakes, potatoes, rice, yams, eddoes	butter, lard, margarine, oil, fatty meat, milk, peanuts	meat, fish, eggs, milk, cheese, peas, beans, tofu

Questions

4.10 Name **two** elements found in proteins that are not found in carbohydrates.

4.11 How many different amino acids are there?

4.12 In what way are protein molecules similar to polysaccharides?

4.13 Give **two** examples of proteins.

4.14 State **three** functions of proteins in living organisms.

◆ Water is needed in the body as a solvent, for transport, for the removal of waste products and for keeping cool.

◆ Carbohydrates include sugars, starch, glycogen and cellulose. They are made of carbon, hydrogen and oxygen.

◆ The simplest carbohydrates are called simple sugars or monosaccharides. They include glucose. Two molecules of simple sugars can join together to form a complex sugar or disaccharide. These include sucrose and maltose. All sugars are soluble in water and taste sweet.

◆ Many simple sugar molecules can link together to form a very long chain. These make up polysaccharides. Examples include cellulose, starch and glycogen. They do not taste sweet and are not soluble in water.

◆ Glucose is the main fuel for most cells. Energy is released from it in respiration. Energy can be stored in the form of starch (in plants) or glycogen (in animals). Cellulose forms plant cell walls.

◆ The Benedict's test is used to test for reducing sugars. Iodine solution is used to test for starch.

◆ Fats are made of carbon, hydrogen and oxygen. Some types of fat molecule are made up of one molecule of glycerol joined to three fatty acids. Fats are insoluble in water.

◆ Fats are used as an energy source. They are especially useful for energy storage, because they contain more than twice as much energy per gram as carbohydrates.

◆ The emulsion test is used to test for fats.

◆ Proteins contain carbon, hydrogen, oxygen, nitrogen and sometimes sulfur. They are made of many amino acids linked in a long chain. Some are soluble in water and some are insoluble.

◆ Proteins are needed for the growth and repair of the body. They are also needed to produce enzymes, antibodies, haemoglobin and many other vital chemicals in the body. They can be used as a source of energy too.

◆ The biuret test is used to test for proteins.

Revision questions

1 For each of these carbohydrates, state: (i) whether it is a monosaccharide, disaccharide or polysaccharide; (ii) whether it is found in plants only, animals only or in both plants and animals; (iii) one function.
   a glucose
   b starch
   c cellulose
   d glycogen

2 Name:
   a an element found in proteins but not in carbohydrates or lipids
   b the small molecules that are linked together to form a protein molecule
   c the reagent used for testing for reducing sugars
   d the substance which the emulsion test detects
   e the form in which carbohydrate is transported in a plant
   f the term that describes all the chemical reactions taking place in an organism.

5 Enzymes

The functions of enzymes

5.1 Enzymes are biological catalysts.

Many chemical reactions can be speeded up by substances called **catalysts**. A catalyst alters the rate of a chemical reaction, without being changed itself.

> **Key definition**
>
> **catalyst** a substance that speeds up a chemical reaction and is not changed by the reaction

Within any living organism, chemical reactions take place all the time. They are sometimes called metabolic reactions. Almost every metabolic reaction is controlled by catalysts called **enzymes**.

> **Key definition**
>
> **enzymes** proteins that function as biological catalysts

For example, inside the alimentary canal, large molecules are broken down to smaller ones in the process of digestion. These reactions are speeded up by enzymes. A different enzyme is needed for each kind of food. For example, starch is digested to the sugar maltose by an enzyme called **amylase**. Protein is digested to amino acids by **protease**.

These enzymes are also found in plants – for example, in germinating seeds, where they digest the food stores for the growing seedling. Many seeds contain stores of starch. As the seed soaks up water, the amylase is activated and breaks down the starch to maltose. The maltose is soluble, and is transported to the embryo in the seed. The embryo uses it to provide energy for growth, and also to provide glucose molecules that can be strung together to make cellulose molecules, for the cell walls of the new cells produced as it grows.

Another enzyme which speeds up the breakdown of a substance is **catalase**. Catalase, however, does not work in the alimentary canal. It works inside the cells of living organisms – both animals and plants – for example, in liver cells or potato cells. It breaks down hydrogen peroxide to water and oxygen. This is necessary because hydrogen peroxide is produced by many of the chemical reactions which take place inside cells. Hydrogen peroxide is a very dangerous substance, and must be broken down immediately.

Not all enzymes help to break things down. Many enzymes help to make large molecules from small ones. One example of this kind of enzyme is starch phosphorylase, which builds starch molecules from glucose molecules inside plant cells.

5.2 Enzymes are given special names.

Enzymes are named according to the reaction that they catalyse. For example, enzymes which catalyse the breakdown of carbohydrates are called **carbohydrases**. If they break down proteins, they are **proteases**. If they break down fats (lipids) they are **lipases**.

Sometimes, they are given more specific names than this. For example, we have seen that the carbohydrase that breaks down starch is called **amylase**. One that breaks down maltose is called **maltase**. One that breaks down sucrose is called **sucrase**.

5.3 Enzymes change substrates to products.

A chemical reaction always involves one substance changing into another. The substance which is present at the beginning of the reaction is called the **substrate**. The substance which is made by the reaction is called the **product**.

For example, in saliva there is an enzyme called amylase. It catalyses the breakdown of starch to the complex sugar maltose. In this reaction, starch is the substrate, and maltose is the product.

$$\text{starch} \xrightarrow{\text{amylase}} \text{maltose}$$

S

Figure 5.1 shows how amylase does this. An amylase molecule has a dent in it called its **active site**. This is exactly the right size and shape for part of a starch molecule to fit in. When the starch molecule is in the active site, the enzyme breaks it apart.

All enzymes have active sites. Each enzyme has an active site that exactly fits its substrate. This means that each

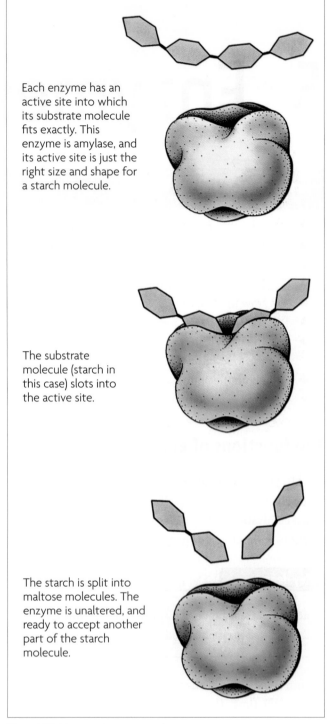

Each enzyme has an active site into which its substrate molecule fits exactly. This enzyme is amylase, and its active site is just the right size and shape for a starch molecule.

The substrate molecule (starch in this case) slots into the active site.

The starch is split into maltose molecules. The enzyme is unaltered, and ready to accept another part of the starch molecule.

Figure 5.1 How an enzyme works.

enzyme can only act on a particular kind of substrate. Amylase, for example, cannot break down protein molecules, because they do not fit into its active site.

The enzyme can be thought of as a lock, and the substrate as a key. Only the correctly shaped key will fit into the lock.

5.4 All enzymes have certain properties.

1 **All enzymes are proteins**   This may seem rather odd, because some enzymes actually digest proteins.

2 **Enzymes are made inactive by high temperature**   This is because they are protein molecules, which are damaged by heat.

3 **Enzymes work best at a particular temperature**   Enzymes which are found in the human body usually work best at about 37 °C.

4 **Enzymes work best at a particular pH**   pH is a measure of how acid or alkaline a solution is. Some enzymes work best in acid conditions (low pH). Others work best in alkaline conditions (high pH).

5 **Enzymes are catalysts**   They are not changed in the chemical reactions which they control. They can be used over and over again, so a small amount of enzyme can change a lot of substrate into a lot of product.

6 **Enzymes are specific**   This means that each kind of enzyme will only catalyse one kind of chemical reaction.

5.5 High temperature denatures enzymes.

Most chemical reactions happen faster at higher temperatures. This is because the molecules have more kinetic energy – they are moving around faster, so they bump into each other more frequently. This means that at higher temperatures an enzyme is likely to bump into its substrate more often than at lower temperatures. They will also hit each other with more energy, so the reaction is more likely to take place (Figure **5.2**).

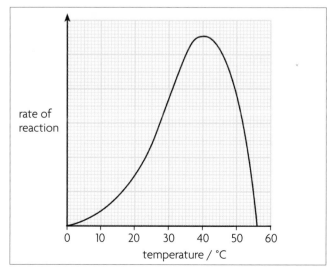

Figure 5.2 How temperature affects enzyme activity.

However, enzymes are damaged by high temperatures. For most human enzymes, this begins to happen from about 40 °C upwards. As the temperature increases beyond this, the enzyme molecules start to lose their shape. The active site no longer fits perfectly with the substrate. The enzyme is said to be **denatured**. It can no longer catalyse the reaction.

The temperature at which an enzyme works fastest is called its **optimum** temperature. Different enzymes have different optimum temperatures. For example, enzymes from the human digestive system generally have an optimum of around 37 °C. Enzymes from plants often have optimums around 28 °C to 30 °C. Enzymes from bacteria that live in hot springs may have optimums as high as 75 °C.

5.6 pH affects enzymes.

The pH of a solution affects the shape of an enzyme. Most enzymes are their correct shape at a pH of about 7 – that is, neutral. If the pH becomes very acidic or very alkaline, then they lose their shape. This means that the active site no longer fits the substrate, so the enzyme can no longer catalyse its reaction (Figure **5.3**).

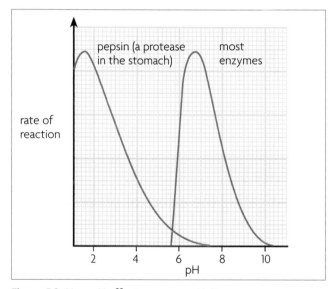

Figure 5.3 How pH affects enzyme activity.

Some enzymes have an optimum pH that is not neutral. For example, there is a protease enzyme in the human stomach that has an optimum pH of about 2. This is because we have hydrochloric acid in the stomach. This protease must be able to work well in these very acidic conditions.

Questions

5.1 What is a catalyst?

5.2 What are the catalysts inside a living organism called?

5.3 Which kinds of reaction inside a living organism are controlled by enzymes?

5.4 What is meant by a carbohydrase?

5.5 Give one example of a carbohydrase.

5.6 Name the substrate and product of a reaction involving a carbohydrase.

5.7 What is meant by an optimum temperature?

5.8 Why are enzymes damaged by high temperatures?

Activity 5.1
The effect of catalase on hydrogen peroxide

skills

C1 Using techniques, apparatus and materials

C2 Observing, measuring and recording

C3 Interpreting and evaluationg

Safety Wear eye protection if available. Hydrogen peroxide is a powerful bleach. Wash it off with plenty of water if you get it on your skin.

Catalase is found in almost every kind of living cell. It catalyses this reaction:

hydrogen peroxide $\xrightarrow{\text{catalase}}$ water + oxygen

1 Read through the instructions. Decide what you will need to observe and measure, and draw a suitable results table.

2 Measure 10 cm³ of hydrogen peroxide into each of five test tubes or boiling tubes.

3 To each tube, add one of the following substances:
   a some chopped raw potato
   b some chopped boiled potato
   c some fruit juice
   d a small piece of liver
   e some yeast suspension.

4 Light a wooden splint, and then blow it out so that it is glowing. Gently push the glowing splint down through the bubbles in your tubes.

5 Record your observations, and explain them as fully as you can.

Activity 5.2
Investigating the effect of pH on the activity of catalase

skills

C1 Using techniques, apparatus and materials

C2 Observing, measuring and recording

C3 Interpreting and evaluationg

Safety Wear eye protection if available. Hydrogen peroxide is a powerful bleach. Wash it off with plenty of water if you get it on your skin.

Catalase is a common enzyme which is the catalyst in the breakdown of hydrogen peroxide, H_2O_2. Catalase is found in almost every kind of living cell. Hydrogen peroxide is a toxic substance formed in cells. The breakdown reaction is as follows:

$$2H_2O_2 \longrightarrow 2H_2O + O_2$$

The rate of the reaction can be determined from the rate of oxygen production.

(continued...)

(*... continued*)

One indirect but simple way to measure rate of oxygen production is to soak up a catalase solution onto a little square of filter paper and then drop it into a beaker containing a solution of H_2O_2.

The paper sinks at first, but as the reaction proceeds, bubbles of oxygen collect on its surface and it floats up. The time between placing the paper in the beaker and it floating to the surface is a measure of the rate of the reaction.

In this investigation, you will test this hypothesis:

Catalase works best at a pH of 7 (neutral).

1   Label five 50 cm³ beakers pH 5.6, 6.2, 6.8, 7.4, 8.0.

2   Measure 5 cm³ of 3% hydrogen peroxide solution into each beaker.

3   Add 10 cm³ of the correct buffer solution to each beaker. (A buffer solution keeps the pH constant at a particular value.)

4   Cut out 20 squares of filter paper exactly 5 mm × 5 mm. Alternatively, use a hole punch to cut out circles of filter paper all exactly the same size. Avoid handling the paper with your fingers, as you may get grease onto it. Use forceps (tweezers) instead.

5   Prepare a leaf extract by grinding the leaves in a pestle and mortar. Add 25 cm³ of water and stir well.

6   Allow the remains of the leaves to settle and then pour the fluid into a beaker. This fluid contains catalase.

7   Prepare a results table like the one below.

	time taken for paper to float in seconds				
pH	5.6	6.2	6.8	7.4	8.0
tests 1					
2					
3					
mean					
boiled extract					

8   Pick up a filter paper square with forceps and dip it into the leaf extract.

9   Make sure you are ready to start timing. Then place the filter paper square at the bottom of the beaker containing H_2O_2 and pH 5.6 buffer solution. (Do not let it fall near the side of the beaker.) As you put the square into the beaker, start a stopwatch. Stop the watch when the paper floats horizontally at the surface.

10   Record the time in your table and repeat steps **8** and **9** twice more.

11   Follow steps **8–10** for each of the other pHs.

12   Pour some of the remaining leaf extract into a test tube and boil for 2 minutes. Cool under a tap.

13   Repeat steps **8–10**, using the boiled extract.

14   Calculate the mean (average) time taken at each pH and enter it into your table.

15   Draw a graph to show time taken for flotation plotted against pH.

Questions

1   Does the enzyme have an optimum pH? If it does, what do your results suggest it to be?

2   Do your results support the hypothesis you were testing, or do they disprove it? Explain your answer.

3   What is the effect of boiling the extract?

4   Why do the filter paper squares have to be exactly the same size?

5   In most experiments in biology, we can never be quite sure that we would get exactly the same results if we did it again. There are always some limitations on the reliability of the data that we collect. Can you think of any reasons why the results you got in your experiment might not be absolutely reliable? For example:

   ● Might there have been any variables that were not controlled and that might have affected the results?

   ● Were you able to measure the volumes and times as accurately as you would have liked?

Activity 5.3
Investigating the effect of temperature on the activity of catalase

Safety Wear eye protection if available. Hydrogen peroxide is a powerful bleach. Wash it off with plenty of water if you get it on your skin.

You are going to plan this investigation yourself. You can use ideas from Activities **5.1** and **5.2** to help you.

You can vary temperature by using a water bath. Your teacher may be able to provide electrically controlled water baths. If not, you can make one by placing a beaker of water on a tripod and gauze over a Bunsen burner. You can make cold temperatures by using ice. Your teacher will show you how to do this.

You need to think about each of the following points carefully. Once you have an idea about how you will do your experiment, write it down as a list of points. Then think through it again, and make improvements to your plan. Once you are fairly happy with it, show your teacher. You must not try to do your experiment until your teacher says that you may begin.

- What is the hypothesis you are going to test? (Hint: use similar wording to the hypothesis in Activity **5.2**. Also look at section **5.4**, point **3**.)

- What apparatus and other materials will you need for your experiment?

- What will you vary in your experiment? How will you vary it?

- What will you keep the same in all the tubes or beakers in your experiment? How will you do this?

- What will you measure in your experiment? How will you measure it? When will you measure it? Will you do repeat measurements and calculate a mean?

- How will you record your results? (You can sketch out a results chart, ready to fill in.)

- How will you display your results? (You can sketch the axes of the graph you plan to draw.)

- What will your results be if your hypothesis is correct? (You can sketch the shape of the graph you think you will get.)

Once you have approval from your teacher, you should do your experiment. Most scientific researchers find that they want to make changes to their experiment once they actually begin doing it. This is a good thing to do. Make careful notes about all the changes that you make.

Finally, write up your experiment in the usual way, including:

- a heading, and the hypothesis that you tested

- a diagram of the apparatus that you used, and a full description of your method

- a neat and carefully headed table of results, including means if you decided to do repeats

- a neat and carefully headed line graph of your results

- a conclusion, in which you say whether or not your results support your hypothesis

- a discussion, in which you use what you know about enzymes to try to explain the pattern in your results

- an evaluation, in which you explain the main limitations that you feel might have affected the reliability of your data.

Making use of enzymes

5.7 Enzymes are used in washing powders.

Biological washing powders contain enzymes, as well as detergents. The detergents help greasy dirt to mix with water, so that it can be washed away. The enzymes help to break down other kinds of substances which can stain clothes. They are especially good at removing dirt which contains coloured substances from animals or plants, like blood or egg stains.

Some of the enzymes are proteases, which catalyse the breakdown of protein molecules. This helps with the removal of stains caused by proteins, such as blood stains. Blood contains the red protein haemoglobin. The proteases in biological washing powders break the haemoglobin molecules into smaller molecules, which are not coloured, and which dissolve easily in water and can be washed away.

Some of the enzymes are lipases, which catalyse the breakdown of fats to fatty acids and glycerol. This is good for removing greasy stains.

The first biological washing powders only worked in warm, rather than hot, water, because the proteases in them had optimum temperatures of about 40 °C. However, proteases have now been developed which can work at much higher temperatures. These proteases have often come from bacteria which naturally live in hot water, in hot springs. This is useful, because the other components of washing powders – which get rid of grease and other kinds of dirt – work best at these higher temperatures.

5.8 Enzymes are often used in the food industry.

Fruit juices are extracted using an enzyme called **pectinase**. Pectin is a substance which helps to stick plant cells together. A fruit such as an apple or orange contains a lot of pectin. If the pectin is broken down, it can be much easier to squeeze juice from the fruit. Pectinase is widely used commercially both in the extraction of juice from fruit, and in making the juice clear rather than cloudy.

Enzymes are sometimes used when making baby foods. Some high-protein foods are treated with proteases, to break down the proteins to polypeptides and amino acids. This makes it easier for young babies to absorb the food.

Activity 5.4
Investigating biological washing powders

skills
C1 Using techniques, apparatus and materials
C2 Observing, measuring and recording
C3 Interpreting and evaluating
C4 Planning

Safety   Wear eye protection if available. Whichever methods you use, do not let the enzymes come into contact with your hands any more than necessary. Remember, you contain a lot of protein and fat! If you do get them on your skin, wash with plenty of water.

Design and carry out an experiment to test one or more of the hypotheses in the box.

a   A biological washing powder removes egg stains from fabric better than a non-biological washing powder.

b   The optimum temperature for biological washing powders is lower than that for non-biological washing powders.

c   Lipases – enzymes which digest fats – help to remove grease stains from fabrics.

If you want to test the protease-containing powders on fabrics, you could first stain the fabrics with a protein-containing stain such as egg.

Your teacher will suggest suitable amounts of enzymes or washing powders to use.

There is a great demand for sugar in the food industry. Not only is it used in making many sweet foods, but it can also be supplied as a food for microorganisms used in making food substances. (You can read about some of these in Chapter 7.) As well as getting sugar directly from sugar cane and sugar beet, it can be made from starch. This is done by crushing starch-containing materials – perhaps potatoes or grain – with water, and then adding amylase. The amylase digests the starch to maltose, making a syrup.

Some sugars are sweeter than others. Fructose, for example – a sugar found in fruits – is sweeter than most other sugars. People who really like sweet things, but are worried about eating too much sugar, may prefer to eat fructose rather than glucose or sucrose, because they can get just as much sweet taste with less sugar. However, most of the sugar we get from plants is either glucose or sucrose. An enzyme called isomerase can be used to convert glucose into fructose.

Activity 5.5
Investigating the use of pectinase in making fruit juice

skills

C1 *Using techniques, apparatus and materials*

C2 *Observing, measuring and recording*

C3 *Interpreting and evaluating*

C4 *Planning*

Safety Take care if using a sharp blade to cut the fruit. Since you are doing this investigation in a laboratory, and since the pectinase you use may not be food grade, you must not taste the fruit juice you make.

Design and carry out an experiment to test one of the hypotheses in the box.

You will find it best to pulp your fruit before trying to extract juice from it. If you are using apples, you can do this by chopping them up roughly, and then blending them with some distilled water in a food processor or other blender. You may then like to heat the blended apple to about 40 °C for around 15 minutes, as this speeds up the juice extraction. You can extract the juice by placing the crushed apple in a funnel lined with filter paper, and collecting the juice which drips through.

Your teacher will suggest suitable amounts of pectinase to use.

a   You can extract more juice if you add pectinase to the fruit than if you do not.

b   Juice is extracted more quickly if pectinase is added to the fruit than if it is not.

c   The effect of pectinase varies on different kinds of fruit – for example, apples and pears.

d   Pectinase has a greater effect on the amount of juice extracted from old fruit than from freshly picked fruit.

e   It is more difficult to extract juice, even when using pectinase, from Golden Delicious apples than from other varieties.

f   People cannot tell the difference between the appearance of juice extracted using pectinase and that of juice extracted without it.

g   Pectinase added to the extracted juice can make it clear.

h   Pectinase has an optimum temperature.

i   Bought fruit juice contains pectinase.

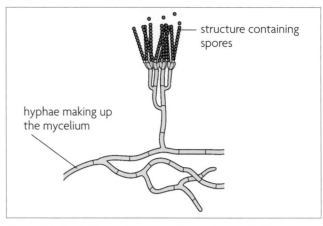

Figure 5.4 *Penicillium*, the fungus that makes penicillin.

Question

5.9 Name the substrate for each of the following enzymes:
  a pectinase
  b isomerase
  c amylase.

5.9 We obtain many enzymes from microorganisms.

The enzymes that are used in industry are usually obtained from microorganisms. These include bacteria and microscopic fungi, such as yeast.

The microorganisms are grown inside large vessels called fermenters. Inside the fermenter, the microorganisms are provided with everything they need to grow and reproduce. This generally includes oxygen, a supply of a nutrients, a suitable pH and a suitable temperature. The microorganisms make the enzymes and release them into the liquid in which they are growing. The liquid can then be collected from the fermenter, and the enzymes purified before use.

5.10 The fungus *Penicillium* makes penicillin.

In many cases, it is easier for us just to use the whole microorganism, rather than extract its enzymes from it. The microorganism is grown in a fermenter, and its enzymes convert a substrate to a desired product.

One example is the production of the antibiotic penicillin. Antibiotics are substances which kill bacteria without harming human cells. We take antibiotics to help to cure bacterial infections.

Penicillin is made by growing the fungus *Penicillium* (Figure 5.4) in a large fermenter (Figure 5.5). It is grown in a culture medium containing carbohydrates and amino acids. The contents of the fermenter look a bit like watery porridge. They are stirred continuously. This not only keeps the fungus in contact with fresh supplies of nutrients, and mixes oxygen into the culture, but also rolls the fungus up into little pellets. This makes it quite easy to separate the liquid part of the culture – which contains the pencillin – from the fungus, at a later stage.

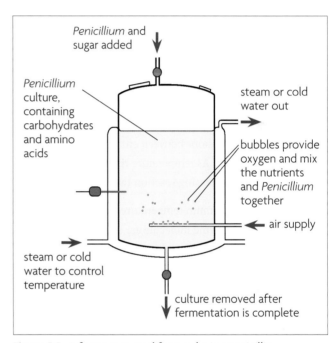

Figure 5.5 A fermenter used for producing penicillin.

To begin with, the fungus just grows. This stage takes about 15–24 hours. After that, it begins to secrete penicillin. The rate at which it produces penicillin partly depends on how much sugar it has available. If there is a lot of sugar, then not much penicillin is made. If there is no sugar at all, then no penicillin is made. So small amounts of sugar have to be fed into the fermenter all the time that the fungus is producing penicillin.

The culture is kept going until it is decided that the rate of penicillin production has slowed down so much that it is not worth waiting any longer. This is often after about a week, although the exact time can vary quite a lot on either side of this. Then the culture is filtered, and the liquid is treated to concentrate the penicillin which it contains.

◆ Enzymes are proteins that work as biological catalysts.

◆ Enzymes are named according to the substrate on which they act. Proteases act on proteins, carbohydrases on carbohydrates and lipases on fats (lipids). The substance that is produced by the reaction is called the product.

◆ An enzyme molecule has a depression called its active site, which is exactly the right shape for the substrate to fit into. The enzyme can be thought of as a lock, and the substrate as the key.

◆ Reactions catalysed by enzymes work faster at higher temperatures, up to an optimum that differs for different enzymes. Above the optimum temperature, reaction rate rapidly decreases.

◆ At low temperatures, molecules have low kinetic energy, so collisions between enzyme and substrate are infrequent. As temperature rises they collide more frequently, increasing reaction rate.

◆ Above the optimum temperature, the vibrations within the enzyme molecule are so great that it begins to lose its shape. The enzyme is said to be denatured. The substrate no longer fits into the active site and the reaction stops.

◆ Reactions catalysed by enzymes work fastest at a particular pH. The optimum pH for most enzymes is around pH7 (neutral), but some have an optimum pH much higher or lower than that.

◆ Extremes of pH cause enzyme molecules to lose their shape, so they no longer bind with their substrate.

◆ Amylase is found in seeds. When the seed begins to germinate, the amylase is activated and catalyses the breakdown of insoluble starch to soluble maltose in the seed. The maltose is used by the growing embryo as an energy source and to make cellulose for new cell walls.

◆ Biological washing powders contain enzymes, often obtained from microorganisms such as bacteria or fungi. The enzymes break down proteins or fats on the fabric, forming water-soluble substances that can be washed away.

◆ Pectinase is used to break down cell walls in fruits, making it easier to extract juice from them.

◆ The antibiotic penicillin is made by cultivating the fungus *Penicillium* in a fermenter. The fermenter is kept at the correct pH and temperature for the enzymes of the fungus to work well.

Revision questions

1 Explain the meaning of each of these terms:
   a enzyme
   b substrate
   c product
   d active site
   e denatured.

2 A protease enzyme is found in the stomachs of humans. It catalyses the breakdown of long chains of amino acids (proteins) into individual amino acid molecules.
   a Suggest the optimum temperature for the activity of this protease enzyme.
   b The stomach contains hydrochloric acid. Suggest the optimum pH for the activity of this protease enzyme.
   c Explain why the rate of an enzyme-controlled reaction is relatively slow at low temperatures.

   d Explain why the rate of the reaction slows down above the enzyme's optimum temperature.

3 Amylase is found in germinating seeds.
   a What are the substrate and product of amylase?
   b Explain the function of amylase in a germinating seed.

4 a Name the organism that is used to make the antibiotic penicillin.
   b What is an antibiotic?
   c The organism is cultured in a fermenter.
      i Explain why air is bubbled through the fermenter.
      ii Explain why water is made to run through a jacket surrounding the fermenter.
      iii Explain why the pH inside the fermenter is controlled.

6 Plant nutrition

In this chapter, you will find out:

- how plants make carbohydrates by photosynthesis
- about the structure of leaves
- the ways in which leaves are adapted for photosynthesis
- how plants use the glucose they produce in photosynthesis
- how to carry out investigations into photosynthesis
- about the factors that affect the rate of photosynthesis
- about the importance of photosynthesis for all living things.

Types of nutrition

6.1 All organisms feed.

All living organisms need to take many different substances into their bodies. Some of these may be used to make new parts, or repair old parts. Others may be used to release energy. Taking in useful substances is called feeding, or **nutrition**.

> **Key definition**
>
> **nutrition** taking in of nutrients which are organic substances and mineral ions, containing raw materials or energy for growth and tissue repair, absorbing and assimilating them

6.2 Animals take complex substances from plants.

Animals and fungi cannot make their own food. They feed on **organic** substances that have originally been made by plants. Some animals eat other animals, but all the substances passing from one animal to another were first made by plants. Animal nutrition is described in Chapter 7.

6.3 Green plants can make complex substances from simple chemicals.

Green plants make their own food. They use simple **inorganic** substances – carbon dioxide, water and minerals – from the air and soil. Plants build these substances into complex materials, making all the carbohydrates, lipids, proteins and vitamins that they need. Substances made by living things are said to be **organic**.

Photosynthesis

6.4 Photosynthesis is a chemical process.

Green plants first make the carbohydrate glucose (section **4.3**) from carbon dioxide and water. At the same time oxygen is produced.

However, if you just mix carbon dioxide and water together, they will not make glucose. They have to be given energy before they will combine. Green plants use the energy of sunlight for this. The reaction is therefore called **photosynthesis** ('photo' means light, and 'synthesis' means manufacture).

6.5 Chlorophyll absorbs sunlight.

However, sunlight shining onto water and carbon dioxide still will not make them react together to make glucose. The sunlight energy has to be trapped, and then used in the reaction. Green plants have a substance which does this. It is called **chlorophyll**.

Chlorophyll is the pigment which makes plants look green. It is kept inside the chloroplasts of plant cells. When sunlight falls on a chlorophyll molecule, the energy is absorbed. The chlorophyll molecule then releases the energy. The energy makes carbon dioxide combine with water, with the help of enzymes inside the chloroplast. The glucose that is made contains energy that was originally in the sunlight. So, in this processs, light energy is converted to chemical energy.

6.6 Photosynthesis can be shown as an equation.

The full equation for photosynthesis is written like this:

$$\text{carbon dioxide} + \text{water} \xrightarrow[\text{chlorophyll}]{\text{sunlight}} \text{glucose} + \text{oxygen}$$

To show the number of molecules involved in the reaction, a balanced equation needs to be written. Carbon dioxide contains two atoms of oxygen, and one of carbon, so its molecular formula is CO_2. Water has the formula H_2O. Glucose has the formula $C_6H_{12}O_6$. Oxygen molecules contain two atoms of oxygen, and so they are written O_2.

The balanced equation for photosynthesis is this:

$$6CO_2 + 6H_2O \xrightarrow[\text{chlorophyll}]{\text{sunlight}} C_6H_{12}O_6 + 6O_2$$

Leaves

6.7 Plant leaves are food factories.

Photosynthesis happens inside chloroplasts. This is where the enzymes and chlorophyll are that catalyse and supply energy for the reaction. In a typical plant, most chloroplasts are in the cells in the leaves. A leaf is a factory for making carbohydrates.

Leaves are therefore specially adapted to allow photosynthesis to take place as quickly and efficiently as possible.

6.8 The structure of leaves.

A leaf consists of a broad, flat part called the **lamina** (Figure **6.1**), which is joined to the rest of the plant by a leaf stalk or **petiole**. Running through the petiole are **vascular bundles** (section **8.26**), which then form the **veins** in the leaf. These contain tubes which carry substances to and from the leaf.

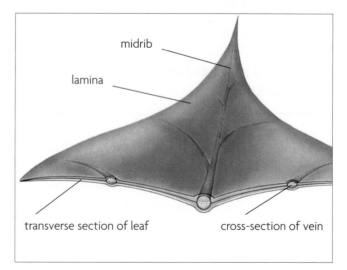

midrib

lamina

transverse section of leaf

cross-section of vein

Figure 6.1 The structure of a leaf.

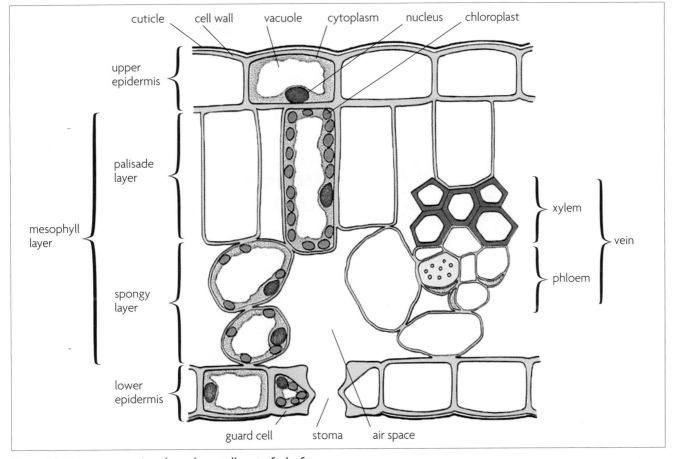

cuticle cell wall vacuole cytoplasm nucleus chloroplast

upper
epidermis

mesophyll
layer

palisade
layer

spongy
layer

lower
epidermis

xylem

phloem

vein

guard cell stoma air space

Figure 6.2 Transverse section through a small part of a leaf.

Although a leaf looks thin, it is in fact made up of several layers of cells. You can see these if you look at a transverse section (TS) of a leaf under a microscope (Figures **6.2** and **6.3**).

The top and bottom of the leaf are covered with a layer of closely fitting cells called the **epidermis** (Figures **6.4** and **6.5**, overleaf). These cells do not contain chloroplasts. Their function is to protect the inner layers of cells in the leaf. The cells of the upper epidermis often secrete a waxy substance, that lies on top of them. It is called the **cuticle**, and it helps to stop water evaporating from the leaf. There is sometimes a cuticle on the underside of the leaf as well.

In the lower epidermis, there are small openings called **stomata** (singular: **stoma**). Each stoma is surrounded by a pair of sausage-shaped **guard cells** (Figures **6.4** and **6.5**, overleaf) which can open or close the hole (Figure **8.35**, page **108**). Guard cells, unlike other cells in the epidermis, do contain chloroplasts.

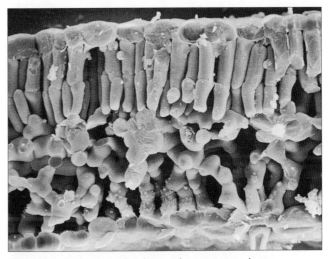

Figure 6.3 A photograph taken with a scanning electron microscope, showing the cells inside a leaf. Notice the many air spaces between the cells (×400).

The middle layers of the leaf are called the **mesophyll** ('meso' means middle, and 'phyll' means leaf). These cells all contain chloroplasts. The cells nearer to the top of the leaf are arranged like a fence or palisade, and they form the **palisade layer**. The cells beneath them are rounder,

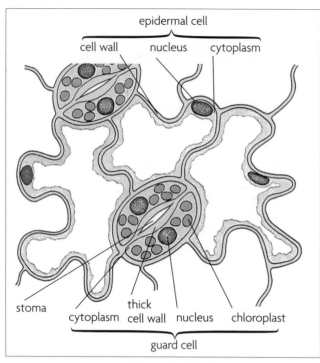

Figure 6.4 Surface view of the lower epidermis of a leaf.

Figure 6.5 The lower surface of a leaf, showing the closely fitting cells of the epidermis. The oval openings are stomata, and the two curved cells around each stoma are guard cells (×450).

and arranged quite loosely, with large **air spaces** between them. They form the **spongy layer** (Figures 6.2 and 6.3).

Running through the mesophyll are veins. Each vein contains large, thick-walled **xylem vessels** (section 8.24) for carrying water, and smaller, thin-walled **phloem tubes** (section 8.25) for carrying away sucrose and other substances that the leaf has made.

Activity 6.1
Looking at the epidermis of a leaf

skills

C2 *Observing, measuring and recording*

Using a piece of epidermis

1   Using forceps, carefully peel a small piece of epidermis from the underside of a leaf.

2   Put the piece of epidermis into a drop of water on a microscope slide.

3   Spread it out carefully, trying not to let any part of it fold over. Cover it with a coverslip.

4   Look at your slide under the microscope, and make a labelled drawing of a few cells.

Making a nail varnish impression

1   Paint the underside of a leaf with transparent nail varnish. Leave to dry thoroughly.

2   Peel off part of the nail varnish, and mount it in a drop of water on a microscope slide.

3   Spread it out carefully, and cover with a coverslip.

4   Look at your slide under the microscope, and make a labelled drawing of the impressions made by a few cells.

5   Repeat with the upper surface of a leaf.

Questions

1   On which surface of the leaf did you find most stomata?

2   Which of these two techniques for examining the epidermis of a leaf do you consider **(a)** is easier, and **(b)** gives you better results?

3   There are two kinds of cell in the lower epidermis of a leaf. What are they, and what are their functions?

Questions

6.5 What is another name for a leaf stalk?

6.6 Which kind of cell makes the cuticle on a leaf?

6.7 What is the function of the cuticle?

6.8 What are stomata?

6.9 What are guard cells?

6.10 List **three** kinds of cell in a leaf which contain chloroplasts, and **one** kind which does not.

6.9 Leaves are adapted to obtain carbon dioxide, water and sunlight.

Carbon dioxide   Carbon dioxide is obtained from the air. There is not very much available, because only about 0.04% of the air is carbon dioxide. Therefore the leaf must be very efficient at absorbing it. The leaf is held out into the air by the stem and the leaf stalk, and its large surface area helps to expose it to as much air as possible.

The cells which need the carbon dioxide are the mesophyll cells, inside the leaf. The carbon dioxide can get into the leaf through the stomata. It does this by diffusion, which is described in Chapter 3. Behind each stoma is an air space (Figure 6.2) which connects up with other air spaces between the spongy mesophyll cells. The carbon dioxide can therefore diffuse to all the cells in the leaf. It can then diffuse through the cell wall and cell membrane of each cell, and into the chloroplasts.

Water   Water is obtained from the soil. It is absorbed by the root hairs (section 8.24), and carried up to the leaf in the xylem vessels. It then travels from the xylem vessels to the mesophyll cells by osmosis, which was described in Chapter 3. The path it takes is shown in Figures 6.6 and 6.7.

Sunlight   The position of a leaf and its broad, flat surface help it to obtain as much sunlight as possible. If you look up through the branches of a tree, you will see that the leaves are arranged so that they do not cut off light from

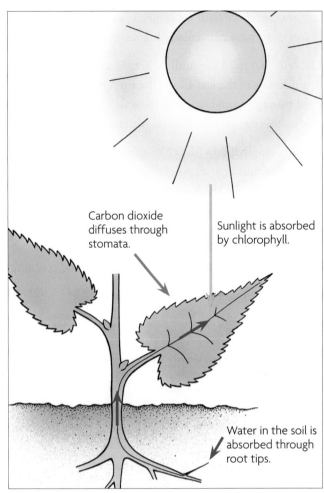

Carbon dioxide diffuses through stomata.

Sunlight is absorbed by chlorophyll.

Water in the soil is absorbed through root tips.

Figure 6.6 How the materials for photosynthesis get into a leaf.

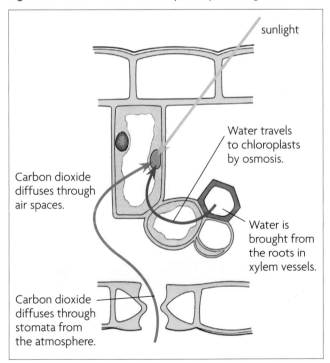

sunlight

Water travels to chloroplasts by osmosis.

Carbon dioxide diffuses through air spaces.

Water is brought from the roots in xylem vessels.

Carbon dioxide diffuses through stomata from the atmosphere.

Figure 6.7 How the raw materials for photosynthesis get into a palisade cell.

one another more than necessary. Plants that live in shady places often have particularly big leaves.

The cells that need the sunlight are the mesophyll cells. The thinness of the leaf allows the sunlight to penetrate right through it, and reach all the cells. To help this the epidermal cells are transparent, with no chloroplasts.

In the mesophyll cells, the chloroplasts are arranged to get as much sunlight as possible, particularly those in the palisade cells. The chloroplasts can lie broadside on to do this, but in strong sunlight, they often arrange themselves end on. This reduces the amount of light absorbed. Inside them, the chlorophyll is arranged on flat membranes (Figure 6.8) to expose as much as possible to the sunlight.

Table 6.1 sums up the adaptations of leaves for photosynthesis.

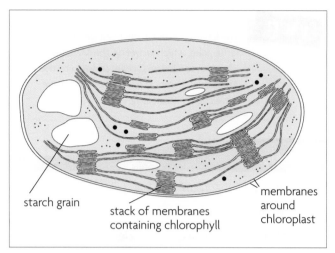

Figure 6.8 The structure of a chloroplast.

Table 6.1 Adaptations of leaves for photosynthesis

Adaptation	Function
supported by stem and petiole	to expose as much of the leaf as possible to the sunlight and air
large surface area	to expose as large an area as possible to the sunlight and air
thin	to allow sunlight to penetrate to all cells; to allow CO_2 to diffuse in and O_2 to diffuse out as quickly as possible
stomata in lower epidermis	to allow CO_2 to diffuse in and O_2 to diffuse out
air spaces in spongy mesophyll	to allow CO_2 and O_2 to diffuse to and from all cells
no chloroplasts in epidermal cells	to allow sunlight to penetrate to the mesophyll layer
chloroplasts containing chlorophyll present in the mesophyll layer	to absorb energy from sunlight, so that CO_2 will combine with H_2O
palisade cells arranged end on	to keep as few cell walls as possible between sunlight and the chloroplasts
chloroplasts inside palisade cells often arranged broadside on	to expose as much chlorophyll as possible to sunlight
chlorophyll arranged on flat membranes inside the chloroplasts	to expose as much chlorophyll as possible to sunlight
xylem vessels within short distance of every mesophyll cell	to supply water to the cells in the leaf, some of which will be used in photosynthesis
phloem tubes within short distance of every mesophyll cell	to take away sucrose and other organic products of photosynthesis

Questions

6.11 What are the raw materials needed for photosynthesis?

6.12 What percentage of the air is carbon dioxide?

6.13 How does carbon dioxide get into a leaf?

6.14 How does a leaf obtain its water?

6.15 Give two reasons why the large surface area of leaves is advantageous to the plant.

6.16 Leaves are thin. What purpose does this serve?

Uses of glucose

6.10 Glucose is used in different ways.

One of the first carbohydrates to be made in photosynthesis is glucose. There are several things that may then happen to it (Figure **6.9**).

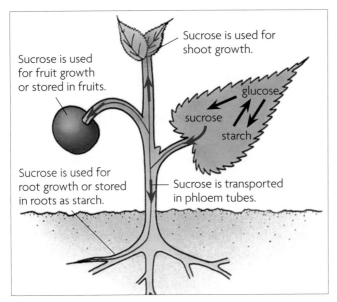

Figure 6.9 The products of photosynthesis.

Used for energy Energy may be released from glucose in the leaf. All cells need energy, which they obtain by the process of respiration (section **9.1**). Some of the glucose which a leaf makes will be broken down by respiration, to release energy.

Stored as starch Glucose may be turned into starch and stored in the leaf. Glucose is a simple sugar (section **4.3**). It is soluble in water, and quite a reactive substance. It is not, therefore, a very good storage molecule. Firstly, being reactive, it might get involved in chemical reactions where it is not wanted. Secondly, it would dissolve in the water in and around the plant cells, and might be lost from the cell. Thirdly, when dissolved, it would increase the concentration of the solution in the cell, which could cause damage.

The glucose is therefore converted into starch to be stored. Starch is a polysaccharide, made of many glucose molecules joined together. Being such a large molecule, it is not very reactive, and not very soluble. It can be made into granules which can be easily stored inside the chloroplasts.

Used to make proteins and other organic substances
Glucose may be used to make other organic substances. The plant can use glucose as a starting point for making all the other organic substances it needs. These include the carbohydrates sucrose and cellulose. Plants also make fats and oils.

Plants can also use the sugars they have made in photosynthesis to make proteins. To do this, they need **nitrogen**. Unfortunately, even though the air around us is 78% nitrogen, this is completely useless to plants because it is very unreactive. Plants have to be supplied with nitrogen in a more reactive form, usually as nitrate ions. They absorb nitrate ions from the soil, through their root hairs, by diffusion and active transport. The nitrate ions combine with glucose to make amino acids. The amino acids are then strung together to form protein molecules.

Another substance that plants make is chlorophyll. Once again, they need nitrogen to do this, and also another element – **magnesium**. The magnesium, like the nitrate ions, is obtained from the soil.

Table **6.2** shows what happens to a plant if it does not have enough of these ions. Figure **6.10** (overleaf) shows what happens when a plant does not have enough nitrogen. Farmers often add extra mineral ions to the soil in which their crops are growing, to make sure that they do not run short of these essential substances. You can read more about this in section **15.11**.

Table 6.2 Mineral ions required by plants

	nitrogen	magnesium
mineral salt	nitrates or ammonium ions	magnesium ions
why needed	to make proteins	to make chlorophyll
deficiency	weak growth, yellow leaves	yellowing between the veins of leaves

Changed to sucrose for transport A molecule has to be small and soluble to be transported easily. Glucose has both of these properties, but it is also rather reactive. It is therefore converted to the complex sugar sucrose to be transported to other parts of the plant. Sucrose molecules

Figure 6.10 These stunted, yellow maize seedlings are suffering from nitrogen deficiency.

are also quite small and soluble, but less reactive than glucose. They dissolve in the sap in the phloem vessels, and can be distributed to whichever parts of the plant need them (Figure **6.9**).

The sucrose may later be turned back into glucose again, to be broken down to release energy, or turned into starch and stored, or used to make other substances which are needed for growth.

Photosynthesis investigations

6.11 Investigations need controls.

If you do Activities **6.3**, **6.4** and **6.5**, you can find out for yourself which substances a plant needs for photosynthesis. In each investigation, the plant is given everything it needs, except for one substance. Another plant is used at the same time. This is a **control**. The control is given everything it needs, including the substance being tested for. Sometimes the control is a leaf, or even a part of a leaf, from the experimental plant. The important thing is that the control has all the substances it needs, while the experimental plant – or leaf – is lacking one substance.

Both plants (or leaves) are then treated in exactly the same way. Any differences between them at the end of the investigation, therefore, must be because of the substance being tested.

At the end of the investigation, test a leaf from your experimental plant and one from your control to see if they have made starch. By comparing them, you can find out which substances are necessary for photosynthesis.

6.12 Plants for photosynthesis investigations must be destarched.

It is very important that the leaves you are testing should not have any starch in them at the beginning of the investigation. If they did, and you found that the leaves contained starch at the end of the investigation, you could not be sure that they had been photosynthesising. The starch might have been made before the investigation began.

So, before doing any of these investigations, you must destarch the plants. The easiest way to do this is to leave them in a dark cupboard for at least 24 hours. The plants cannot photosynthesise while they are in the cupboard because there is no light. So they use up their stores of starch. To be certain that they are thoroughly destarched, test a leaf for starch before you begin your investigation.

6.13 Iodine solution can stain starch in leaves.

Iodine solution is used to test for starch. A blue-black colour shows that starch is present. However, if you put iodine solution onto a leaf which contains starch, it will not immediately turn black. This is because the starch is inside the chloroplasts in the cells (Figure **6.8**, page **54**). The iodine solution cannot get through the cell membranes to reach the starch and react with it.

Another difficulty is that the green colour of the leaf and the brown iodine solution can look black together.

Therefore before testing a leaf for starch, you must break down the cell membranes, and get rid of the green colour (chlorophyll). The way this is done is described in Activity **6.2**. The cell membranes are first broken down by boiling water, and then the chlorophyll is removed by dissolving it out with alcohol.

Activity 6.2
Testing a leaf for starch

skills

C1 Using techniques, apparatus and materials

Safety Wear eye protection if available.
Take care with the boiling water.
Alcohol is **very flammable**. Turn out your
Bunsen flame **before** putting the tube of
alcohol into the hot water.
Use forceps to handle the leaf.

1 Take a leaf from a healthy plant, and drop it into
boiling water in a water bath. Leave for about 30 s.
Turn out the Bunsen flame.

2 Remove the leaf, which will be very soft, and drop
it into a tube of alcohol in the water bath. Leave it
until all the chlorophyll has come out of the leaf.

3 The leaf will now be brittle. Remove it from the
alcohol, and dip it into hot water again to soften it.

4 Spread out the leaf on a white tile, and cover it
with iodine solution. A blue-black colour shows
that the leaf contains starch.

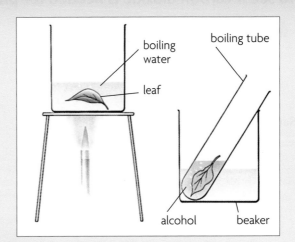

Questions

1 Why was the leaf put into boiling water?

2 Why did the alcohol become green?

3 Why was the leaf put into alcohol *after* being put into
boiling water?

Activity 6.3
To see if chlorophyll is needed for photosynthesis

skills

C1 Using techniques, apparatus and materials

C2 Observing, measuring and recording

C3 Interpreting and evaluating

Safety Wear eye protection if available.
Take care with the boiling water.
Alcohol is **very flammable**. Turn out your
Bunsen flame **before** putting the tube of
alcohol into the hot water.
Use forceps to handle the leaf.

1 Destarch a plant with variegated (green and white)
leaves.

2 Leave your plant in a warm, sunny spot for
a few days.

3 Test one of the leaves for starch (Activity **6.2**).

4 Make a drawing of your leaf before and
after testing.

Questions

1 What was the control in this investigation?

2 What do your results tell you about chlorophyll and
photosynthesis?

Activity 6.4
To see if carbon dioxide is needed for photosynthesis

C1 *Using techniques, apparatus and materials*

C2 *Observing, measuring and recording*

C3 *Interpreting and evaluating*

Safety Wear eye protection if available.
Take care with the boiling water.
Alcohol is **very flammable**. Turn out your
Bunsen flame **before** putting the tube of
alcohol into the hot water.
Use forceps to handle the leaf.

1   Destarch a plant.

2   Set up your apparatus as shown in the diagram.
Take special care that no air can get into the flasks.
Leave the plant in a warm sunny window for
a few days.

3   Test each treated leaf for starch.

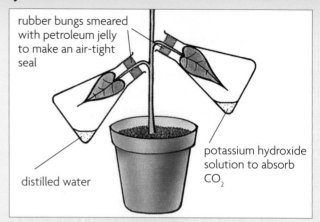

rubber bungs smeared with petroleum jelly to make an air-tight seal

potassium hydroxide solution to absorb CO_2

distilled water

Questions

1   Why was potassium hydroxide put in with one leaf, and water with the other?

2   Which was the control?

3   Why was petroleum jelly put around the bungs?

4   What do your results suggest about carbon dioxide and photosynthesis?

Activity 6.5
To see if light is needed for photosynthesis

C1 *Using techniques, apparatus and materials*

C2 *Observing, measuring and recording*

C3 *Interpreting and evaluating*

Safety Wear eye protection if available.
Take care with the boiling water.
Alcohol is **very flammable**. Turn out your
Bunsen flame **before** putting the tube of
alcohol into the hot water.
Use forceps to handle the leaf.

1   Take a healthy bean or *Pelargonium* plant, growing
in a pot. Leave it in a cupboard for a few days, to
destarch it.

2   Test one of its leaves for starch, to check that it
does not contain any.

3   Using a folded piece of black paper or aluminium
foil, a little larger than a leaf, cut out a shape (see
diagram). Fasten the paper or foil over both sides
of a leaf on your plant, making sure that the edges
are held firmly together. Don't take the leaf off
the plant!

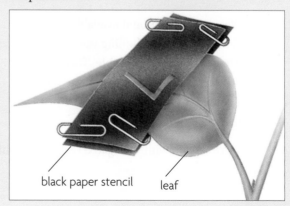

black paper stencil   leaf

(*continued...*)

(... continued)

4   Leave the plant near a warm, sunny window for a few days.

5   Remove the cover from your leaf, and test the leaf for starch.

6   Make a labelled drawing of the appearance of your leaf after testing for starch.

Questions

1   Why was the plant destarched before the beginning of the experiment?

2   Why was part of the leaf left uncovered?

3   What do your results tell you about light and photosynthesis?

Activity 6.6
To show that oxygen is produced in photosynthesis

C1   *Using techniques, apparatus and materials*

C2   *Observing, measuring and recording*

1   Set up the apparatus shown in the diagram. Make sure that the test tube is completely full of water.

2   Leave the apparatus near a warm, sunny window for a few days.

3   Carefully remove the test tube from the top of the funnel, allowing the water to run out, but not allowing the gas to escape.

4   Light a wooden splint, and then blow it out so that it is just glowing. Carefully put it into the gas in the test tube. If it bursts into flame, then the gas is oxygen.

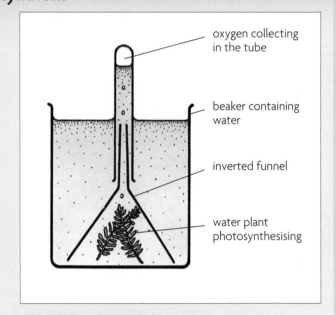

- oxygen collecting in the tube
- beaker containing water
- inverted funnel
- water plant photosynthesising

Questions

1   Why was this investigation done under water?

2   This investigation has no control. Try to design one.

Activity 6.7
Investigating the effect of light intensity on photosynthesis

Safety If you use an electric lamp, keep water well away from it.

If you did Activity **6.6**, you may have noticed that the plant seemed to produce more bubbles in bright sunlight than when it was in the shade. This could mean that the rate of photosynthesis is affected by light intensity.

1 Write down a hypothesis that you will investigate. The hypothesis should be one sentence, and it should describe the relationship that you think exists between light intensity and the rate of photosynthesis.

  You can vary light intensity by moving a light source closer to the plant. The shorter the distance between the light and the plant, the greater the light intensity.

  You can use a water plant in your investigation.

2 Once you have an idea about how you will do your experiment, write it down as a list of points. Then think through it again, and make improvements to your plan. Once you are fairly happy with it, show your teacher. You must not try to do your experiment until your teacher says that you may begin.

- What apparatus and other materials will you need for your experiment?

- What will you vary in your experiment? How will you vary it?

- What will you keep the same in all the tubes or beakers in your experiment? How will you do this?

- What will you measure in your experiment? How will you measure it? When will you measure it? Will you do repeat measurements and calculate a mean?

- How will you record your results? (You can sketch out a results chart, ready to fill in.)

- How will you display your results? (You can sketch the axes of the graph you plan to draw.)

- What will your results be if your hypothesis is correct? (You can sketch the shape of the graph you think you will get.)

Once you have approval from your teacher, you should do your experiment. Most scientific researchers find that they want to make changes to their experiment once they actually begin doing it. This is a good thing to do. Make careful notes about all the changes that you make.

Finally, write up your experiment in the usual way, including:

- a heading, and the hypothesis that you tested

- a diagram of the apparatus that you used, and a full description of your method

- a neat and carefully headed table of results, including means if you decided to do repeats

- a neat and carefully headed line graph of your results

- a conclusion, in which you say whether or not your results support your hypothesis

- a discussion, in which you use what you know about photosynthesis to try to explain the pattern in your results

- an evaluation, in which you explain the main limitations that you feel might have affected the reliability of your data.

Limiting factors

6.14 Many factors affect photosynthesis.

If a plant is given plenty of sunlight, carbon dioxide and water, the limit on the rate at which it can photosynthesise is its own ability to absorb these materials, and make them react. However, quite often plants do not have unlimited supplies of these materials, and so their rate of photosynthesis is not as high as it might be.

Sunlight   In the dark, a plant cannot photosynthesise at all. In dim light, it can photosynthesise slowly. As light intensity increases, the rate of photosynthesis will increase, until the plant is photosynthesising as fast as it can. At this point, even if the light becomes brighter, the plant cannot photosynthesise any faster (Figure **6.11**).

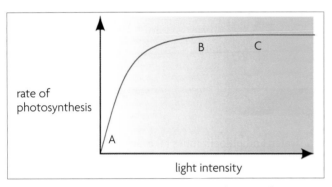

Figure 6.11 The effect of light intensity on the rate of photosynthesis.

Over the first part of the curve in Figure **6.11**, between A and B, light is a **limiting factor**. The plant is limited in how fast it can photosynthesise because it does not have enough light. You can see this because when the plant is given more light it photosynthesises faster.

> **Key definition**
>
> **limiting factor**   something present in the environment in such short supply that it restricts life processes

Between B and C, however, light is not a limiting factor. You can show this because, even if more light is shone on the plant, it still cannot photosynthesise any faster. It already has as much light as it can use.

Carbon dioxide   Carbon dioxide can also be a limiting factor (Figure **6.12**). The more carbon dioxide a plant is given, the faster it can photosynthesise up to a point, but then a maximum is reached.

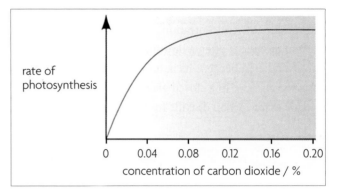

Figure 6.12 The effect of carbon dioxide concentration on the rate of photosynthesis.

Temperature   The chemical reactions of photosynthesis can only take place very slowly at low temperatures (section **5.5**), so a plant can photosynthesise faster on a warm day than on a cold one.

Stomata   The carbon dioxide which a plant uses diffuses into the leaf through the stomata. If the stomata are closed, then photosynthesis cannot take place. Stomata often close if the weather is very hot and sunny, to prevent too much water being lost (Figure **8.35**, page **108**). This means that on a really hot day photosynthesis may slow down.

6.15 Conditions for growing crops can be controlled in glasshouses.

When plants are growing outside, we cannot do much about changing the conditions that they need for photosynthesis. If a field of sorghum does not get enough sunshine, or is short of carbon dioxide, then it just has to stay that way. But if crops are grown in glasshouses, then it is possible to control the conditions so that they are photosynthesising as fast as possible.

For example, in parts of the world where it is often too cold for good growth of some crop plants, they can be grown in heated glasshouses. This is done, for example, with tomatoes. The temperature in the glasshouse can be kept at the optimum level to encourage the tomatoes to grow fast and strongly, and to produce a large yield of fruit that ripens quickly.

Light can also be controlled. In cloudy or dark conditions, extra lighting can be provided, so that light is not limiting the rate of photosynthesis. The kind of lights that are used can be chosen carefully so that they provide just the right wavelengths that the plants need.

Carbon dioxide concentration can also be controlled. Carbon dioxide is often a limiting factor for photosynthesis, because its natural concentration in the air is so very low. In a closed glasshouse, it is possible to provide extra carbon dioxide for the plants.

Questions

6.21 What is meant by a limiting factor?

6.22 Name **two** factors which may limit the rate of photosynthesis of a healthy plant.

6.23 Why do plants sometimes stop photosynthesising on a very hot, dry day?

6.16 The process of photosynthesis is vital to all living things.

Photosynthesis is of importance, not only to green plants, but to all living organisms. It is the basic reaction which brings the energy of the Sun into ecosystems (section **15.3**). The flow of energy in ecosystems is one-way. So there is a constant need for replenishment from the energy source, and therefore a constant need for photosynthesis.

Photosynthesis is also essential for maintaining a constant global level of oxygen and carbon dioxide (section **15.10**). The oxygen given off is available for respiration. Carbon dioxide produced by respiration and from the combustion of fuels is used in photosynthesis, which helps to stop the levels of carbon dioxide in the atmosphere from rising too high.

Key ideas

- Photosynthesis takes place in chloroplasts in the leaves of plants.

- The word equation for photosynthesis is:

$$\text{carbon dioxide} + \text{water} \xrightarrow[\text{chlorophyll}]{\text{sunlight}} \text{simple sugars} + \text{oxygen}$$

The balanced equation is:

$$6CO_2 + 6H_2O \xrightarrow[\text{chlorophyll}]{\text{sunlight}} C_6H_{12}O_6 + 6O_2$$

- Chlorophyll traps energy from light. In photosynthesis, this energy is converted to chemical energy in carbohydrates.

- Photosynthesis takes place in the cells of the mesophyll layer, especially the palisade mesophyll. Leaves are thin and have a large surface area, to speed up the supply of carbon dioxide to the palisade cells and to maximise the amount of sunlight that hits the leaf and can be absorbed by chlorophyll. Stomata and air spaces allow carbon dioxide to diffuse quickly from the air to the chloroplasts. Xylem vessels bring water, and phloem tubes take away the products of photosynthesis.

- Some of the glucose that is made is used in respiration, to provide energy to the plant cells. Some is stored as starch. Some is used to make cellulose for cell walls. Some is transported around the plant in the form of sucrose, in the phloem tubes. Some is combined with nitrate or ammonium ions to make proteins. Some is used to make other substances such as fats. With the addition of magnesium ions, chlorophyll can be made.

- When testing a leaf for starch, it must first be boiled to break down cell membranes and allow iodine solution to make contact with any starch inside the cells. Hot alcohol will remove chlorophyll from the leaf, making it easier to see any colour changes.

- Plants need light and carbon dioxide for photosynthesis.

- If either light or carbon dioxide are in short supply, they limit the rate of photosynthesis and are said to be limiting factors. The rate of photosynthesis is also affected by temperature.

Revision questions

1 Copy and complete this table to show how, and for what purpose, plants obtain these substances.

	obtained from	used for
nitrates		
water		
magnesium		
carbon dioxide		

2 Explain how each of the following helps a leaf to photosynthesise.
   a There is an air space behind each stoma.
   b The epidermal cells of a leaf do not have chloroplasts.
   c Leaves have a large surface area.
   d The veins in a leaf branch repeatedly.
   e A living leaf containing starch does not turn blue-black when you put iodine solution onto it.
   f Chloroplasts have many membranes in them.

3 Which carbohydrate does a plant use for each of these purposes? Explain why.
   a transport
   b storage

4 Describe how a carbon atom in a carbon dioxide molecule in the air could become part of a starch molecule in a carrot root. Mention all the structures it would pass through, and what would happen to it at each stage.

5 Read the following passage carefully, and then answer the questions, using both the information in the passage and your own knowledge.

White light is made up of all the colours of the rainbow. Sea water acts as a light filter which screens off some of the light energy, starting at the red end of the spectrum. As sunlight travels downwards through the water, first the red light is lost, then green and yellow and finally blue. In very clear water, the blue light can penetrate to a maximum of 1000 m. Below this, all is dark.

The upper layers of the sea contain a large community of microscopic floating organisms called plankton, many of which are tiny plants known as phytoplankton. These act as a gigantic solar cell, which feeds all the animals of the sea and supplies both them and the atmosphere above with oxygen.

Nearer the shore, larger plants are found. Seaweeds grow on rocky shores, brown and green ones high on the shore, and red ones lower down, where they are covered with deep water when the tide is in. The colours of the seaweeds are due to their light-absorbing pigments, not all of which are chlorophyll.

   a Why are no green plants found below the upper few hundred metres of the sea?
   b Some living organisms are found in the permanently dark depths of the oceans. What might they feed on?
   c What are phytoplankton?
   d Explain as fully as you can the last sentence of paragraph 2, 'These act as ... with oxygen.'
   e Chlorophyll is a green pigment. Which colours of light does it (i) absorb, and (ii) reflect?
   f What colour light would you expect the pigment of red seaweeds to absorb?
   g Why are red seaweeds normally found lower down the shore than green ones?

7 Animal nutrition

In this chapter, you will find out:

◆ what is meant by a balanced diet
◆ the different types of nutrients we should eat, and foods that are good sources of them
◆ that different people need different amounts of energy in their diet
◆ some problems that can be caused by eating an unbalanced diet
◆ about producing food by agriculture
◆ why we need to digest the food that we eat
◆ the structure of the alimentary canal, and the functions of each of its parts
◆ how digested food is absorbed and assimilated.

Diet

7.1 Animals need food made by plants.

Animals get their food from other organisms – from plants or other animals. They cannot make their own food as plants do.

The food an animal eats every day is called its **diet**. Most animals need seven types of nutrients in their diets. These are:

carbohydrates	minerals
proteins	water
fats	fibre
vitamins	

A diet which contains all of these things, in the correct amounts and proportions, is called a **balanced diet**.

7.2 A balanced diet provides the right amount of energy.

Every day, a person uses up energy. The amount you use partly depends on how old you are, which sex you are and what job you do. A few examples are shown in Figure 7.1.

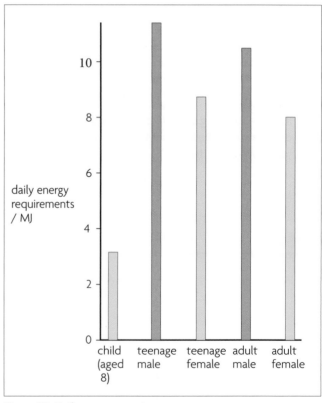

Figure 7.1 Daily energy requirements.

The energy you use each day comes from the food you eat. If you eat too much food, some of the extra will probably be stored as fat. If you eat too little, you may not be able to obtain as much energy as you need. This will make you feel tired.

All food contains some energy. Scientists have worked out how much energy there is in particular kinds of food. You can look up this information. A few examples are given in Table 7.1. You may remember that one gram of fat contains about twice as much energy as one gram of protein or carbohydrate (section 4.9). This is why fried foods should be avoided if you are worried about putting on weight.

7.3 Diets should contain a variety of food.

As well as providing you with energy, food is needed for many other reasons. To make sure that you eat a balanced diet you must eat foods containing carbohydrate, fat and protein. You also need each kind of vitamin and mineral, fibre and water. If you miss out on any of these things, your body will not be able to work properly.

7.4 Vitamins and minerals are needed in small amounts.

Vitamins are organic substances which are only needed in tiny amounts. If you do not have enough of a vitamin, you may get a deficiency disease. Table 7.2 (overleaf) provides information about vitamins C and D.

Minerals are inorganic substances. Once again, only small amounts of them are needed in the diet. Table 7.3 shows two of the most important ones.

7.5 Fibre prevents constipation.

Fibre helps to keep the alimentary canal working properly. Food moves through the alimentary canal (section 7.23) because the muscles contract and relax to squeeze it along. This is called **peristalsis** (Figure 7.18, page 78). The muscles are stimulated to do this when there is food in the alimentary canal. Soft foods do not stimulate the muscles very much. The muscles work more strongly when there is harder, less digestible food, like fibre, in the alimentary canal. Fibre keeps the digestive system in good working order, and helps to prevent constipation.

All plant foods, such as fruits and vegetables, contain fibre. This is because the plant cells have cellulose cell walls. Humans cannot digest cellulose.

Table 7.1 Energy content of some different kinds of food

food	kJ / 100 g	food	kJ / 100 g
Breakfast foods		**Main meals**	
cornflakes	1567	stewed steak	932
oatmeal	1698	roast chicken	599
boiled egg	612	fish (fresh)	340
brown bread	948	fish (dried, salt)	1016
white bread	991	sardines	906
milk	272	fried liver	1016
sugar	1682	cheddar cheese	1682
marmalade	1035		
unsweetened fruit juice	143	cottage cheese	402
		baked beans	270
Desserts		cabbage	66
pawpaw	160	carrots	98
bananas	326	lettuce	36
melon	96	peas	161
oranges	150	boiled white (Irish) potatoes	339
canned peaches	373		
		french fries	1065
ice cream	698	tomatoes	60
custard	496	rice	1536
Snacks		spaghetti	1612
chocolate	2214	lentils	1293
fruit yoghurt	405		
plain biscuits	1925		
chocolate biscuits	2197		
roast peanuts	2364		

One common form of fibre is the outer husk of cereal grains, such as oats, wheat and barley. This is called **bran**. Some of this husk is found in wholemeal bread. Brown or unpolished rice is also a good source of fibre.

Table 7.2 Vitamins

vitamin	foods that contain it	why it is needed	deficiency disease
C	citrus fruits (such as oranges, limes), raw vegetables	to make the stretchy protein collagen, found in skin and other tissues; keeps tissues in good repair	scurvy, which causes pain in joints and muscles, and bleeding from gums and other places; this used to be a common disease of sailors, who had no fresh vegetables for long voyages
D	butter, egg yolk (and can be made by the skin when sunlight falls on it)	helps calcium to be absorbed, for making bones and teeth	rickets, in which the bones become soft and deformed; this disease was common in young children in industrial areas, who rarely got out into the sunshine

Table 7.3 Minerals

mineral element	foods that contain it	why it is needed	deficiency disease
calcium, Ca	milk and other dairy products, bread	for bones and teeth; for blood clotting	brittle bones and teeth; poor blood clotting
iron, Fe	liver, red meat, egg yolk, dark green vegetables	for making haemoglobin, the red pigment in blood which carries oxygen	anaemia, in which there are not enough red blood cells so the tissues do not get enough oxygen delivered to them

7.6 Too much saturated fat increases the risk of heart disease.

The kind of fat found in animal foods is called **saturated fat**. These foods also contain **cholesterol**. Some research suggests that people who eat a lot of saturated fat and cholesterol are more likely to get heart disease than people who do not. This is because fat deposits build up on the inside of arteries, making them stiffer and narrower. If this happens in the coronary arteries supplying the heart muscle with blood, then not enough blood can get through. The heart muscles run short of oxygen and cannot work properly. This is called **coronary heart disease**. The deposits can also cause a blood clot, which results in a **heart attack** (section 8.4).

Dairy products such as milk, cream, butter and cheese contain a lot of saturated fat. So do red meat and eggs (Figure 7.2). But vegetable oils are usually unsaturated fats. These, and also oils from fish (Figure 7.3), do not increase the risk of heart disease, so it is sensible to use these instead of animal fats when possible.

Figure 7.2 These foods are high in fat. Small amounts in a diet are fine, but too much could increase the risk of coronary heart disease.

Figure 7.3 Oils from fish and plants are good for you.

Vegetable oil can be used for frying instead of butter or lard. Polyunsaturated spreads can be used instead of butter.

Fish and white meat such as chicken do not contain much saturated fat, so eating more of these and less red meat may help to cut down the risk of heart disease.

7.7 Obesity causes health problems.

People who take in more energy than they use up get fat. Being very fat is called **obesity** (Figure 7.4). Obesity is dangerous to health. Obese people are more likely to get heart disease, strokes, diabetes and many other problems.

Figure 7.4 Being very overweight increases the risk of many different, and serious, health problems. Weight around your middle has been shown to be clearly linked to heart disease.

Most people can control their weight by eating normal, well-balanced meals and taking regular exercise. Crash diets are not a good idea, except for someone who is very overweight. Although a person may manage to lose a lot of weight quickly, he or she will almost certainly put it on again once he or she stops dieting.

7.8 Starvation and malnutrition are different.

In many countries in the world, there is no danger of people suffering from obesity. In some parts of Africa, for example, several years of drought can mean that the harvests do not provide enough food to feed all the people. Despite help from other countries, many people have died from **starvation**.

Even if there is enough food to keep people alive, they may suffer from **malnutrition**. Malnutrition is caused by not eating a balanced diet. One common form of malnutrition is **kwashiorkor** (Figure 7.5). This is caused by a lack of protein in the diet. It is most common in children between the ages of nine months and two years, after they have stopped feeding on breast milk.

Figure 7.5 The older boy is thin, but has a swollen abdomen, suggesting he is suffering from kwashiorkor. This photo was taken at a refugee camp in Ethiopa.

Kwashiorkor is often caused by poverty, because the child's carers do not have any high-protein food to give to the child. But sometimes it is caused by a lack of knowledge about the right kinds of food that should be eaten.

Children suffering from kwashiorkor are always underweight for their age. But they may often look quite fat, because their diet may contain a lot of carbohydrate. If they are put onto a high-protein diet, they usually begin to grow normally again.

Questions

7.1 A balanced diet contains these nutrients:

carbohydrates   fats   proteins
vitamins   minerals   water

  a Which of these nutrients are organic, and which are inorganic?
  b Which of these nutrients can provide energy?
  c What is the role of fibre in the diet?

7.2 List **three** health problems associated with obesity.

7.3 What is coronary heart disease?

7.4 What is the difference between starvation and malnutrition?

7.5 What is meant by a deficiency disease?

7.6 Give **two** examples of deficiency diseases.

Food production

7.9 There are problems with world food supplies.

It has been calculated that more than enough food is produced on Earth to provide every single person with more than enough for their needs. Yet many people do not get enough food. Each year, many people – both children and adults – die because they have an inadequate diet.

The fundamental problem is that, while some parts of the world produce more than enough food for the people that live there, in other parts of the world nowhere near enough food is produced. Food is distributed unequally on our planet. Although large amounts of food are transported from one area to another, this is still not sufficient to supply enough food to everybody. Also, if food prices rise too high, then even if there is plenty of food around, many people may not be able to afford to buy it.

Famines can occur for many different reasons. Often, the main cause is the weather. If an area suffers drought for several years in succession, then it becomes impossible for the people to grow crops. Their animals die, too. Sometimes, however, the problem is exactly the opposite – so much rain falls that it causes flooding, again preventing crops from growing (Figure 7.6). Sometimes, even though the weather remains normal, the human population may grow so large that the land on which they live can no longer provide enough food for them. Sometimes, wars raging in an area prevent people from working the land and harvesting their crops.

Figure 7.6 These people, living near the village of Muzaffarpur in northern India, had no food for four days after flooding swept away their houses and drowned their farmland. These floods happened in 2007 and were the worst in living memory.

When the world becomes aware that an area is suffering from famine, other countries are usually very willing to donate food supplies to the people. Hopefully, this will only need to happen for a relatively short time, until things improve and people can plant their crops and become self-sufficient again. Most people would much prefer this, rather than having to rely on hand-outs of food.

7.10 Agriculture produces most of our food.

Most of the world's supply of food is produced by growing crops or by keeping animals. During the last century, the quantity of food produced has increased greatly. Much of this increase is due to the development of modern technology which allows more crops to be grown, or animals to be kept, on the same area of land (Figures 7.7 and 7.8). The improvements are brought about through the use of modern agricultural machinery, chemical fertilisers, pesticides and herbicides. You can read about their benefits, and some of the problems that they can cause, in Chapter **16**.

Figure 7.7 This farmer in the Philippines is preparing land for planting rice. His work is labour intensive, as people in this region cannot afford much machinery.

Figure 7.8 Here in the US, farmers have tractors and other machinery to work the land. These farmers are preparing the soil for sowing seeds.

We have also made changes to the crops and animals themselves. Selective breeding has been used for centuries, and has produced varieties of plants and animals that have features we desire, such as high yields of grain or milk. You can find out about selective breeding in Chapter 14.

7.11 Microorganisms are used to make yoghurt.

In many parts of the world, milk from cattle, sheep, goats or buffalo is used to make yoghurt.

Yoghurt is produced when a bacterium called *Lactobacillus bulgaricus* is added to milk. *Lactobacillus* uses sugar from milk as its energy source. This sugar is called **lactose**. *Lactobacillus* converts the lactose to lactic acid.

lactose ⟶ lactic acid + energy

Lactic acid, like all acids, tastes sour. However, most people like its taste. The presence of the lactic acid lowers the pH of the milk. This affects the proteins in the milk. They coagulate, forming clumps. The milk separates out into these clumps, called curds, and a liquid, called whey.

A culture of the bacterium is simply added to warm milk, and left for a few hours. Usually, the milk is heated to around 70 °C, and then cooled, before the *L. bulgaricus* is added. The heating is done to kill any other microorganisms in the milk, which might also ferment it, making different, unpleasant-tasting substances.

Activity 7.1
Making yoghurt

skills
C1 *Using techniques, apparatus and materials*
C2 *Observing, measuring and recording*
C3 *Interpreting and evaluating*
C4 *Planning*

Safety If you are able to do this practical in a room used for food preparation, then you could taste your yoghurt. However, if you are working in a science laboratory, then it is not safe to do this. You should not eat your yoghurt if you have added anything to your milk and starter culture mixture, such as enzymes, just in case they make you ill.

Here is a basic yoghurt recipe. You will need to make sure that all the apparatus you use has been sterilised. You can do this by washing it with very hot water. If not, then other microorganisms will act on the milk, producing substances which you do not want.

1   Collect some 'live' yoghurt. This is yoghurt which has not been heat treated, so it still contains living *Lactobacillus*. This is your 'starter culture'. You will need roughly 1 cm³ of starter culture for each sample of yoghurt you make.

2   Measure 10 cm³ of milk into a sterile container, such as a test tube.

3   Add the starter culture, and mix gently. Cover the tube with cling film, to stop any other microorganisms getting in.

4   Stand your tube in a water bath or incubator at about 40 °C, and leave for approximately two hours for the bacteria to turn the milk into yoghurt.

Many factors affect the speed at which the yoghurt is formed, and the kind of yoghurt which is made. You can tell what is happening just by looking at your milk/yoghurt, or you could test its pH (because the bacteria are producing lactic acid). Choose two of the following factors, and investigate how they affect the rate of action of the bacteria, and/or the final properties of the yoghurt which is produced.

- the type of milk used
- the type of starter culture used
- whether air can get to the milk or not
- the temperature at which the milk is kept
- adding lactase to the milk

7.12 Microorganisms can be used as food.

In the production of yoghurt, we use microorganisms to change one substance into another, which we use as food. But we can also use the microorganisms themselves as food.

There are many good arguments for doing this. In many parts of the world, there is a shortage of food, especially protein-rich food. Microorganisms could provide a good source of protein in these areas. Microorganisms do not need soil to grow in. They can use many different substances as food sources, including wastes from other processes – so they could be grown very cheaply. Producing food in this way wastes less energy than producing meat, because it 'taps in' to an earlier stage in the food chain (section **15.5**).

The first attempts to make microorganisms into food used yeast. In Germany during World War I, yeast was cultured in large vats, using molasses as a food source for the yeast, to produce a protein supplement for people. More recently, different kinds of protoctists (usually single-celled photosynthetic ones) and bacteria have been grown for food production. The food made from all of these microorganisms is known as **single cell protein** or SCP.

However, there have been big problems in selling SCP as food for people. People are very suspicious of eating microorganisms – even though they like eating yoghurt! The first SCPs also tasted rather unpleasant, partly because they contained a lot of DNA and RNA, which taste bitter. Most SCPs are now marketed as animal feed.

One SCP which has found a market as human food, however, is **mycoprotein**. This is made from a fungus. In Britain, the fungus which is used is called *Fusarium*. Its structure is rather like that of *Penicillium* (Figure **5.4**, page **47**), so it is made of hyphae rather than single cells. But mycoprotein is still often called SCP.

The *Fusarium* is grown in large vats, using carbohydrates as a food source, with other nutrients such as ammonium nitrate added as well. The carbohydrates often come from waste left over from making flour. *Fusarium* reproduces quickly and makes a mass of mycelium, which is harvested and treated to remove a lot of the RNA it contains. Then it is dried, and shaped into chunks or cakes, ready for eating as it is, or for making into pies or

Figure 7.9 Mycoprotein can be made to look like pieces of meat. It is quite bland and can be cooked with other ingredients.

other foods. Some people think that mycoprotein looks and tastes a bit like chicken (Figure **7.9**). If you have not seen any, have a look for some next time you are shopping in a supermarket. In Britain it is sold as Quorn®.

Mycoprotein is an excellent food. It has a high protein content, very little fat, no cholesterol, and a lot of fibre. Because the mycelium of the fungus is made up of long thread-like hyphae, mycoprotein has a fibrous texture which many people like – because it is a bit like meat. It has quite a bland taste, and can easily be flavoured to make a pleasant-tasting food.

7.13 Food additives can be helpful.

A **food additive** is something which is added to the food for reasons other than nutrition. The main kinds of food additives, and the reasons for adding them to food, are listed in Table **7.4**.

In Europe, each permitted food additive is given an E number. This is meant to reassure people, because it shows that each additive has been tested and passed as safe. However, many people do not like the idea of anything 'unnatural' being added to food, and so E numbers have come to be viewed with suspicion.

In some cases, this suspicion may be justified. The orange food colouring tartrazine, for example, does appear to cause behavioural problems in some children. Food colourings do not improve the food in any way.

However, many food additives are really very good for us. Ascorbic acid, for example, added to many foods to help it to keep well and not go brown, is vitamin C. And without preservatives, there would be many more cases of food poisoning each year.

Table 7.4 Some examples of food additives

type of additive	example	types of food	function of additive	notes
flavourings	monosodium glutamate	soups, stock cubes, many convenience foods	enhances the flavour of savoury foods; provides umame flavour	an amino acid that occurs naturally in many foods and has been used in Chinese cooking for centuries
	vanillin	desserts, chocolate, cakes	gives vanilla taste	natural vanillin comes from the fruits of the vanilla orchid, but most is now made artificially
colourings	tartrazine	sweets, drinks	gives yellow or orange colour	can cause hyperactivity in some children, and is no longer used in many countries
	caramel	sweets, drinks, soups	gives brown colour	caramel is made by heating sugar
preservatives	sulfur dioxide	fruit juice, dried fruit	kills bacteria and preserves the vitamin C content of the food	
	sodium nitrite	meat products, for example sausages	stops growth of harmful bacteria	there is some evidence that very large amounts of nitrite in the diet may increase the risk of cancer
	ascorbic acid	fruits, meat	stops browning – it is an antioxidant	ascorbic acid is vitamin C; it is also used as a flour improver, as it helps bread dough to rise
emulsifiers	lecithin	powdered milk	stops oil and water separating out into different layers	

Questions

7.7 What kind of microorganism is used to make yoghurt?

7.8 What gives yoghurt its sour taste?

7.9 What is mycoprotein?

7.10 Mycoprotein is sometimes known as single cell protein. Why is this not a very good name?

7.11 Outline two benefits and two potential hazards of food additives.

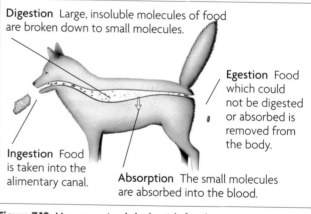

Digestion Large, insoluble molecules of food are broken down to small molecules.

Egestion Food which could not be digested or absorbed is removed from the body.

Ingestion Food is taken into the alimentary canal.

Absorption The small molecules are absorbed into the blood.

Figure 7.10 How an animal deals with food.

Digestion

7.14 Digestion makes nutrients easier to absorb.

The alimentary canal of a mammal is a long tube running from one end of its body to the other (Figure 7.10). Before food can be of any use to the animal, it has to get out of the alimentary canal and into the bloodstream. This is called **absorption**. To be absorbed, molecules of food have to get through the walls of the alimentary canal. They need to be quite small to be able to do this.

The food that is eaten by mammals usually contains some large molecules (Chapter 4). Before these molecules can be absorbed, they must be broken down into small ones. This is called **digestion**.

> **Key definition**
>
> **digestion** the break-down of large, insoluble food molecules into small, water-soluble molecules using mechanical and chemical processes

7.15 Not all foods need digesting.

Figure 7.11 shows what happens to the three kinds of nutrients that need to be digested – fats, proteins and carbohydrates. Look at one column at a time, and work down it, to follow what happens to that type of food as it passes through the alimentary canal.

Large carbohydrate molecules, such as polysaccharides, have to be broken down into simple sugars (monosaccharides). Proteins are broken down to amino acids. Fats are broken down to fatty acids and glycerol.

Simple sugars, water, vitamins and minerals are small molecules, and they can be absorbed just as they are. They do not need to be digested.

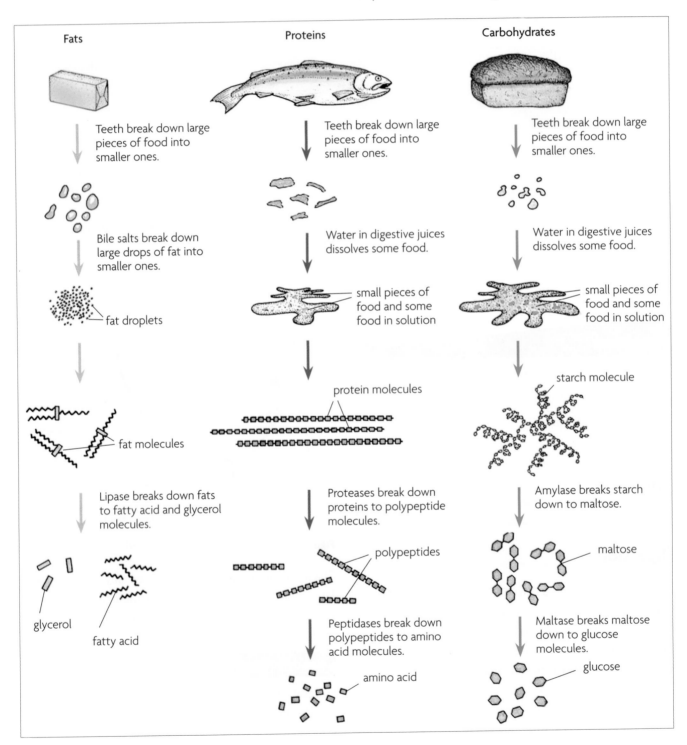

Figure 7.11 Digestion.

7.16 Digestion may be mechanical and chemical.

Often the food an animal eats is in quite large pieces. These need to be broken up by teeth, and by churning movements of the alimentary canal. This is called **mechanical digestion**.

Once pieces of food have been ground up, the large molecules present are then broken down into small ones. This is called **chemical digestion**. It involves a chemical change from one sort of molecule to another. Enzymes are involved in this process (Chapter 5). Figure **7.11** summarises how mechanical and chemical digestion work together to produce small molecules the body can use.

Questions

7.12 What is digestion?

7.13 Name **two** groups of food that do not need to be digested.

7.14 What does digestion change each of these kinds of food into: **(a)** polysaccharides, **(b)** proteins, **(c)** fats?

7.15 What is meant by chemical digestion?

Digestion in humans – teeth

7.17 Teeth are important for the ingestion and mechanical digestion of food.

Teeth help with the **ingestion** and mechanical digestion of the food we eat.

Key definition

ingestion   taking substances (e.g. food, drink) into the body through the mouth

Teeth can be used to bite off pieces of food. They then chop, crush or grind them into smaller pieces. This gives the food a larger surface area, which makes it easier for enzymes to work on the food in the digestive system. It also helps to dissolve soluble parts of the food.

The structure of a tooth is shown in Figure **7.12**. The part of the tooth which is embedded in the gum is called the **root**. The part which can be seen is the **crown**. The crown

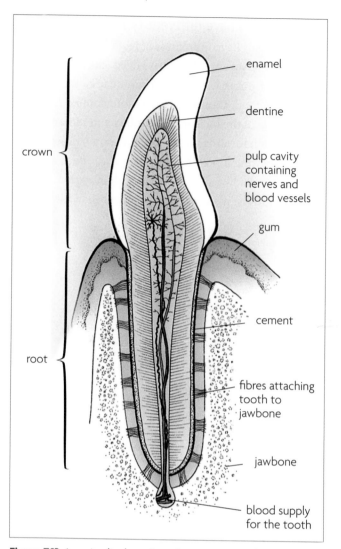

Figure 7.12 Longitudinal section of an incisor tooth.

is covered with **enamel**. Enamel is the hardest substance made by animals. It is very difficult to break or chip it. However, it can be dissolved by acids. Bacteria feed on sweet foods left on the teeth. This makes acids, which dissolve the enamel and decay sets in.

Under the enamel is a layer of **dentine**, which is rather like bone. Dentine is quite hard, but not as hard as enamel. It has channels in it which contain living cytoplasm.

In the middle of the tooth is the **pulp cavity**. It contains nerves and blood vessels. These supply the cytoplasm in the dentine with food and oxygen.

The root of the tooth is covered with **cement**. This has **fibres** growing out of it. These attach the tooth to the jawbone, but allow it to move slightly when biting or chewing.

7.18 Mammals have different types of teeth.

Most mammals have four kinds of teeth (Figures 7.13 and 7.14). **Incisors** are the sharp-edged, chisel-shaped teeth at the front of the mouth. They are used for biting off pieces of food. **Canines** are the more pointed teeth at either side of the incisors. **Premolars** and **molars** are the large teeth towards the back of the mouth. They are used for chewing food. The ones right at the back are sometimes called wisdom teeth. They do not grow until much later in the mammal's development than the others.

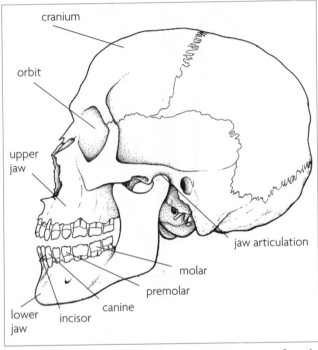

Figure 7.13 A human skull, showing the different types of teeth.

7.19 Mammals have two sets of teeth in their life.

Mammals also differ from other animals in having two sets of teeth. The first set is called the **milk teeth** or **deciduous teeth**. In humans, these start to grow through the gum, one or two at a time, when a child is about five months old. By the age of 24 to 30 months, most children have a set of 20 teeth.

This first set of teeth begins to fall out when the child is about seven years old. Twenty teeth to replace the ones which fall out, plus 12 new teeth, make up the complete set of **permanent teeth**. There are 32 altogether. Most people have all their permanent teeth by about 17 years of age.

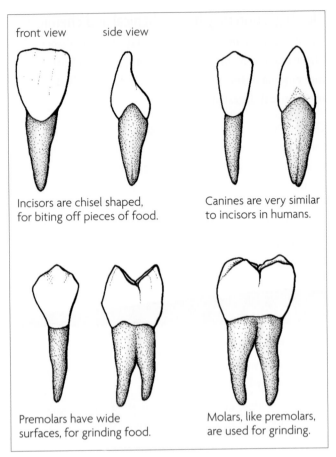

Figure 7.14 Types of human teeth.

7.20 Plaque causes tooth decay.

Tooth decay and gum disease are common problems. Both are caused by bacteria. You have large numbers of bacteria living in your mouth, most of which are harmless. However, some of these bacteria, together with substances from your saliva, form a sticky film over your teeth, especially next to the gums and in between the teeth. This is called **plaque**.

Plaque is soft and easy to remove at first, but if it is left it hardens to form **tartar**, which cannot be removed by brushing.

Gum disease If plaque is not removed, the bacteria in it may infect the gums. The gums swell, become inflamed, and may bleed when you brush your teeth. This is usually painless, but if the bacteria are allowed to spread they may work down around the root of the tooth. The tooth becomes loose, and needs removing (Figure 7.15).

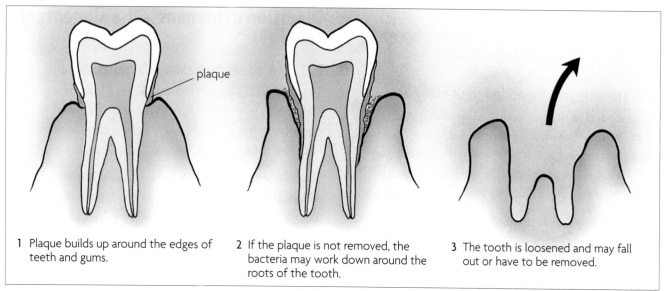

1 Plaque builds up around the edges of teeth and gums.

2 If the plaque is not removed, the bacteria may work down around the roots of the tooth.

3 The tooth is loosened and may fall out or have to be removed.

Figure 7.15 Gum disease.

Tooth decay If sugar is left on the teeth, bacteria in the plaque will feed on it, changing it into acid. The acid gradually dissolves the enamel covering the tooth, and works its way into the dentine (Figure **7.16**). Dentine is dissolved away more rapidly than the enamel. If nothing is done about it, the tooth will eventually have to be taken out.

7.21 Tooth decay and gum disease can be prevented.

There are several easy things which you can do to keep your teeth and gums healthy and free from pain.

1 Don't eat too much sugar. If you never eat any sugar, you will not have tooth decay. But nearly everyone enjoys sweet foods, and if you are careful you can still eat them without damaging your teeth. The rule is to eat sweet things only once or twice a day, preferably with your meals. The worst thing you can do is to suck or chew sweet things all day long. And don't forget that many drinks also contain a lot of sugar.

2 Use a fluoride toothpaste regularly. Fluoride makes your teeth more resistant to decay. Drinking water which contains fluoride, or brushing teeth with a fluoride toothpaste, makes it much less likely that you will have to have teeth filled or extracted.

Regular and thorough brushing also helps to remove plaque, which will prevent gum disease and reduce decay.

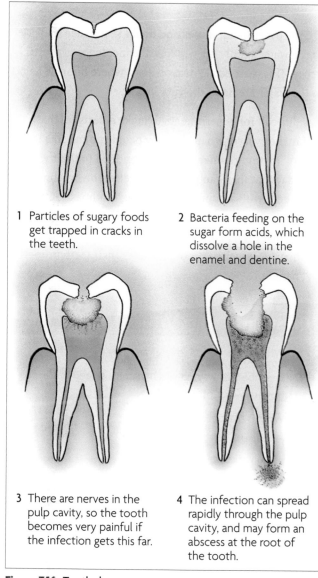

1 Particles of sugary foods get trapped in cracks in the teeth.

2 Bacteria feeding on the sugar form acids, which dissolve a hole in the enamel and dentine.

3 There are nerves in the pulp cavity, so the tooth becomes very painful if the infection gets this far.

4 The infection can spread rapidly through the pulp cavity, and may form an abscess at the root of the tooth.

Figure 7.16 Tooth decay.

3 Make regular visits to a dentist. Regular dental check-ups will make sure that any gum disease or tooth decay is stopped before it really gets a hold.

7.22 Fluoride may be added to drinking water.

In some parts of the world, fluoride is added to drinking water. This is done because fluoride helps to reduce tooth decay. If a person has fluoride ions in their saliva, the fluoride helps any damage to the tooth to be repaired. It does this by forming a hard mineral in the tooth enamel.

However, some people do not like the idea of adding fluoride to drinking water. They feel that they ought to be able to make the choice about whether the water they drink contains extra fluoride or not. Also, too much fluoride can cause problems of its own. It may make the teeth go black.

On the other hand, there is no doubt that added fluoride in the water supply does reduce the incidence of tooth decay. This is especially true in areas where fluoride levels in the water are naturally very low.

It is difficult for governments to make a decision that will make everyone happy. Many people do not like the idea of anything unnecessary being added to their water. They say that they can get the same effect by brushing their teeth with a fluoride-containing toothpaste. Others think that it is a good idea to add fluoride, because this means that even people who cannot afford expensive toothpastes, or who do not look after their teeth properly, will benefit.

Questions

7.16 What are incisors, and what are they used for?

7.17 Describe **two** ways in which mammals' teeth differ from those of other animals.

7.18 What is plaque?

7.19 Explain how plaque can cause:
 a gum disease
 b tooth decay.

Digestion in humans – the alimentary canal

7.23 The alimentary canal is a muscular tube.

The **alimentary canal** (Figure 7.17) is a long tube which runs from the mouth to the anus. It is part of the **digestive system**. The digestive system also includes the **liver** and the **pancreas**.

The wall of the alimentary canal contains muscles, which contract and relax to make food move along. This movement is called **peristalsis** (Figure 7.18, page 78).

Sometimes, it is necessary to keep the food in one part of the alimentary canal for a while, before it is allowed to move to the next part. Special muscles can close the tube completely in certain places. They are called **sphincter muscles**.

To help the food to slide easily through the alimentary canal, it is lubricated with **mucus**. Mucus is made in goblet cells which occur along the alimentary canal.

Each section of the alimentary canal has its own part to play in the digestion, absorption, and egestion of food.

7.24 In the mouth, food is mixed with saliva.

Food is ingested using the teeth, lips and tongue. The teeth then bite or grind the food into smaller pieces, increasing its surface area.

The tongue mixes the food with saliva, and forms it into a **bolus**. The bolus is then swallowed.

Saliva is made in the salivary glands. It is a mixture of water, mucus and the enzyme **amylase**. The water helps to dissolve substances in the food, allowing us to taste them. The mucus helps the chewed food to bind together to form a bolus, and lubricates it so that it slides easily down the oesophagus when it is swallowed. Amylase begins to digest starch in the food to maltose. Usually, it does not have time to finish this because the food is not kept in the mouth for very long. However, if you chew something starchy (such as a piece of bread) for a long time, you may be able to taste the sweet maltose that is produced.

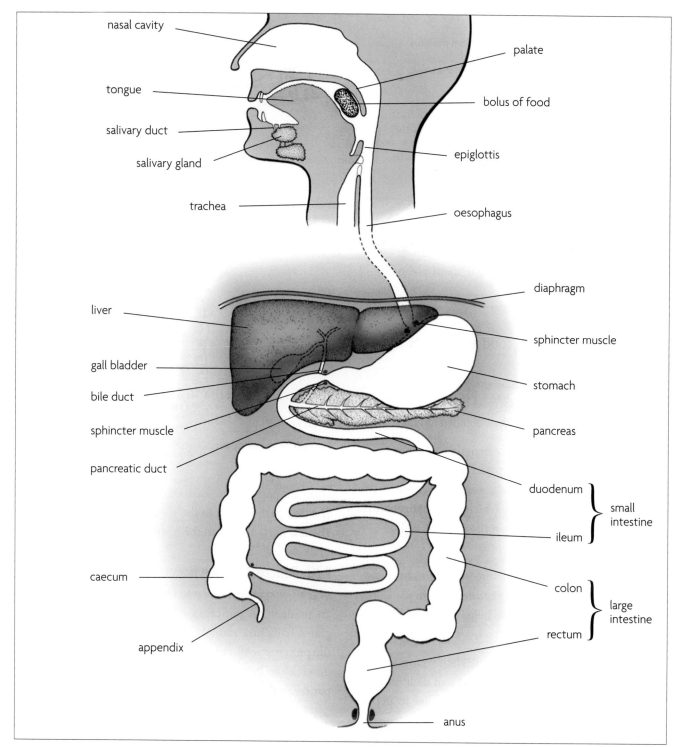

Figure 7.17 The human digestive system.

7.25 The oesophagus carries food to the stomach.

There are two tubes leading down from the back of the mouth. The one in front is the **trachea** or windpipe, which takes air down to the lungs. Behind the trachea is the **oesophagus**, which takes food down to the stomach.

When you swallow, a piece of cartilage covers the entrance to the trachea. It is called the **epiglottis**, and it stops food from going down into the lungs.

The entrance to the stomach from the oesophagus is guarded by a ring of muscle called a **sphincter**. This muscle relaxes to let the food pass into the stomach.

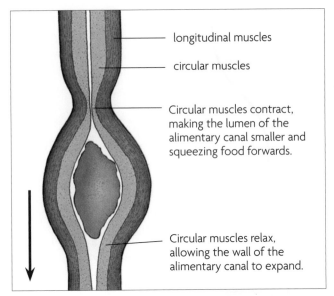

- longitudinal muscles
- circular muscles
- Circular muscles contract, making the lumen of the alimentary canal smaller and squeezing food forwards.
- Circular muscles relax, allowing the wall of the alimentary canal to expand.

Figure 7.18 Peristalsis.

7.26 The stomach stores food and digests proteins.

The **stomach** has strong, muscular walls. The muscles contract and relax to churn the food and mix it with the enzymes and mucus. The mixture is called **chyme**.

Like all parts of the alimentary canal, the stomach wall contains **goblet cells** which secrete mucus. It also contains other cells which produce enzymes called **pepsin** and **rennin**, and others which make **hydrochloric acid**. These are situated in pits in the stomach wall (Figure 7.19).

Pepsin is a protease. It begins to digest proteins by breaking them down into polypeptides. Pepsin works best in acid conditions. The acid also helps to kill any bacteria in the food.

Rennin is only produced in the stomach of young mammals. It causes milk that they get from their mothers to clot. The milk proteins are then broken down by pepsin.

The stomach can store food for quite a long time. After one or two hours, the sphincter at the bottom of the stomach opens and lets the chyme into the duodenum.

7.27 The small intestine is very long.

The **small intestine** is the part of the alimentary canal between the stomach and the colon. It is about 5 m long. It is called the small intestine because it is quite narrow.

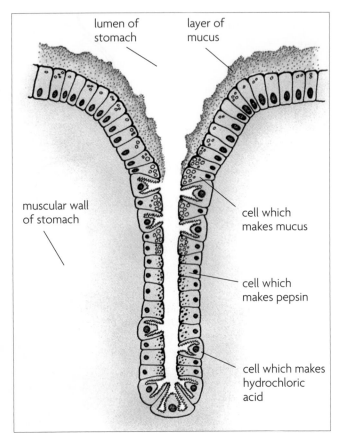

- lumen of stomach
- layer of mucus
- muscular wall of stomach
- cell which makes mucus
- cell which makes pepsin
- cell which makes hydrochloric acid

Figure 7.19 A gastric pit. 'Gastric' means 'to do with the stomach'.

Different parts of the small intestine have different names. The first part, nearest to the stomach, is the **duodenum**. The last part, nearest to the colon, is the **ileum**.

7.28 Pancreatic juice flows into the duodenum.

Several enzymes are secreted into the duodenum. They are made in the **pancreas**, which is a cream-coloured gland, lying just underneath the stomach. A tube called the **pancreatic duct** leads from the pancreas into the duodenum. **Pancreatic juice**, which is a fluid made by the pancreas, flows along this tube.

This fluid contains many enzymes. One is **amylase**, which breaks down starch to maltose. Another is **trypsin**, which is a protease and breaks down proteins to polypeptides. Another is **lipase**, which breaks down fats (lipids) to fatty acids and glycerol.

These enzymes do not work well in acid environments, but the chyme which has come from the stomach contains hydrochloric acid. Pancreatic juice contains **sodium hydrogencarbonate** which partially neutralises the acid.

7.29 Bile helps digest fats.

As well as pancreatic juice, another fluid flows into the duodenum. It is called **bile**. Bile is a yellowish green, watery liquid. It is made in the liver, and then stored in the **gall bladder**. It flows to the duodenum along the **bile duct**.

Bile does not contain any enzymes. It does, however, help to digest fats. It does this by breaking up the large drops of fat into very small ones, making it easier for the lipase in the pancreatic juice to digest them. This is called **emulsification**, and is done by salts in the bile called **bile salts**. Emulsification is a type of mechanical digestion.

Bile also contains yellowish **bile pigments**. These are made by the liver when it breaks down old red blood cells. The bile pigments are made from haemoglobin. The pigments are not needed by the body, so they are eventually excreted in the faeces.

7.30 Digestion is completed in the small intestine.

As well as receiving enzymes made in the pancreas, the small intestine makes some enzymes itself. They are made by cells in its walls.

The inner wall of all parts of the small intestine – the duodenum and ileum – is covered with millions of tiny projections. They are called **villi** (singular: **villus**). Each villus is about 1 mm long (Figures **7.20**, **7.21**, **7.22** and **7.23**, overleaf). Cells covering the villi make enzymes. The enzymes do not come out into the lumen of the small intestine, but stay close to the cells which make them. These enzymes complete the digestion of food.

Maltase breaks down maltose to glucose. **Sucrase** breaks down sucrose to glucose and fructose. **Lactase** breaks down lactose to glucose and galactose. These three enzymes are all **carbohydrases**. There are also **proteases**, which finish breaking down any polypeptides into amino acids. **Lipase** completes the breakdown of fats to fatty acids and glycerol.

7.31 Digested food is absorbed in the small intestine.

By now, most carbohydrates have been broken down to simple sugars, proteins to amino acids, and fats to fatty acids and glycerol.

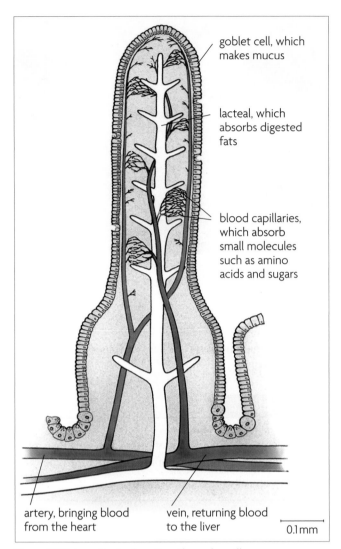

Figure 7.20 Longitudinal section through a villus.

goblet cell, which makes mucus

lacteal, which absorbs digested fats

blood capillaries, which absorb small molecules such as amino acids and sugars

artery, bringing blood from the heart

vein, returning blood to the liver

0.1 mm

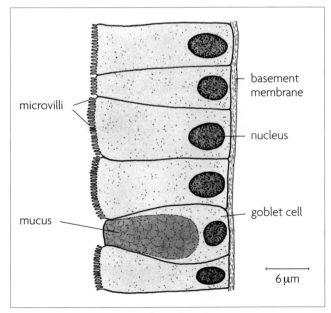

Figure 7.21 Detail of the surface of a villus.

microvilli

mucus

basement membrane

nucleus

goblet cell

6 μm

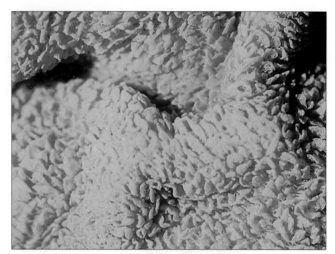

Figure 7.22 This micrograph shows thousands of villi covering the inner wall of the small intestine. It is magnified about 20 times.

These molecules are small enough to pass through the wall of the small intestine and into the blood. This is called **absorption**.

The small intestine is especially adapted to allow absorption to take place very efficiently. Some of its features are listed in Table 7.5.

Water, mineral salts and vitamins are also absorbed in the small intestine.

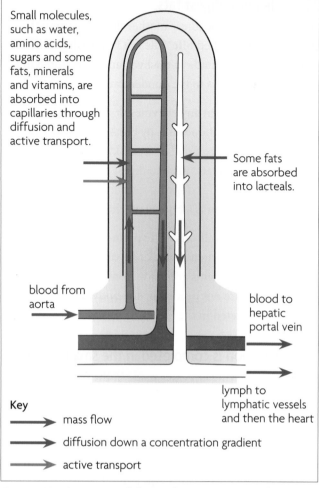

Figure 7.23 Absorption of digested nutrients into a villus.

The small intestine absorbs between 5 and 10 dm³ of water each day.

Table 7.6 (page 82) gives a summary of digestion in the human alimentary canal.

Table 7.5 How the small intestine is adapted for absorbing digested nutrients

feature	how this helps absorption take place
It is very long, about 5 m in an adult human.	This gives plenty of time for digestion to be completed, and for digested food to be absorbed as it slowly passes through.
It has villi. Each villus is covered with cells which have even smaller projections on them, called microvilli.	This gives the inner surface of the small intestine a very large surface area. The larger the surface area, the faster nutrients can be absorbed.
Villi contain blood capillaries.	Monosaccharides, amino acids, water, minerals and vitamins, and some fats, pass into the blood, to be taken to the liver and then round the body.
Villi contain lacteals, which are part of the lymphatic system.	Fats are absorbed into lacteals.
Villi have walls only one cell thick.	The digested nutrients can easily cross the wall to reach the blood capillaries and lacteals.

7.32 The colon absorbs water.

Not all the food that is eaten can be digested, and this undigested food cannot be absorbed in the small intestine. It travels on, through the caecum, past the appendix and into the **colon**. In humans, the caecum and appendix have no function. In the colon, more water and salt are absorbed.

However, the colon absorbs much less water than the small intestine, generally around 0.3 to 0.5 dm³ per day.

The colon and rectum are sometimes called the **large intestine,** because they are wider tubes than the duodenum and ileum.

7.33 The rectum temporarily stores undigested food.

By the time the food reaches the **rectum**, most of the substances which can be absorbed have gone into the blood. All that remains is indigestible food (fibre, or roughage), bacteria, and some dead cells from the inside of the alimentary canal. This mixture forms the **faeces**, which are passed out at intervals through the **anus**. This process is called **egestion**.

Key definition

egestion   the passing out of food that has not been digested, as faeces, through the anus

7.34 Most absorbed food goes straight to the liver.

After they have been absorbed into the blood, the nutrients are taken to the liver, in the **hepatic portal vein** (Figure 7.24). The liver processes some of it, before it goes any further (Table 11.1, page 153). Some of these nutrients can be broken down, some converted into other substances, some stored and the remainder left unchanged.

The nutrients, dissolved in the blood plasma, are then taken to other parts of the body where they may become assimilated as part of a cell.

Key definition

assimilation   the movement of digested food molecules into the cells of the body where they are used, becoming part of the cells

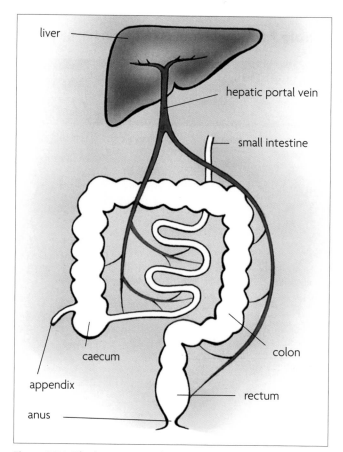

Figure 7.24 The hepatic portal vein transports absorbed nutrients from the small intestine to the liver.

The liver has an especially important role in the metabolism of glucose. If there is more glucose than necessary in the blood, the liver will convert some of it to the polysaccharide **glycogen**, and store it. You can find out more about this in section **11.10**.

The liver also deals with excess amino acids. Our bodies are not able to store amino acids or proteins. Rather than waste the energy in them, the liver converts them to other substances that we can use or store in the body, such as fats.

This involves removing the nitrogen from them, a process called **deamination**. The nitrogen-containing part is made into **urea**, which is transported to the kidneys and excreted. The rest of the amino acid molecule is used to provide energy for the liver cells.

The liver also has the role of breaking down any toxins (harmful substances) that have been absorbed from the alimentary canal. These include drugs such as alcohol. Cells in the liver are able to convert many toxins to harmless substances that can be transported in the blood and excreted from the body.

Table 7.6 Summary of digestion in the human alimentary canal

part of the canal	juices secreted	where made	enzymes in juice	substrate	product	other substances in juice	functions of other substances
mouth	saliva	salivary glands	amylase	starch	maltose		
oesophagus	none						
stomach	gastric juice	in pits in wall of stomach	pepsin	proteins	polypeptides	hydrochloric acid	acid environment for pepsin; kills bacteria in food
			rennin (only in young mammals)	milk protein	curdled milk protein		
duodenum	pancreatic juice	pancreas	amylase	starch	maltose	sodium hydro-gencarbonate	reduces acidity of chyme
			trypsin	proteins	polypeptides		
			lipase	fats	fatty acids and glycerol		
	bile	liver, stored in gall bladder	none			bile salts	emulsify fats
						bile pigments	excretory products
ileum	no juice secreted; enzymes remain in or on the cells covering the villi	by cells covering the villi	maltase	maltose	glucose		
			sucrase	sucrose	glucose and fructose		
			lactase	lactose	glucose and galactose		
			peptidase	polypeptides	amino acids		
			lipase	fats	fatty acids and glycerol		

All of the digestive juices contain water and mucus. The water is used for the digestion of large molecules to small ones. It is also a solvent for the nutrients and enzymes. Mucus acts as a lubricant. It also forms a covering over the inner surface of the alimentary canal, preventing enzymes from digesting the cells.

Questions

7.20 What is a sphincter muscle?

7.21 Name **two** places in the alimentary canal where sphincter muscles are found.

7.22 In which parts of the alimentary canal is mucus secreted? Explain why.

7.23 Name **two** parts of the alimentary canal where amylase is secreted. What does it do?

7.24 What is the epiglottis?

7.25 Why do the walls of the stomach secrete hydrochloric acid?

7.26 Which **two** parts of the alimentary canal make up the small intestine?

7.27 Which **two** digestive juices are secreted into the duodenum?

7.28 How do bile salts help in digestion?

7.29 What do faeces contain?

◆ A balanced diet contains suitable proportions of each group of nutrients – carbohydrates, fats, proteins, minerals, vitamins, water and fibre – and the correct amount of energy.

◆ Eating food containing more energy than you can use up causes weight increase, which can lead to obesity. Children who do not get enough food may suffer from energy protein malnutrition, in which they do not grow properly and have little energy.

◆ Digestion is the breakdown of large molecules of food into small ones, so that they can be absorbed through the wall of the alimentary canal.

◆ Mechanical digestion breaks down large pieces of food to small ones. It is done by the teeth, the muscles in the wall of the alimentary canal and bile salts. Chemical digestion breaks down large molecules to small ones. It is done by enzymes.

◆ Mammals have four types of teeth – incisors, canines, premolars and molars – each with their own functions.

◆ Digestion begins in the mouth, as teeth grind food into smaller pieces, and amylase digests starch to maltose.

◆ Protein digestion begins in the stomach, where pepsin digests proteins to polypeptides. Rennin is present in young mammals, and clots milk protein. Hydrochloric acid kills bacteria and provides a low pH for the action of pepsin.

◆ Pancreatic juice flows into the duodenum. It contains enzymes that digest starch, proteins and lipids, and also sodium hydrogencarbonate to partly neutralise the acidity of food coming from the stomach.

◆ Bile also flows into the duodenum. It contains bile salts which emulsify fats, making it easier for lipase to digest them.

◆ The lining of the small intestine is covered with villi, giving it a very large surface area which helps to speed up absorption. Cells on the surface of the villi make enzymes, which complete the digestion of food. The villi contain blood capillaries to absorb glucose, amino acids, water, vitamins and minerals, and lacteals to absorb fatty acids and glycerol.

◆ The absorbed nutrients are carried to the liver in the hepatic portal vein. Some are used in the liver, some are stored, and some are sent on in the blood to be delivered to cells all over the body.

◆ The colon absorbs more water from the food. In the rectum, the undigested food is formed into faeces, which are eventually egested through the anus.

Revision questions

1 With the aid of examples wherever possible, explain the differences between each of the following pairs of terms.
   a enamel, dentine
   b digestion, absorption
   c small intestine, large intestine
   d bile, pancreatic juice

2 a What is meant by a balanced diet?
   b Using Table 7.1 and Figure 7.1, plan menus for one day which would provide a balanced diet for (i) a teenage boy, and (ii) a pregnant woman. For each food you include, state how much energy, and which types of nutrients it contains.

3 Describe what would happen to a piece of steak, containing only protein, and a chip, containing starch and fat, as they passed through a person's alimentary canal.

4 a Name three enzymes made by the cells covering the villi in the small intestine and explain what they do.
   b From your knowledge of the properties of enzymes, suggest why pepsin stops working when it passes out of the stomach and enters the duodenum.
   c Explain how the small intestine is adapted for absorption.

S

8 Transport

In this chapter, you will find out:

◆   why large organisms need transport systems
◆   which substances are transported in animals and plants
◆   about the structure and function of the human circulatory system, including the heart and blood vessels
◆   about the structure and functions of blood
◆   about the structure and functions of xylem vessels and phloem tubes in plants
◆   about transpiration and how to measure its rate, and the factors that affect it.

Transport in humans

8.1 Mammals have double circulatory systems.

The main transport system of all mammals, including humans, is the blood system, also known as the **circulatory system**. It is a network of tubes, called **blood vessels**. A pump, the **heart**, keeps blood flowing through the vessels.

Figure **8.1** illustrates the general layout of the human blood system. The arrows show the direction of blood flow. If you follow the arrows, beginning at the lungs, you can see that blood flows into the left-hand side of the heart, and then out to the rest of the body. It is brought back to the right-hand side of the heart, before going back to the lungs again.

This is called a **double circulatory system**, because the blood travels through the heart twice on one complete journey around the body.

8.2 Blood becomes oxygenated and deoxygenated.

The blood in the left-hand side of the heart has come from the lungs. It contains oxygen, which was picked up by the capillaries surrounding the alveoli. It is called **oxygenated blood**.

This oxygenated blood is then sent around the body. Some of the oxygen in it is taken up by the body cells,

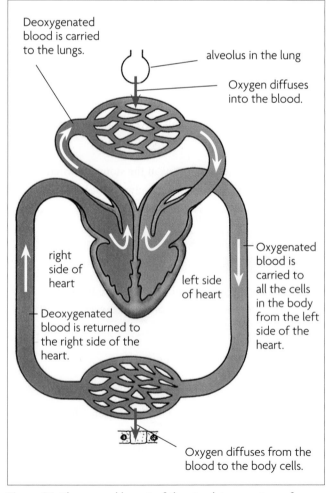

Figure 8.1 The general layout of the circulatory system of a human, as seen from the front.

Within the figure:
- Deoxygenated blood is carried to the lungs.
- alveolus in the lung
- Oxygen diffuses into the blood.
- right side of heart
- left side of heart
- Oxygenated blood is carried to all the cells in the body from the left side of the heart.
- Deoxygenated blood is returned to the right side of the heart.
- Oxygen diffuses from the blood to the body cells.

which need oxygen for respiration (Chapter **9**). When this happens the blood becomes **deoxygenated**. The deoxygenated blood is brought back to the right-hand side of the heart. It then goes to the lungs, where it becomes oxygenated once more.

The heart

8.3 The structure of the heart.

The function of the heart is to pump blood around the body. It is made of a special type of muscle called **cardiac muscle**. This muscle contracts and relaxes regularly, throughout life.

Figure **8.2** is a section through a heart. It is divided into four chambers. The two upper chambers are called **atria**.

The two lower chambers are **ventricles**. The chambers on the left-hand side are completely separated from the ones on the right-hand side by a **septum**.

If you look at Figures **8.1** and **8.2**, you will see that blood flows into the heart at the top, into the atria. Both of the atria receive blood. The left atrium receives blood from the **pulmonary veins**, which come from the lungs. The right atrium receives blood from the rest of the body, arriving through the **venae cavae** (singular: vena cava).

From the atria, the blood flows into the ventricles. The ventricles then pump it out of the heart. They do this by contracting the muscle in their walls. The strong cardiac muscle contracts with considerable force, squeezing inwards on the blood inside the heart and pushing it out. The blood in the left ventricle is pumped into the **aorta**, which takes the blood around the body. The right ventricle pumps blood into the **pulmonary artery**, which takes it to the lungs.

The function of the ventricles is quite different from the function of the atria. The atria simply receive blood, from either the lungs or the body, and supply it to the ventricles. The ventricles pump blood out of the heart and

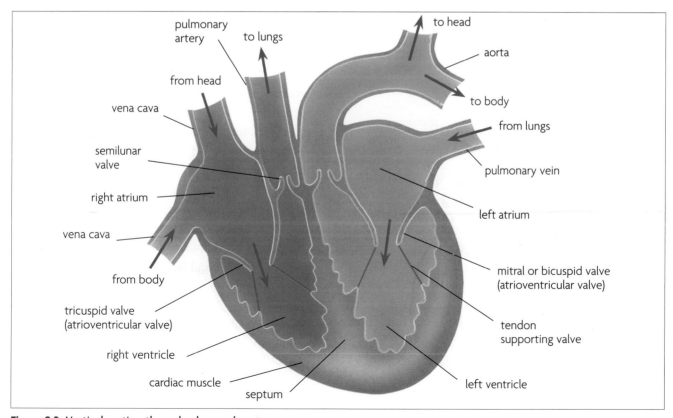

Figure 8.2 Vertical section through a human heart.

all around the body. To help them do this, the ventricles have much thicker, more muscular walls than the atria.

There is also a difference in the thickness of the walls of the right and left ventricles. The right ventricle pumps blood to the lungs, which are very close to the heart. The left ventricle, however, pumps blood all around the body. The left ventricle has an especially thick wall of muscle to enable it to do this. The blood flowing to the lungs in the pulmonary artery has a much lower pressure than the blood in the aorta.

8.4 Coronary arteries supply heart muscle.

In Figure 8.3, you can see that there are blood vessels on the outside of the heart. They are called the **coronary arteries**. These vessels supply blood to the heart muscles.

It may seem odd that this is necessary, when the heart is full of blood. However, the muscles of the heart are so thick that the nutrients and oxygen in the blood inside the heart would not be able to diffuse to all the muscles quickly enough. The heart muscle needs a constant supply of nutrients and oxygen, so that it can keep contracting and relaxing. The coronary arteries supply this.

If a coronary artery gets blocked – for example, by a blood clot – the cardiac muscles run short of oxygen. They cannot respire, so they cannot obtain energy to

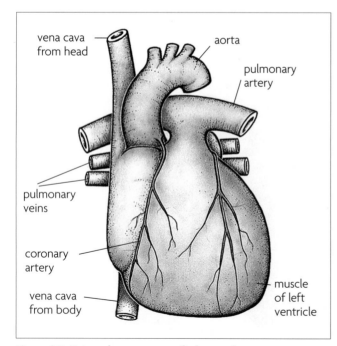

pulmonary veins

coronary artery

vena cava from body

vena cava from head

aorta

pulmonary artery

muscle of left ventricle

Figure 8.3 External appearance of a human heart.

Questions

8.5 What kind of muscle is found in the heart?

8.6 Which parts of the heart receive blood from **(a)** the lungs, and **(b)** the body?

8.7 Which parts of the heart pump blood into **(a)** the pulmonary artery, and **(b)** the aorta?

8.8 Why do the ventricles have thicker walls than the atria?

8.9 Why does the left ventricle have a thicker wall than the right ventricle?

8.10 What is the function of the coronary arteries?

allow them to contract. The heart therefore stops beating. This is called a **heart attack** or **cardiac arrest**.

Blockage of the coronary arteries is called coronary heart disease. It is a very common cause of illness and death, especially in developed countries. We know several factors that increase a person's risk of getting coronary heart disease.

- **Smoking cigarettes** Several components of cigarette smoke, including nicotine, cause damage to the circulatory system. Stopping smoking is the single most important thing a smoker can do in order to reduce their chances of getting coronary heart disease.
- **Diet** There is evidence that a diet high in salt, saturated fats (fats from animals) or cholesterol increases the chances of getting coronary heart disease. To reduce the risk, it is good to eat a diet containing a very wide variety of foods, with not too many fats in it (though we do need some fat in the diet to stay healthy). Oils from plants and fish, on the other hand, can help to prevent heart disease.
- **Obesity** Being very overweight increases the risk of coronary heart disease. Keeping your body weight at a suitable level, and taking plenty of exercise, helps to maintain the coronary arteries in a healthy condition.
- **Stress** We all need some stress in our lives, or they would be very dull. However, unmanageable or long-term stress appears to increase the risk of developing heart disease. Avoiding severe or long-term stress is a good idea, if you can manage it. Otherwise, it is important to find ways to manage stress.

- **Genes** Some people have genes that make it more likely they will get coronary heart disease. There is not really anything you can do about this. However, if several people in your family have had problems with their hearts, then this could mean that you have these genes. In that case, it is important to try hard to reduce the other risk factors by having a healthy life-style.

A type of drug called **statin** has been found to reduce the risk of the coronary arteries (and other arteries) becoming blocked. A doctor may prescribe statins to anyone who has a high risk of developing heart disease.

8.5 The pacemaker controls the rate of heart beat.

The heart beats as the cardiac muscles in its walls contract and relax. When they contract, the heart becomes smaller, squeezing blood out. This is called **systole**. When they relax, the heart becomes larger, allowing blood to flow into the atria and ventricles. This is called **diastole**. Figure 8.4 illustrates this.

The rate at which the heart beats is controlled by a patch of muscle in the right atrium called the **pacemaker**. The pacemaker sends electrical signals through the walls of the heart at regular intervals, which make the muscle

contract. The pacemaker's rate, and therefore the rate of heart beat, changes according to the needs of the body. For example, during exercise, when extra oxygen is needed by the muscles, the brain sends messages along nerves to the pacemaker, to make the heart beat faster.

Sometimes, the pacemaker stops working properly. An artificial pacemaker can then be placed in the person's heart. It produces an electrical impulse at a regular rate of about one impulse per second. Artificial pacemakers last for up to ten years before they have to be replaced.

8.6 Blood flows one way through heart valves.

There is a valve between the left atrium and the left ventricle, and another between the right atrium and ventricle. These are called **atrioventricular valves** (Figure 8.4).

The valve on the left-hand side of the heart is made of two parts and is called the **bicuspid valve**, or the **mitral valve**. The valve on the right-hand side has three parts, and is called the **tricuspid valve**.

The function of these valves is to stop blood flowing from the ventricles back to the atria. This is important, so that when the ventricles contract, the blood is pushed up

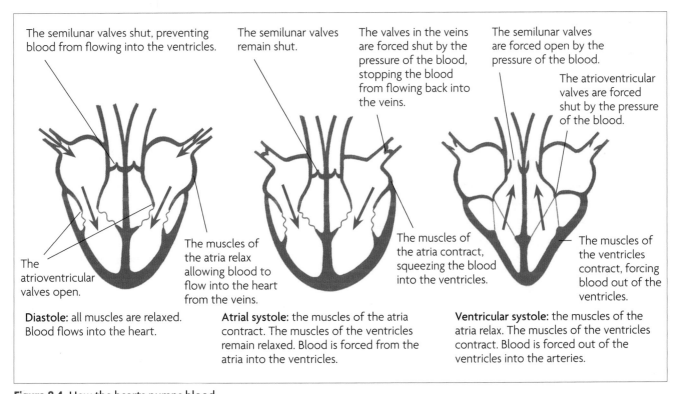

The semilunar valves shut, preventing blood from flowing into the ventricles.

The semilunar valves remain shut.

The valves in the veins are forced shut by the pressure of the blood, stopping the blood from flowing back into the veins.

The semilunar valves are forced open by the pressure of the blood.

The atrioventricular valves are forced shut by the pressure of the blood.

The atrioventricular valves open.

The muscles of the atria relax allowing blood to flow into the heart from the veins.

The muscles of the atria contract, squeezing the blood into the ventricles.

The muscles of the ventricles contract, forcing blood out of the ventricles.

Diastole: all muscles are relaxed. Blood flows into the heart.

Atrial systole: the muscles of the atria contract. The muscles of the ventricles remain relaxed. Blood is forced from the atria into the ventricles.

Ventricular systole: the muscles of the atria relax. The muscles of the ventricles contract. Blood is forced out of the ventricles into the arteries.

Figure 8.4 How the hearts pumps blood.

into the arteries, not back into the atria. As the ventricles contract, the pressure of the blood pushes the valves upwards. The **tendons** attached to them stop them from going up too far.

Blood vessels

8.7 There are three kinds of blood vessels.

There are three main kinds of blood vessels: **arteries**, **capillaries** and **veins** (Figure **8.5**). Arteries carry blood away from the heart. They divide again and again, and eventually form very tiny vessels called capillaries. The capillaries gradually join up with one another to form large vessels called veins. Veins carry blood towards the heart. These vessels are compared in Table **8.1**, page **90**.

8.8 Arteries have thick elastic walls.

When blood flows out of the heart, it enters the arteries. The blood is then at very high pressure, because it has been forced out of the heart by the contraction of the

Activity 8.1
To find the effect of exercise on the rate of heart beat

skills

C2 *Observing, measuring and recording*

C3 *Interpreting and evaluating*

C4 *Planning*

Safety Don't do vigorous exercise if you know it could harm your health.

The best way to measure the rate of your heart beat is to take your pulse. Use the first two fingers of your right hand and rest them on the inside of your left wrist. Feel for the tendon near the outside of your wrist. If you rest your fingers lightly just over this tendon, you can feel the artery in your wrist pulsing as your heart pumps blood through it.

1 Construct a results chart, ready to fill in your results. You will need to read through the whole set of instructions first, so that you can see exactly what you need to have in your results chart.

2 Sit quietly for two minutes, to make sure you are completely relaxed.

3 Count the number of pulses in one minute. Record it in your table.

4 Wait one minute, then count your pulse again, and record.

5 Now do some vigorous exercise, such as stepping up and down onto a chair, for exactly two minutes. At the end of this time, sit down. Immediately count your pulse in the next minute, and record.

6 Continue to record your pulse rate every other minute, until it has returned to near the level before you started to exercise.

7 Draw a graph of your results, putting time on the bottom axis (*x*-axis).

8 Compare your results with those of other people in your class. How much variation is there in pulse rate when resting? How much variation is there in pulse rate after exercise? How much variation is there in the time taken for pulse rate to return to normal after exercise?

9 Design an experiment to test the hypothesis that training reduces the time taken for the pulse rate to return to normal after exercise. Remember to think hard about controlling variables. This will be very difficult for this experiment, but do the best you can.

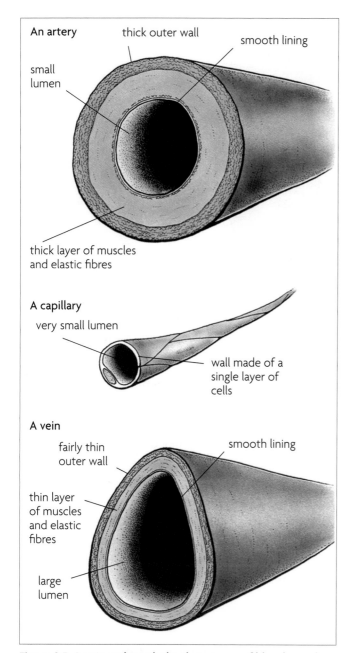

An artery
- thick outer wall
- smooth lining
- small lumen
- thick layer of muscles and elastic fibres

A capillary
- very small lumen
- wall made of a single layer of cells

A vein
- fairly thin outer wall
- smooth lining
- thin layer of muscles and elastic fibres
- large lumen

Figure 8.5 Sections through the three types of blood vessels.

muscular ventricles. Arteries therefore need very strong walls to withstand the high pressure of the blood flowing through them.

The blood does not flow smoothly through the arteries. It pulses through, as the ventricles contract and relax. The arteries have elastic tissue in their walls which can stretch and recoil with the force of the blood. This helps to make the flow of blood smoother. You can feel your arteries stretch and recoil when you feel your pulse in your wrist.

The blood pressure in the arteries of your arm can be measured using a **sphygmomanometer** (Figure 8.6).

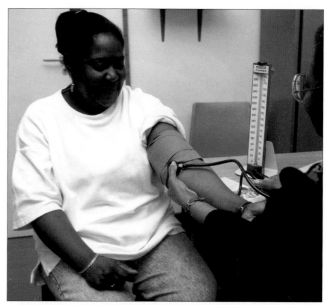

Figure 8.6 A sphygmomanometer being used to measure blood pressure.

8.9 Capillaries are very narrow, with thin walls.

The arteries gradually divide to form smaller and smaller vessels (Figures **8.7** and **8.8**, overleaf). These are the capillaries. The capillaries are very small and penetrate to every part of the body. No cell is very far away from a capillary.

The function of the capillaries is to take nutrients, oxygen and other materials to all the cells in the body, and to take away their waste materials. To do this, their walls must be very thin so that substances can get in and out of them easily. The walls of the smallest capillaries are only one cell thick (Figure **8.5**).

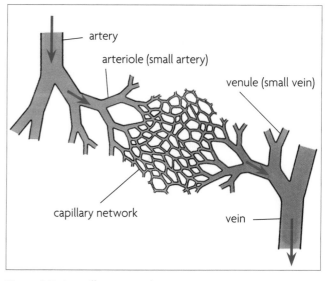

- artery
- arteriole (small artery)
- venule (small vein)
- capillary network
- vein

Figure 8.7 A capillary network.

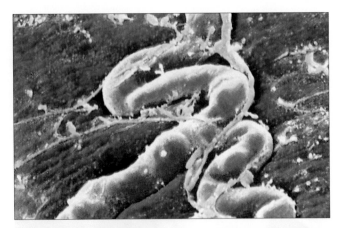

Figure 8.8 A capillary, shown in blue, snakes its way through muscle tissue (× 600).

8.10 Veins have one-way valves.

The capillaries gradually join up again to form veins. By the time the blood gets to the veins, it is at a much lower pressure than it was in the arteries. The blood flows more slowly and smoothly now. There is no need for veins to have such thick, strong, elastic walls.

If the veins were narrow, this would slow down the blood even more. To help keep the blood moving easily through them, the space inside the veins, called the **lumen**, is much wider than the lumen of the arteries.

Veins have valves in them to stop the blood flowing backwards (Figure **8.9**). Valves are not needed in the arteries, because the force of the heart beat keeps blood moving forwards through them.

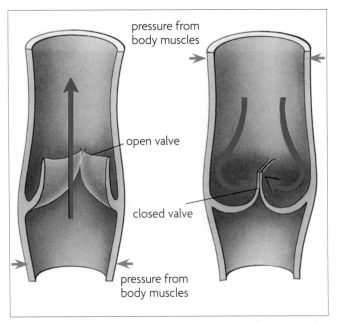

Figure 8.9 Valves in a vein: the valves are like pockets set in the wall of the vein.

Blood is also kept moving in the veins by the contraction of muscles around them (Figure **8.9**). The large veins in your legs are squeezed by your leg muscles when you walk. This helps to push the blood back up to your heart. If a person is confined to bed for a long time, then there is a danger that the blood in these veins will not be kept moving. A clot may form in them, called a **thrombosis**. If the clot is carried to the lungs, it could get stuck in the arterioles. This is called a **pulmonary embolism**, and it may prevent the circulation reaching part of the lungs. In serious cases this can cause death.

Table 8.1 Arteries, veins and capillaries

	function	structure of wall	width of lumen	S	how structure fits function
arteries	carry blood away from the heart	thick and strong, containing muscles and elastic tissues	relatively narrow; it varies with heart beat, as it can stretch and recoil		strength and elasticity needed to withstand the pulsing of the blood as it is pumped through the heart
capillaries	supply all cells with their requirements, and take away waste products	very thin, only one cell thick	very narrow, just wide enough for a red blood cell to pass through		no need for strong walls, as most of the blood pressure has been lost; thin walls and narrow lumen bring blood into close contact with body tissues
veins	return blood to the heart	quite thin, containing far less muscle and elastic tissue than arteries	wide; contains valves		no need for strong walls, as most of the blood pressure has been lost; wide lumen offers less resistance to blood flow; valves prevent backflow

8.11 Each organ has its own blood supply.

Figures **8.10** and **8.11** (overleaf) illustrate the positions of the main arteries and veins in the body.

Each organ of the body, except the lungs, is supplied with oxygenated blood from an artery. Deoxygenated blood is taken away by a vein. The artery and vein are named according to the organ with which they are connected. For example, the blood vessels of the kidneys are the **renal** artery and vein.

All arteries, other than the pulmonary artery, branch from the aorta. All the veins, except the pulmonary veins and hepatic portal vein, join up to one of the two venae cavae.

The liver has two blood vessels supplying it with blood. The first is the **hepatic artery**, which supplies oxygen. The second is the **hepatic portal vein**. This vein brings blood from the digestive system (Figure **8.10**), so that the liver can process the food which has been absorbed, before it travels to other parts of the body. All the blood leaves the liver in the **hepatic vein**.

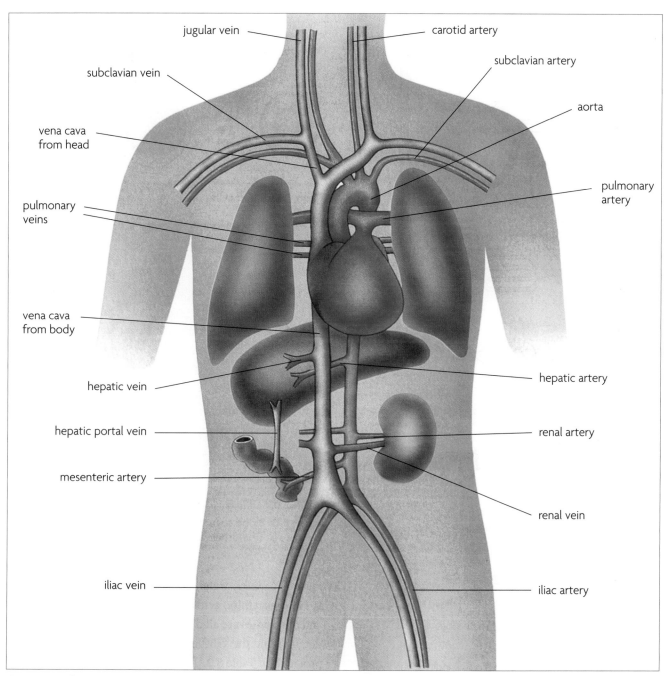

jugular vein — carotid artery
subclavian artery
subclavian vein
aorta
vena cava from head
pulmonary artery
pulmonary veins
vena cava from body
hepatic vein — hepatic artery
hepatic portal vein — renal artery
mesenteric artery
renal vein
iliac vein — iliac artery

Figure 8.10 The main arteries and veins in the human body.

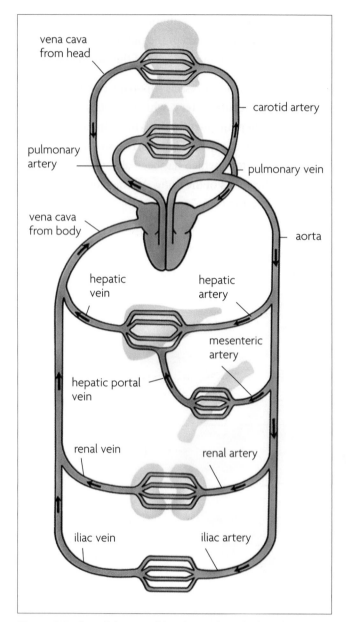

Figure 8.11 Plan of the main blood vessels in the human body.

Labels on Figure 8.11:
vena cava from head, carotid artery, pulmonary artery, pulmonary vein, vena cava from body, aorta, hepatic vein, hepatic artery, mesenteric artery, hepatic portal vein, renal vein, renal artery, iliac vein, iliac artery

Questions

8.16 Which blood vessels carry blood (a) away from, and (b) towards the heart?

8.17 Why do arteries need strong walls?

8.18 Why do arteries have elastic walls?

8.19 What is the function of capillaries?

8.20 Why do veins have a large lumen?

8.21 How is blood kept moving in the large veins of the legs?

8.22 What is unusual about the blood supply to the liver?

Blood

8.12 Blood consists of cells floating in plasma.

The liquid part of blood is called **plasma**. Floating in the plasma are cells. Most of these are **red blood cells**. A much smaller number are **white blood cells**. There are also small fragments formed from special cells in the bone marrow, called **platelets** (Figures **8.12** and **8.13**).

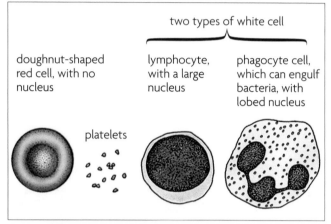

two types of white cell

doughnut-shaped red cell, with no nucleus

lymphocyte, with a large nucleus

phagocyte cell, which can engulf bacteria, with lobed nucleus

platelets

Figure 8.12 Blood cells.

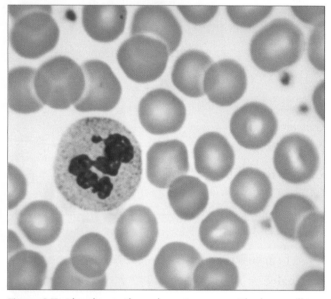

Figure 8.13 Blood seen through a microscope. The large cell is a white cell. The others are all red cells. There are also a few platelets (×1700).

Plasma is mostly water. Many substances are dissolved in it. Glucose, amino acids, salts, hormones, blood proteins, and antibodies are all dissolved in the plasma. More details about the substances carried in blood plasma are provided in Table **8.2**. The functions of components of blood are summarised in Table **8.3** (page **96**).

Table 8.2 Some of the main components of blood plasma

	source	destination	notes
water	Absorbed in small intestine and colon.	All cells.	Excess is removed by the kidneys.
plasma proteins (including fibrinogen and antibodies)	Fibrinogen is made in the liver. Antibodies are made by lymphocytes.	Remain in the blood.	Fibrinogen helps in blood clotting. Antibodies kill invading pathogens.
lipids including cholesterol and fatty acids	Absorbed in the ileum. Also derived from fat reserves in the body.	To the liver, for breakdown. To adipose tissue, for storage. To respiring cells, as an energy source.	Breakdown of fats yields energy – heart muscle depends largely on fatty acids for its energy supply. High cholesterol levels in the blood increase the risk of developing heart disease.
carbohydrates, especially glucose	Absorbed in the ileum. Also produced by the breakdown of glycogen in the liver.	To all cells, for energy release by respiration.	Excess glucose is converted to glycogen and stored in the liver.
excretory substances, e.g. urea	Produced by amino acid deamination in the liver.	To kidneys for excretion.	
mineral ions, e.g. Na^+, Cl^-	Absorbed in the ileum and colon.	To all cells.	Excess ions are excreted by the kidneys.
hormones	Secreted into the blood by endocrine glands.	To all parts of the body.	Hormones only affect their target cells. Hormones are broken down by the liver, and their remains are excreted by the kidneys.
dissolved gases, e.g. carbon dioxide	Carbon dioxide is released by all cells as a waste product of respiration.	To the lungs for excretion.	Most carbon dioxide is carried as hydrogencarbonate ions (HCO_3^-) in the blood plasma.

8.13 Red blood cells carry oxygen.

Red blood cells are made in the bone marrow of some bones, including the ribs, vertebrae and some limb bones. They are produced at a very fast rate – about 9000 million per hour!

Red cells have to be made so quickly because they do not live for very long. Each red cell only lives for about four months. One reason for this is that they do not have a nucleus (Figure **8.12**).

Red cells are red because they contain the pigment **haemoglobin**. This carries oxygen. Haemoglobin is a protein, and contains iron. It is this iron that readily combines with oxygen where the gas is in good supply. It just as readily gives it up where the oxygen supply is low, as in active tissues.

The lack of a nucleus in a red blood cell means that there is more space for packing in millions of molecules of haemoglobin.

Another unusual feature of red blood cells is their shape.

They are biconcave discs – like a flat disc that has been pinched in on both sides. This, together with their small size, gives them a relatively large surface area compared with their volume.

This high surface area to volume ratio speeds up the rate at which oxygen can diffuse in and out of the red blood cell. The small size of the red blood cell is also useful in enabling it to squeeze through even the tiniest capillaries. This means that oxygen can be taken very close to every cell in the body.

Old red blood cells are broken down in the liver, spleen and bone marrow. Some of the iron from the haemoglobin is stored, and used for making new haemoglobin. Some of it is turned into bile pigment and excreted (section **7.29**).

8.14 White blood cells fight infection.

White cells are made in the bone marrow and in the lymph nodes (section **8.21**). White cells do have a nucleus, which is often quite large and lobed (Figure **8.12**). They can move around and can squeeze out through the walls of blood capillaries into all parts of the body. Their function is to fight **pathogens** (disease-causing bacteria and viruses), and to clear up any dead body cells.

There are many different kinds of white blood cells. They all have the function of destroying pathogens in your body, but they do it in different ways.

Phagocytes are cells which can move around the body, engulfing and destroying pathogens (Figure **8.14**). They also destroy any of your own cells that are damaged or worn out. Phagocytes often have lobed nuclei. If you damage your skin, perhaps with a cut or graze, phagocytes will collect at the site of the damage, to 'mop up' any microorganisms which might possibly get in.

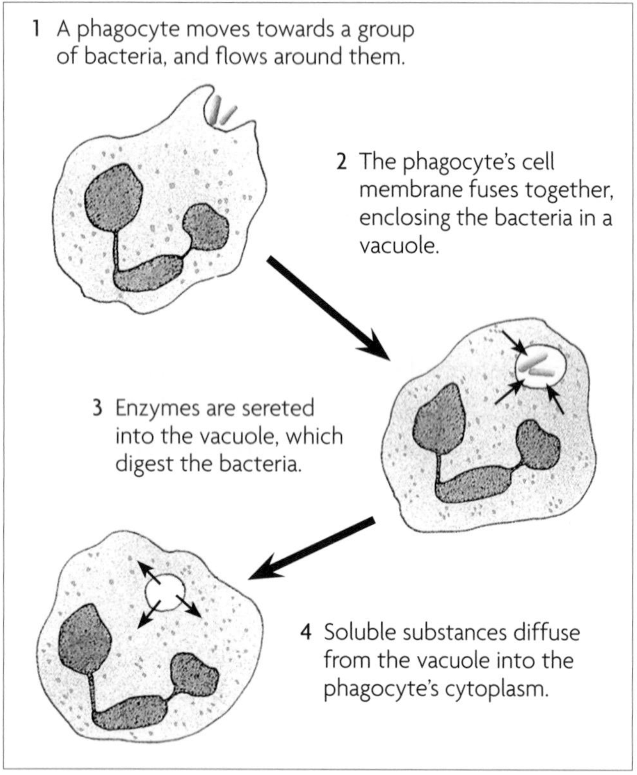

1 A phagocyte moves towards a group of bacteria, and flows around them.

2 The phagocyte's cell membrane fuses together, enclosing the bacteria in a vacuole.

3 Enzymes are sereted into the vacuole, which digest the bacteria.

4 Soluble substances diffuse from the vacuole into the phagocyte's cytoplasm.

Figure 8.14 Phagocytosis.

Lymphocytes have a quite different method of attacking pathogens. They produce chemicals called **antibodies**, which are carried in the blood and tissue fluid to almost every part of the body.

8.15 Antibodies are specific.

In your body, you have thousands of different kinds of lymphocytes. Each kind is able to produce a different sort of antibody.

An antibody is a protein molecule with a particular shape. Rather like an enzyme molecule, this shape is just right to fit into another molecule. To destroy a particular pathogen, antibody molecules must be made which are just the right shape to fit into molecules on the outside of the pathogen. These pathogen molecules are called **antigens**.

When antibody molecules lock onto the pathogen, they kill the pathogen. There are several ways in which they do this. One way is simply to alert phagocytes to the presence of the pathogen, so that the phagocytes will come and destroy them. Or the antibodies may start off a series of reactions in the blood which produce enzymes to digest the pathogens.

8.16 Lymphocytes multiply when 'their' pathogen is present.

Most of the time, most of your lymphocytes do not produce antibodies. It would be a waste of energy and materials if they did. Instead, each lymphocyte waits for a signal that a pathogen which can be destroyed by its particular antibody is in your body.

If a pathogen enters the body, it is likely to meet a large number of lymphocytes. One of these may recognise the pathogen as being something that its antibody can destroy. This lymphocyte will start to divide rapidly by mitosis, making a clone of lymphocytes just like itself. These lymphocytes then secrete their antibody, destroying the pathogen (Figure **8.15**).

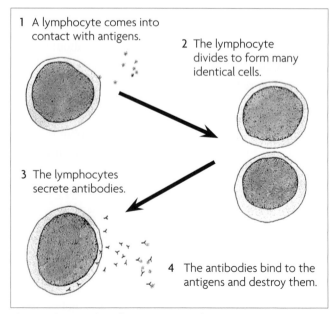

1 A lymphocyte comes into contact with antigens.

2 The lymphocyte divides to form many identical cells.

3 The lymphocytes secrete antibodies.

4 The antibodies bind to the antigens and destroy them.

Figure 8.15 How lymphocytes respond to antigens.

This takes time. It may take a while for the 'right' lymphocyte to recognise the pathogen, and then a few days for it to produce a big enough clone to make enough antibody to kill the pathogen. In the meantime, the pathogen breeds, making you ill. Eventually, however, the lymphocytes get the upper hand, and you get better.

Lymphocytes are a very important part of your immune system. The way in which they respond to pathogens, by producing antibodies, is called the **immune response**.

Lymphocytes and phagocytes will respond to any foreign cells in your body, even if they are not pathogens. This is a real problem in transplant surgery (section **11.22**). If a person's kidneys fail, they can be given a new kidney taken from another person. However, the recipient's immune system will recognise the cells in the new kidney as 'foreign', and will attack and destroy them. This is called **tissue rejection**. To lessen the chance of it happening, the donor must have very similar antigens to the recipient. The best way of achieving this is to take the organ from a close relative. However, this is not usually possible, so surgeons generally have to make do with the best match they can find. The recipient is then given immunosupressant drugs, which 'tone down' the immune system.

8.17 Platelets help blood clot.

Platelets are small fragments of cells, with no nucleus. They are made in the red bone marrow, and they are involved in blood clotting.

Blood clotting stops pathogens getting into the body through breaks in the skin. Normally, your skin provides a very effective barrier against the entry of bacteria and viruses. Blood clotting also prevents too much blood loss.

Figure **8.16** shows how blood clotting happens. Platelets are very important in this process. Normally, blood vessel walls are very smooth. When a blood vessel is cut, the platelets bump into the rough edges of the cut, and react by releasing a chemical. The damaged tissues around the blood vessel also release chemicals.

In the blood plasma, there is a soluble protein called **fibrinogen**. The chemicals released by the platelets and the damaged tissues set off a chain of reactions, which cause the fibrinogen to change into **fibrin**.

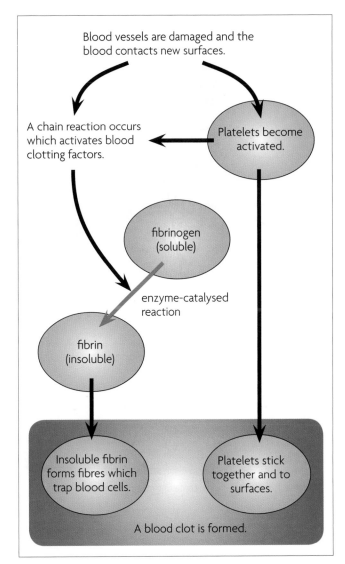

Figure 8.16 How blood clots.

Fibrin is insoluble. As its name suggests, it forms fibres. These form a mesh across the wound. Red blood cells and platelets get trapped in the tangle of fibrin fibres, forming a blood clot (Figures **8.17** and **8.18**, overleaf).

Questions

8.23 List **five** components of plasma.

8.24 Where are red blood cells made?

8.25 What is unusual about red blood cells?

8.26 What is haemoglobin?

8.27 Where are white blood cells made?

8.28 What are platelets?

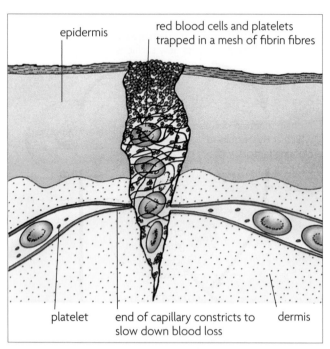

Figure 8.17 Vertical section through a blood clot.

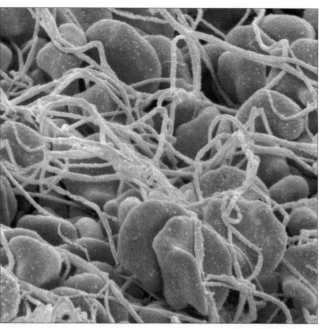

Figure 8.18 A scanning electron micrograph showing red cells tangled up in fibrin fibres (× 3600).

Table 8.3 Components of blood

	structure	functions
plasma	water, containing many substances in solution	1 liquid medium in which cells and platelets can float 2 transports CO_2 in solution 3 transports nutrients in solution 4 transports urea in solution 5 transports hormones in solution 6 transports heat 7 transports substances needed for blood clotting 8 transports antibodies
red cells	biconcave discs with no nucleus, containing haemoglobin	1 transport oxygen 2 transport small amount of CO_2
white cells	variable shapes, with nucleus	1 engulf and destroy pathogens (phagocytosis) 2 make antibodies
platelets	small fragments of cells, with no nucleus	help in blood clotting

8.18 Many substances are transported by blood.

Transport of oxygen In the lungs, oxygen diffuses from the alveoli into the blood (section 9.8). We have seen that the doughnut shape of the red blood cells (Figure 8.19) increases the surface area for diffusion, so that oxygen can diffuse into and out of the cells very rapidly. In the lungs, oxygen diffuses into the red blood cells, where it combines with the haemoglobin (Hb) to form **oxyhaemoglobin** (oxyHb).

The blood is then taken to the heart in the pulmonary veins and pumped out of the heart in the aorta.

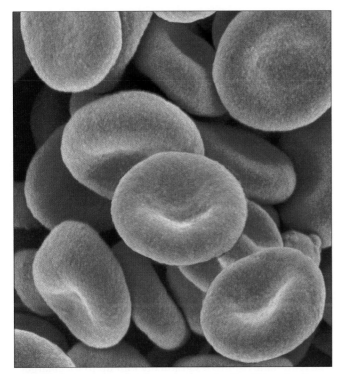

Figure 8.19 Scanning electron micrograph of red blood cells (×4500).

Arteries branch from the aorta to supply all parts of the body with oxygenated blood. When it reaches a tissue which needs oxygen, the oxyHb gives up its oxygen, to become Hb again.

Because capillaries are so narrow, the oxyHb in the red blood cells is taken very close to the tissues which need the oxygen. The oxygen only has a very short distance to diffuse. OxyHb is bright red, whereas Hb is purplish-red. The blood in arteries is therefore a brighter red colour than the blood in veins.

Transport of carbon dioxide   Carbon dioxide is made by all the cells in the body as they respire. The carbon dioxide diffuses through the walls of the capillaries into the blood.

Most of the carbon dioxide is carried by the blood plasma in the form of hydrogencarbonate ions, HCO_3^-. A small amount is carried by Hb in the red cells.

Blood containing carbon dioxide is returned to the heart in the veins, and then to the lungs in the pulmonary arteries. The carbon dioxide diffuses out of the blood and is passed out of the body on expiration.

Transport of food materials   Digested food is absorbed in the ileum (section **7.31**). It includes nutrients such as amino acids, fatty acids and glycerol, monosaccharides (such as glucose), water, vitamins and minerals. These all dissolve in the plasma in the blood capillaries in the villi.

These capillaries join up to form the hepatic portal vein. This takes the dissolved nutrients to the liver. The liver processes each nutrient and returns some of it to the blood.

The nutrients are then carried, dissolved in the blood, to all parts of the body.

Transport of urea   Urea, a waste substance (section **11.14**), is made in the liver. It dissolves in the blood plasma, and is carried to the kidneys. The kidneys excrete it in the urine.

Transport of hormones   Hormones are made in **endocrine** glands (section **10.17**). The hormones dissolve in the blood plasma, and are transported all over the body.

Transport of heat   Some parts of the body, such as the muscles, make a great deal of heat. The blood transports the heat to all parts of the body. This helps to keep the rest of the body warm.

Transport of plasma proteins   Several different proteins are dissolved in plasma. They are called **plasma proteins**. Fibrinogen (page **95**) is an example of a plasma protein.

Questions

8.29 Why is blood in arteries a brighter red than the blood in veins?

8.30 Which vessel transports digested food to the liver?

8.31 How is urea transported?

8.32 Outline **two** functions of blood other than transport.

Lymph and tissue fluid

8.19 Tissue fluid is leaked plasma.

Capillaries leak! The cells in their walls do not fit together exactly, so there are small gaps between them. Plasma can therefore leak out from the blood.

White blood cells can also get through these gaps. They are able to move and can squeeze through, out of the capillaries. Red blood cells cannot get out. They are too large and cannot change their shape very much.

So plasma and white cells are continually leaking out of the blood capillaries. The fluid formed in this way is called **tissue fluid**. It surrounds all the cells in the body (Figure **8.20**).

8.20 The functions of tissue fluid.

Tissue fluid is very important. It supplies cells with all their requirements. These requirements, such as oxygen and nutrients, diffuse from the blood, through the tissue fluid, to the cells. Waste products, such as carbon dioxide, diffuse in the opposite direction.

The tissue fluid is the immediate environment of every cell in your body. It is easier for a cell to carry out its functions properly if its environment stays constant. For example, this means it should stay at the same temperature, and at the same osmotic concentration.

Several organs in the body work to keep the composition and temperature of the blood constant, and therefore the tissue fluid as well. This process is called **homeostasis**, and is described in section **11.1**.

8.21 Lymph is drained tissue fluid.

The plasma and white cells that leak out of the blood capillaries must eventually be returned to the blood. In the tissues, as well as blood capillaries, are other small vessels. They are lymphatic capillaries (Figure **8.21**). The tissue fluid slowly drains into them. The fluid is now called **lymph**.

The lymphatic capillaries gradually join up to form larger lymphatic vessels (Figure **8.21**). These carry the lymph to the **subclavian veins** which bring blood back from the arms (Figure **8.22**). Here the lymph enters the blood again.

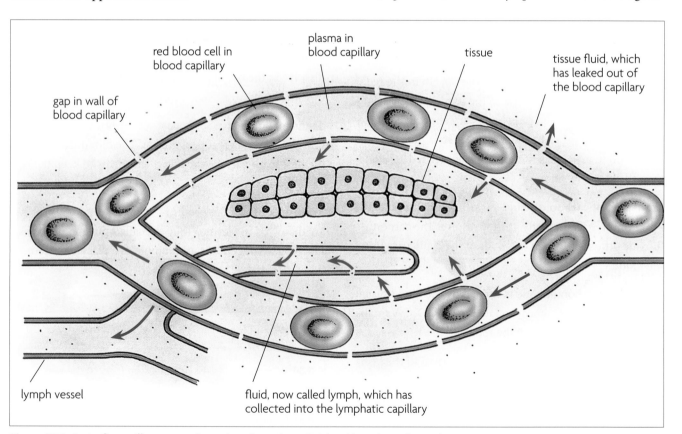

red blood cell in blood capillary

plasma in blood capillary

tissue

tissue fluid, which has leaked out of the blood capillary

gap in wall of blood capillary

lymph vessel

fluid, now called lymph, which has collected into the lymphatic capillary

Figure 8.20 Part of a capillary network, to show how tissue fluid and lymph are formed.

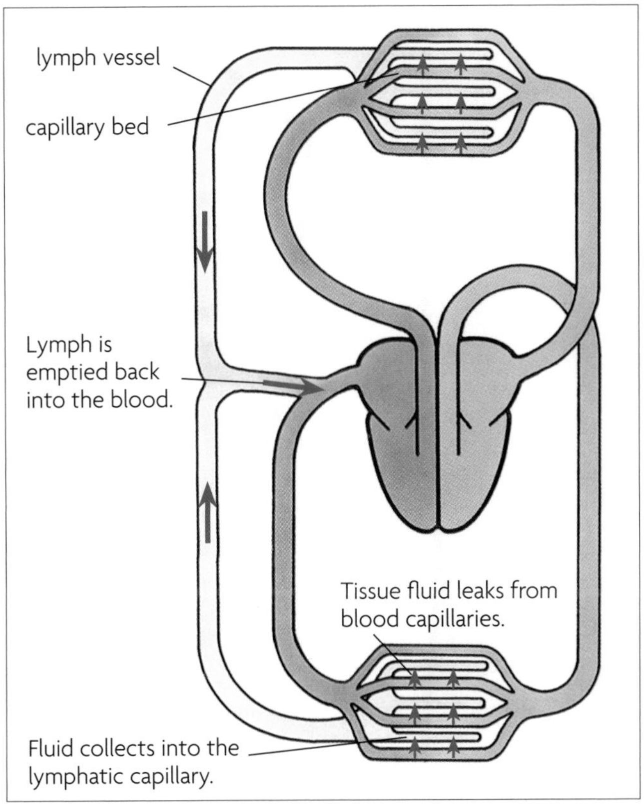

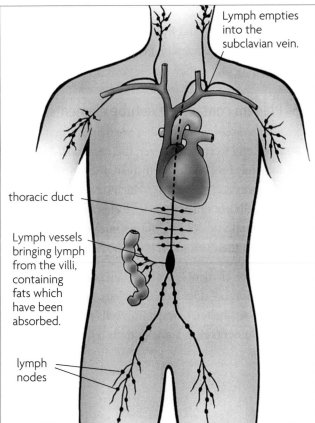

lymph vessel

capillary bed

Lymph is emptied back into the blood.

Tissue fluid leaks from blood capillaries.

Fluid collects into the lymphatic capillary.

Figure 8.21 The relationship between the blood circulation and the lymph circulation.

Lymph empties into the subclavian vein.

thoracic duct

Lymph vessels bringing lymph from the villi, containing fats which have been absorbed.

lymph nodes

Figure 8.22 The main lymph vessels and lymph nodes.

The lymphatic system has no pump to make the lymph flow. Lymph vessels do have valves in them, however, to make sure that movement is only in one direction. Lymph flows much more slowly than blood. Many of the larger lymph vessels run within or very close to muscles, and when the muscles contract they squeeze inwards on the lymph and force it to move along the vessels.

8.22 Lymph nodes contain white blood cells.

On its way from the tissues to the subclavian vein, lymph flows through several **lymph nodes**. Some of these are shown in Figure **8.22**.

Lymph nodes contain large numbers of white cells. Most bacteria or toxins in the lymph can be destroyed by these cells.

Questions

8.33 What is tissue fluid?

8.34 Give **two** functions of tissue fluid.

8.35 What is lymph?

8.36 Why do lymphatic capillaries have valves in them?

8.37 Name **two** places where lymph nodes are found.

8.38 What happens inside lymph nodes?

FACT! A human being contains about 70 cm³ of blood per kilogram of bodyweight. For an adult, this is about 4 or 5 litres of blood, in total.

Transport in flowering plants

8.23 Plants have two transport systems – phloem and xylem.

Transport systems in plants are less elaborate than in mammals. Plants are less active than mammals, and so their cells do not need to be supplied with materials so quickly. Also, the branching shape of a plant means that all the cells can get their oxygen for respiration, and carbon dioxide for photosynthesis, directly from the air, by diffusion.

Plants have two transport systems. The **xylem vessels** carry water and minerals, while the **phloem tubes** carry organic nutrients which the plant has made.

8.24 Xylem helps to support plants.

A xylem vessel is like a long drainpipe (Figures **8.23** and **8.24**). It is made of many hollow, dead cells, joined end to end. The end walls of the cells have disappeared, so a long, open tube is formed. Xylem vessels run from the roots of the plant, right up through the stem. They branch out into every leaf.

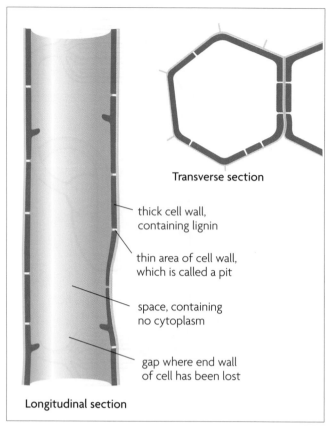

Transverse section

thick cell wall, containing lignin

thin area of cell wall, which is called a pit

space, containing no cytoplasm

gap where end wall of cell has been lost

Longitudinal section

Figure 8.24 Xylem vessels.

Xylem vessels contain no cytoplasm or nuclei. Their walls are made of cellulose and **lignin**. Lignin is very strong, so xylem vessels help to keep plants upright. Wood is made almost entirely of lignified xylem vessels.

8.25 Phloem contains sieve tube elements.

Like xylem vessels, phloem tubes are made of many cells joined end to end. However, their end walls have not completely broken down. Instead, they form **sieve plates** (Figures 8.25 and 8.26), which have small holes in them. The cells are called sieve tube elements. Sieve tube elements contain cytoplasm, but no nucleus. They do not have lignin in their cell walls.

Each sieve tube element has a companion cell next to it. The companion cell does have a nucleus, and also contains many other organelles. Companion cells probably supply sieve tube elements with some of their requirements.

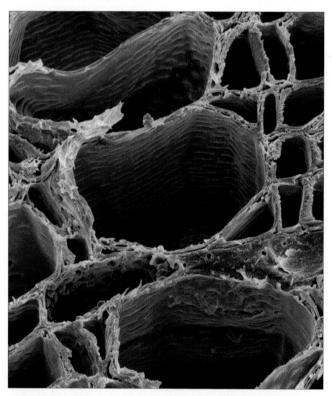

Figure 8.23 This is a scanning electron micrograph of xylem vessels (×1800).

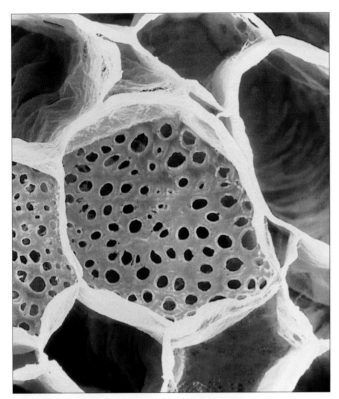

Figure 8.25 This scanning electron micrograph shows a sieve plate in a phloem sieve tube (×1300).

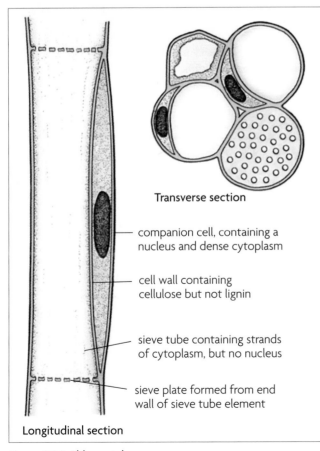

Transverse section

— companion cell, containing a nucleus and dense cytoplasm

— cell wall containing cellulose but not lignin

— sieve tube containing strands of cytoplasm, but no nucleus

— sieve plate formed from end wall of sieve tube element

Longitudinal section

Figure 8.26 Phloem tubes.

8.26 Vascular bundles contain xylem and phloem.

Xylem vessels and phloem tubes are usually found close together. A group of xylem vessels and phloem tubes is called a **vascular bundle**.

The positions of vascular bundles in roots and shoots are shown in Figures **8.27** and **8.28**. In a root, vascular tissue is found at the centre, whereas in a shoot vascular bundles are arranged in a ring near the outside edge. Vascular bundles are also found in leaves (Figure **6.1**). They help to support the plant.

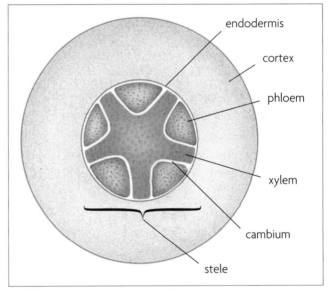

Figure 8.27 Transverse section of a root.

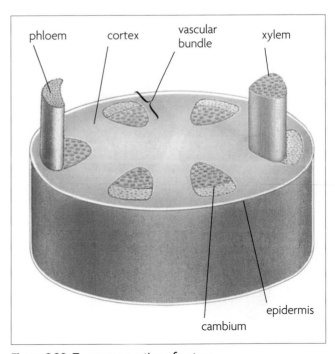

Figure 8.28 Transverse section of a stem.

FACT!

The longest roots measured were produced by a rye plant (a grass). It had over 600 km of roots in only 0.051 m³ of soil.

The deepest roots which have ever been measured belonged to a fig tree growing in southern Africa. They went down to a depth of 120 m.

8.27 The structure of a root.

Plants take in water from the soil, through their **root hairs**. The water is carried by the xylem vessels to all parts of the plant. Figure **8.29** shows the end of a root, magnified. At the very tip is a **root cap**. This is a layer of cells which protects the root as it grows through the soil. The rest of the root is covered by a layer of cells called the **epidermis**.

The root hairs are a little way up from the root tip. Each root hair is a long epidermal cell (Figure **8.30**). Root hairs do not live for very long. As the root grows, they are replaced by new ones.

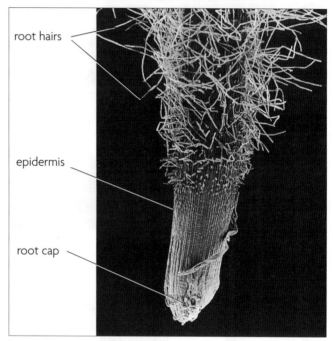

Figure 8.29 A root tip (× 50).

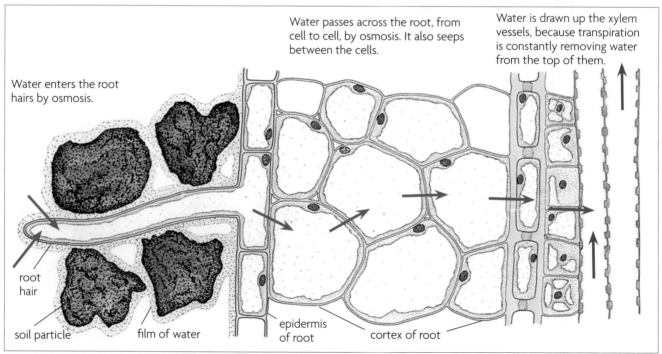

Water passes across the root, from cell to cell, by osmosis. It also seeps between the cells.

Water is drawn up the xylem vessels, because transpiration is constantly removing water from the top of them.

Water enters the root hairs by osmosis.

root hair

soil particle

film of water

epidermis of root

cortex of root

Figure 8.30 How water is absorbed by a plant.

8.28 Root hairs absorb water by osmosis.

The function of root hairs is to absorb water and minerals from the soil. Water gets into a root hair by **osmosis**. The cytoplasm and cell sap inside it are quite concentrated solutions. The water in the soil is normally a more dilute solution. Water therefore diffuses into the root hair, down its concentration gradient, through the partially permeable cell surface membrane (section **3.3**).

8.29 Absorbed water enters the xylem.

The root hairs are on the edge of the root. The xylem vessels are in the centre. Before the water can be taken to the rest of the plant, it must travel to these xylem vessels.

The path it takes is shown in Figure **8.30**. It travels by osmosis through the cortex, from cell to cell. Some of it may also just seep through the spaces between the cells, or through the cell walls, never actually entering a cell at all.

8.30 Water is sucked up the xylem.

Once water reaches the xylem, it moves up xylem vessels in the same way that a drink moves up a straw when you suck it. When you suck a straw, you are reducing the pressure at the top of the straw. The liquid at the bottom of the straw is at a higher pressure, so it flows up the straw into your mouth.

The same thing happens with the water in xylem vessels. The pressure at the top of the vessels is lowered, while the pressure at the bottom stays high. Water therefore flows up the xylem vessels.

How is the pressure at the top of the xylem vessels reduced? It happens because of **transpiration**.

8.31 Transpiration is evaporation from the parts of the plant above the ground.

Transpiration is the evaporation of water from a plant. Most of this evaporation takes place from the leaves.

> ### Key definition
>
> **transpiration**   evaporation of water at the surfaces of the mesophyll cells followed by loss of water vapour from plant leaves, through the stomata

If you look back at Figure **6.4** (page 52), you will see that there are openings on the surface of the leaf called **stomata**. There are usually more stomata on the underside of the leaf, in the lower epidermis. The cells inside the leaf are each covered with a thin film of moisture.

Some of this film of moisture evaporates from the cells, and this water vapour diffuses out of the leaf through the stomata. Water from the xylem vessels in the leaf will travel to the cells by osmosis to replace it.

Water is constantly being taken from the top of the xylem vessels, to supply the cells in the leaves. This reduces the effective pressure at the top of the xylem vessels, so that water flows up them. This process is known as the **transpiration stream** (Figure **8.31**).

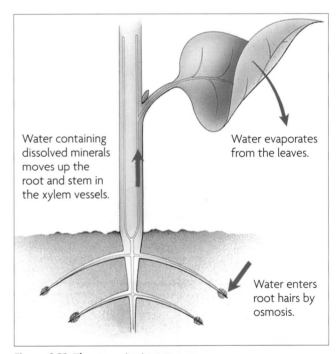

Water containing dissolved minerals moves up the root and stem in the xylem vessels.

Water evaporates from the leaves.

Water enters root hairs by osmosis.

Figure 8.31 The transpiration stream.

8.32 Water moves down a water potential gradient.

You can think of the way that water moves into a root hair, across to the xylem vessels, up to the leaves and then out into the air in terms of water potential.

You may remember that water moves down a water potential gradient, from a high water potential to a low water potential (section **3.3**). All along this pathway, the water is moving down a water potential gradient from one place to another. The highest water potential is in the solution in the soil, and the lower water potential is in the air.

The low water potential in the leaves is caused by the loss of water vapour from the leaves by transpiration. This produces a 'pull' from above, drawing water up the plant.

Water molecules have a strong tendency to stick together. This is called cohesion. When the water is 'pulled' up the xylem vessels, the whole column of water stays together. Without cohesion, the water column would break apart and the whole system would not work.

8.33 A potometer compares transpiration rates.

It is not easy to measure how much water is lost from the leaves of a plant. It is much easier to measure how fast the plant takes up water. The rate at which a plant takes up water depends on the rate of transpiration – the faster a plant transpires, the faster it takes up water.

Figure 8.32 illustrates apparatus which can be used to compare the rate of transpiration in different conditions. It is called a **potometer**. By recording how fast the air/water meniscus moves along the capillary tube you can compare how fast the plant takes up water in different conditions.

There are many different kinds of potometer, so yours may not look like this. The simplest kind is just a long glass tube which you can fill with water. A piece of rubber tubing slid over one end allows you to fix the cut end of a shoot into it, making an air-tight connection. This works just as well as the one in Figure 8.32, but is much harder to refill with water.

8.34 Conditions that affect transpiration rate.

Temperature On a hot day, water will evaporate quickly from the leaves of a plant. Transpiration increases as temperature increases.

Humidity Humidity means the moisture content of the air. The higher the humidity, the less water will evaporate from the leaves. This is because there is not much of a diffusion gradient for the water between the air spaces inside the leaf, and the wet air outside it. Transpiration decreases as humidity increases.

Wind speed On a windy day, water evaporates more quickly than on a still day. Transpiration increases as wind speed increases.

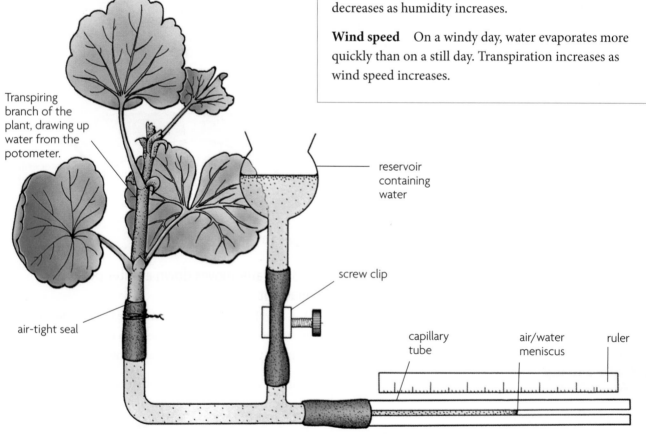

Transpiring branch of the plant, drawing up water from the potometer.

reservoir containing water

screw clip

air-tight seal

capillary tube

air/water meniscus

ruler

Figure 8.32 A potometer.

Light intensity   In bright sunlight, a plant may open its stomata to supply plenty of carbon dioxide for photosynthesis. More water can therefore evaporate from the leaves.

Water supply   If water is in short supply, then the plant will close its stomata. This will cut down the rate of transpiration. Transpiration decreases when water supply decreases below a certain level.

Transpiration is useful to plants, because it keeps water moving up the xylem vessels and evaporation helps to cool the leaves. But if the leaves lose too much water, the roots may not be able to take up enough to replace it. If this happens, the plants **wilts**, because the cells lose water by osmosis and become **flaccid** (section **3.6**).

Activity 8.2
To see which part of a stem transports water and solutes

skills

C1 *Using techniques, apparatus and materials*

C2 *Observing, measuring and recording*

C3 *Interpreting and evaluating*

C4 *Planning*

Safety Take care with the sharp blade when cutting the stem sections.

1 Take a plant, such as *Impatiens*, with a root system intact. Wash the roots thoroughly.

2 Put the roots of the plant into eosin solution. Leave overnight.

3 Set up a microscope.

4 Remove the plant from the eosin solution, and wash the roots thoroughly.

5 Use a razor blade to cut across the stem of the plant about half-way up. Take great care when using a razor blade and do not touch its edges.

6 Now cut very thin sections across the stem. Try to get them so thin that you can see through them. It does not matter if your section is not a complete circle.

7 Choose your thinnest section, and mount it in a drop of water on a microscope slide. Cover with a coverslip.

8 Observe the section under a microscope. Compare what you can see with Figure **8.28** (page **101**). Make a labelled drawing of your section.

Questions

1 Which part of the stem contained the dye? What does this tell you about the transport of water and solutes (substances dissolved in water) up a stem?

2 Why was it important to wash the roots of the plant:

   a before putting it into the eosin solution, and

   b before cutting sections?

3 Design an experiment to investigate the effect of **one** factor (for example, light intensity, temperature, wind speed) on the rate at which the dye is transported up the stem. Remember to write down your hypothesis, and to think about variables. When you have completed your plan, ask your teacher to check it for you. Then carry out your experiment and record and display your results. Write down your conclusions, and discuss them in the light of your knowledge about transport in plants. You should also evaluate the reliability of your results and suggest how you could improve your experiment if you were able to do it again.

Activity 8.3
To see which surface of a leaf loses most water

C2 *Observing, measuring and recording*

C3 *Interpreting and evaluating*

Cobalt chloride paper is blue when dry and pink when wet. Use forceps to handle it.

1   Use a healthy, well-watered potted plant, with leaves which are not too hairy. Fix a small square of blue cobalt chloride paper onto each surface of one leaf, using clear sticky tape. Make sure there are no air spaces around the paper.

2   Leave the paper on the leaf for a few minutes.

Questions

1   Which piece of cobalt chloride paper turned pink first? What does this tell you about the loss of water from a leaf?

2   Why does this surface lose water faster than the other?

3   Why is it important to use forceps, not fingers, for handling cobalt chloride paper?

Activity 8.4
To measure the rate of transpiration of a potted plant

C2 *Observing, measuring and recording*

C3 *Interpreting and evaluating*

1   Use two similar well-watered potted plants. Enclose one plant entirely in a polythene bag, including its pot. This is the control.

2   Enclose only the pot of the second plant in a polythene bag. Fix the bag firmly around the stem of the plant, as in the diagram, and seal with petroleum jelly.

3   Place both plants on balances, and record their masses.

4   Record the mass of each plant every day, at the same time, for at least a week.

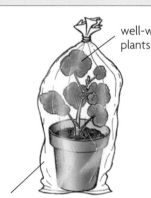 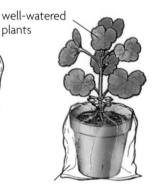

well-watered plants

tightly sealed polythene bag enclosing the entire plant and pot

tightly sealed polythene bag enclosing just the pot and soil

Questions

1   Which plant lost mass? Why?

2   Do you think this is a good method of measuring transpiration rate? How could it be improved?

Activity 8.5
Using a potometer to compare rates of transpiration under different conditions

C1 *Using techniques, apparatus and materials*

C2 *Observing, measuring and recording*

C3 *Interpreting and evaluating*

1 Set up the potometer as in Figure **8.32** (page **104**). The stem of the plant must fit exactly into the rubber tubing, with no air gaps. Petroleum jelly will help to make an air-tight seal.

2 Fill the apparatus with water, by opening the clip.

3 Close the clip again, and leave the apparatus in a light, airy place. As the plant transpires, the water it loses is replaced by water taken up the stem. Air will be drawn in at the end of the capillary tube.

4 When the air/water meniscus reaches the scale, begin to record the position of the meniscus every two minutes.

5 When the meniscus reaches the end of the scale, refill the apparatus with water from the reservoir as before.

6 Now repeat the investigation, but with the apparatus in a different situation. You could try each of these:
   ● blowing it with a fan
   ● putting it in a cupboard
   ● putting it in a refrigerator.

7 Draw graphs of your results.

Questions

1 Under which conditions did the plant transpire **(a)** most quickly, and **(b)** most slowly?

2 You have been using the potometer to compare the rate of uptake of water under different conditions. Does this really give you a good measurement of the rate of transpiration? Explain your answer.

8.35 Desert plants must cut down water loss.

Plants that live in deserts can easily run short of water, especially if the temperatures are hot. Desert plants, such as cacti (Figure **8.33**) and succulents (Figure **8.34**), must be well adapted to survive in these difficult conditions. They all have several ways of cutting down the rate of water loss by transpiration, including those listed on page **108**.

Figure 8.33 *Ferocactus* – a plant adapted to live in deserts.

Figure 8.34 *Pachypodium* – a succulent living in Madagascar.

Closing stomata Plants lose most water through their stomata. If they close their stomata, then transpiration will slow right down. Figure 8.35 shows how they do this.

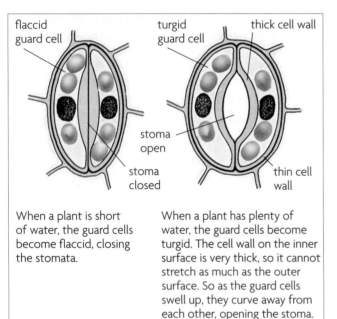

flaccid guard cell

turgid guard cell

thick cell wall

stoma open

stoma closed

thin cell wall

When a plant is short of water, the guard cells become flaccid, closing the stomata.

When a plant has plenty of water, the guard cells become turgid. The cell wall on the inner surface is very thick, so it cannot stretch as much as the outer surface. So as the guard cells swell up, they curve away from each other, opening the stoma.

Figure 8.35 How stomata open and close.

However, if its stomata are closed, then the plant cannot photosynthesise, because carbon dioxide cannot diffuse into the leaf. Stomata close when it is very hot and dry, or when they could not photosynthesise anyway, such as at night.

Waxy cuticle The leaves of desert plants are often covered with a waxy cuticle, made by the cells in the epidermis. The wax makes the leaf waterproof.

Hairy leaves Some plants have hairs on their leaves. These hairs trap a layer of moist air next to the leaf.

Stomata on underside of leaves In most leaves, there are more stomata on the lower surface than on the upper surface. The lower surface is usually cooler than the upper one, so less water will evaporate. In desert plants, there may be fewer stomata than usual, and they may be sunk into deep pits in the leaf.

Cutting down on the surface area The smaller the surface area of the leaf, the less water will evaporate from it. Plants like cacti (Figure 8.33) have leaves with a small surface area, to help them to conserve water. However, this slows down photosynthesis, because it means less light and carbon dioxide can be absorbed.

Having deep or spreading roots Desert plants may have to seek water very deep down in the soil, or across a wide area. They usually have either very deep roots, or roots that spread a long way sideways from where the plant is growing.

In fact, many plants – even those that do not live in deserts – have at least some of these adaptations. For example, a plant growing in your garden may have to cope with hot, dry conditions at least some of the time. Most plants have stomata only on the undersides of their leaves, which close when the need arises. Most of them have waxy cuticles on their leaves, to cut down water loss. Desert plants, though, show these adaptations to a much greater extent.

8.36 Water plants may have stomata on the tops of their leaves.

Water plants have no problem of water shortage. They do not need adaptations to conserve water, as desert plants do.

The water hyacinth, *Eichhornia crassipes*, is an example of a plant adapted to live in water (Figure 8.36). The roots of water hyacinths do not attach to the bed of the river or pond where they grow, but just float freely in the water. The stems and leaf stalks have hollow spaces in them, filled with air, which help them to float on the top of the water where they can get plenty of light for photosynthesis.

Figure 8.36 These big clumps of water hyacinth are floating freely in the water. Water hyacinth has become a very serious weed in many parts of the world, clogging up waterways and preventing light, oxygen and carbon dioxide reaching other plants growing in the water.

Water hyacinth leaves have stomata on both surfaces, not just on the underside as in most plants. This allows them to absorb carbon dioxide from the air, for photosynthesis. The cuticle on the upper and lower surfaces of the leaves is much thinner than in plants that don't live in water. There is no need for a thick cuticle, because there is no need to prevent water loss from the leaves.

Uptake of mineral salts

8.37 Root hairs absorb minerals by active transport.

As well as absorbing water by osmosis, root hairs absorb mineral salts. These are in the form of ions dissolved in the water in the soil. They travel to the xylem vessels along with the water which is absorbed, and are transported to all parts of the plant.

These minerals are usually present in the soil in quite low concentrations. The concentration inside the root hairs is higher. In this situation the mineral ions would normally diffuse out of the root hair into the soil. Root hairs can, however, take up mineral salts against their concentration gradient. It is the cell surface membrane which does this. Special carrier molecules in the cell membrane of the root hair carry the mineral ions across the cell membrane into the cell, against their concentration gradient. This is called **active transport**, and is described in section 3.7.

Transport of manufactured food

8.38 Phloem translocates organic foods.

Leaves make carbohydrates by photosynthesis. They also use some of these carbohydrates to make amino acids, proteins, oils and other organic substances.

Some of the organic food material, especially sugar, that the plant makes is transported in the phloem tubes. It is carried from the leaves to whichever part of the plant needs it. This is called **translocation**. The sap inside the phloem tubes therefore contains a lot of sugar, particularly **sucrose**.

Key definition

translocation   the movement of sucrose and amino acids in phloem, from regions of production to regions of storage, or to regions of utilisation in respiration or growth

Questions

8.45 What is the function of a root cap?

8.46 Explain how water goes into root hairs. How does this process differ from the way in which minerals enter?

8.47 What is transpiration?

8.48 What are stomata?

8.49 What is a potometer used for?

8.50 Explain how (a) temperature, and (b) light intensity affect the rate of transpiration.

8.39 Systemic pesticides are translocated in phloem.

People who grow crops for food sometimes need to use chemicals called pesticides. Pests, such as insects that eat the crop plants, or fungi that grow on them, can greatly reduce the yield of the crop. Pesticides are used to kill the insects or fungi.

Some pesticides kill only the insects or fungus that the spray touches. They are called contact pesticides. They can be very effective if they are applied properly, but sometimes it is better to use a systemic pesticide. A systemic pesticide is absorbed into the plant and carried all the way through it in its phloem tissue. So any insect feeding on the plant, even if it was hidden under the leaf where the spray could not reach it, will eventually end up feeding on pesticide. The same is true for fungi; no matter where they are growing on the plant, the pesticide will eventually reach them. You can read more about the use of pesticides in Chapter 16.

8.40 Sources and sinks vary with the seasons.

The part of a plant from which sucrose and amino acids are being translocated is called a **source**. The part of the plant to which they are being translocated is called a **sink**.

When a plant is actively photosynthesising and growing, the leaves are generally the major sources of translocated material. They are constantly producing sucrose, which is carried in the phloem to all other parts of the plant. These parts – the sinks – include the roots and flowers. The roots may change some of the sucrose to starch and store it. The flowers use the sucrose to make fructose (an especially sweet-tasting sugar found in nectar). Later, when the fruits are developing, quite large amounts of sucrose may be used to produce sweet, juicy fruits ready to attract animals (section **13.39**).

But many plants have a time of year when they become **dormant**. During this stage, they wait out harsh conditions in a state of reduced activity. In a hot climate, this may be during the hottest, driest season. In temperate countries, it may be during the winter.

Dormant plants do not photosynthesise, but survive on their stored starch, oils and other materials. When the seasons change, they begin to grow again. Now the stored materials are converted to sucrose and transported to the growing regions.

For example, potato plants (Figure **8.37**) grow in temperate regions, and are not able to survive the cold frosts of winter. During the summer, the leaves photosynthesise and send sucrose down into underground stems. Here,

swellings called stem **tubers** develop. The cells in the root tubers change the sucrose to starch and store it.

In autumn, the leaves die. Nothing is left of the potato plant above ground – just the stem tubers beneath the soil. In spring, they begin to grow new shoots and leaves. The starch in the tubers is changed back to sucrose, and transported in the phloem to the growing stems and leaves. This will continue until the leaves are above ground and photosynthesising.

So, in summer the leaves are sources and the growing stem tubers are sinks. In spring, the stem tubers are sources and the growing leaves are sinks.

You can see from this example that phloem can transfer sucrose in either direction – up or down the plant. This isn't true for the transport of water in the xylem vessels. That can only go upwards, because transpiration always happens at the leaf surface, and it is this that provides the 'pull' to draw water up the plant.

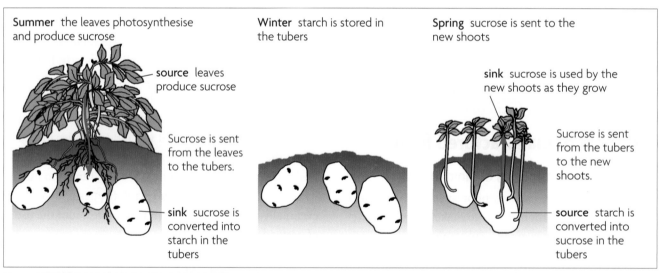

Figure 8.37 Potato plants in summer and spring.

Key ideas

- Mammals have a double circulatory system, in which blood is moved through vessels by the regular contraction and relaxation of cardiac muscles in the wall of the heart.

- Blood enters the atria of the heart, flows through open valves into the ventricles, and is then forced out into the arteries during systole.

- The ventricles have thicker walls than the atria, and the left ventricle has a thicker wall than the right ventricle, to allow them to produce a greater force when the muscles contract, necessary so that they can push the blood further.

(continued ...)

(... continued)

◆ In coronary heart disease, the coronary arteries become blocked, so oxygen is not delivered to the heart muscles and they stop contracting. Smoking, stress and a diet high in saturated fats increase the risk.

◆ Arteries are thick-walled, elastic vessels that carry pulsing, high-pressure blood away from the heart. They split into capillaries, which are tiny vessels with walls only one cell thick. Capillaries take blood close to every cell in the body, so that the cells are supplied with oxygen and nutrients and have their waste products removed. Capillaries join up to form veins. Veins are thin-walled vessels with valves, which carry low-pressure blood back to the heart.

◆ Blood contains red cells, white cells and platelets floating in plasma. Plasma transports many different substances in solution. Red cells contain the iron-containing protein haemoglobin, which transports oxygen. White cells fight against bacteria and viruses. Platelets help the blood to clot.

◆ Fluid leaks out of capillaries to fill the spaces between all the body cells, where it is called tissue fluid. It is collected into lymph vessels which carry it back to the bloodstream.

◆ In plants, xylem vessels transport water and mineral ions from the roots upwards to the leaves. Phloem tubes transport sucrose and other organic nutrients from the leaves where they are made to all parts of the plant. This is called translocation.

◆ Xylem vessels are made of dead, empty cells with strong lignin in their walls. As well as transporting water, they help to support the plant.

◆ Water is drawn up xylem vessels by the evaporation of water from the leaves, called transpiration. Transpiration happens fastest when it is hot, dry, windy and sunny.

◆ Water enters root hairs by osmosis, and then moves across the cortex of the root into the xylem.

◆ Root hairs take up mineral ions by active transport, using energy supplied by respiration to move them against their concentration gradient.

◆ Phloem is made of living cells with sieve plates at their ends. A companion cell is associated with each phloem sieve tube element.

◆ Systemic pesticides are translocated in phloem.

◆ Sucrose is translocated from sources to sinks. Different parts of a plant may become sources and sinks in different seasons.

Revision questions

1 Using Figure 8.11 to help you, list in order the blood vessels and parts of the heart which:
   a a glucose molecule would travel through on its way from your digestive system to a muscle in your leg
   b a carbon dioxide molecule would travel through on its way from the leg muscle to your lungs.

2 Explain the difference between each of the following pairs.
   a blood, lymph
   b diastole, systole
   c artery, vein
   d deoxygenated blood, oxygenated blood
   e atrium, ventricle
   f hepatic vein, hepatic portal vein
   g red blood cell, white blood cell
   h xylem, phloem
   i diffusion, active transport
   j source, sink

3 Arteries, veins, capillaries, xylem vessels and phloem tubes are all tubes used for transporting substances in mammals and flowering plants. Describe how each of these tubes is adapted for its particular function.

9 Respiration

In this chapter, you will find out:

- ◆ how respiration in all cells releases useful energy from food
- ◆ the equations for aerobic respiration and anaerobic respiration
- ◆ how yeast is used in brewing and baking
- ◆ what a gas exchange surface is, and the features it must have
- ◆ how gas exchange occurs in human lungs
- ◆ the roles of cilia and goblet cells
- ◆ how breathing takes place
- ◆ the differences between inspired and expired air, and how to measure them
- ◆ the effect of exercise on rate and depth of breathing
- ◆ how a fall in blood pH increases rate and depth of breathing
- ◆ about oxygen debt.

Energy for cells

9.1 Respiration releases energy from food.

Every living cell needs energy. In humans, our cells need energy for:

- contracting muscles, so that we can move parts of the body
- making protein molecules by linking together amino acids into long chains
- cell division, so that we can repair damaged tissues and can grow
- active transport, so that we can move substances across cell membranes up their concentration gradients
- transmitting nerve impulses, so that we can transfer information quickly from one part of the body to another
- producing heat inside the body, to keep the body temperature constant even if the environment is cold.

All of this energy comes from the food that we eat. The food is digested – that is, broken down into smaller molecules – which are absorbed from the intestine into the blood. The blood transports the nutrients to all the cells in the body. The cells take up the nutrients that they need.

The main nutrient used to provide energy in cells is **glucose**. Glucose contains a lot of chemical energy. In order to make use of this energy, cells have to break down the glucose molecules and release the energy from them. They do this in a metabolic reaction called **respiration**.

> **Key definition**
>
> **respiration** the chemical reactions that break down nutrient molecules in living cells to release energy

9.2 Aerobic respiration involves oxygen.

Most of the time, our cells release energy from glucose by combining it with oxygen. This is called **aerobic respiration**.

This happens in a series of small steps, each one controlled by enzymes. We can summarise the reactions of aerobic respiration as an equation.

> glucose + oxygen $\longrightarrow$ carbon dioxide + water

> $C_6H_{12}O_6 + 6O_2 \longrightarrow 6CO_2 + 6H_2O$

Key definition

aerobic respiration   the release of a relatively large amount of energy in cells by the breakdown of food substances in the presence of oxygen

Activity 9.1
Investigating heat production by germinating peas

skills

C2 *Observing, measuring and recording*

C3 *Interpreting and evaluating*

1   Soak some peas (or beans) in water for a day, so that they begin to germinate.

2   Boil a second set of peas, to kill them.

3   Wash both sets of peas in dilute disinfectant, so that any bacteria and fungi on them are killed.

4   Put each set of peas into a vacuum flask as shown in the diagram. Do not fill the flasks completely.

5   Note the temperature of each flask.

6   Support each flask upside down, and leave them for a few days.

7   Note the temperature of each flask at the end of your experiment.

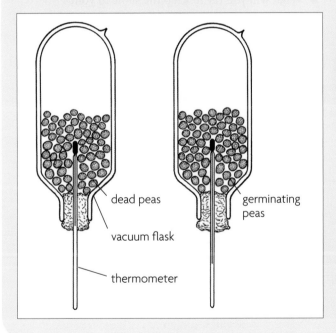

dead peas

germinating peas

vacuum flask

thermometer

Questions

1   Which flask showed the higher temperature at the end of the experiment? Explain your answer.

2   Why is it important to kill any bacteria and fungi on the peas?

3   Why should the flasks not be completely filled with peas?

4   Carbon dioxide is a heavy gas. Why were the flasks left upside down, with porous cotton wool plugs in them?

5   Not all of the energy produced by the respiring peas is given off as heat. What happens to the rest of it?

Activity 9.2
Comparing the energy content of two kinds of food

skills

C1 *Using techniques, apparatus and materials*

C2 *Observing, measuring and recording*

C3 *Interpreting and evaluating*

C4 *Planning*

Safety Wear eye protection if available.

You know that food contains energy. We can change this energy into heat energy by burning the food. We can measure the amount of heat energy that is produced by measuring the temperature change in a known volume of water.

The diagram shows the apparatus you can use. You will also need a thermometer.

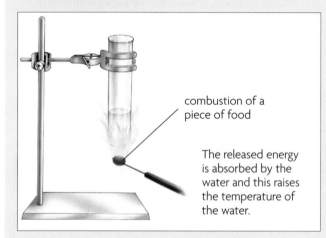

combustion of a piece of food

The released energy is absorbed by the water and this raises the temperature of the water.

In order to calculate the energy that is released as heat when you burn the food, you need to know:
- the volume of water in the tube
- the initial temperature of water in the tube
- the final temperature of water in the tube.

You can then calculate the amount of heat energy that went into the water using this formula:

$$\text{heat energy in J} = \text{temperature change in °C} \times \text{volume of water in cm}^3 \times 4.2$$

Your task is to design and carry out an investigation to compare the amount of energy in two kinds of food.

Suitable foods could be: plain popcorn and popcorn soaked in oil; white bread and brown bread; a peanut and a cashew nut. Your teacher will suggest which foods you can use.

1   Decide on a hypothesis you will investigate. The hypothesis should be one sentence, and should state which of the two foods you predict contains more energy.

2   Plan how you will carry out your investigation. Then think through it again, and make improvements to your plan. Once you are fairly happy with it, show your teacher. You must not try to do your experiment until your teacher says that you may begin.

- What apparatus and other materials will you need for your experiment?
- What will you vary in your experiment? How will you vary it?
- What will you try to keep the same in your experiment? How will you do this?
- What will you measure in your experiment? How will you measure it? When will you measure it? Will you do repeat measurements and calculate a mean?
- How will you record your results? (You can sketch out a results chart, ready to fill in.)
- How will you use your results to calculate the amount of energy in the food?
- What will your results be if your hypothesis is correct?

Once you have approval from your teacher, you should do your experiment. Most scientific researchers find that they want to make changes to their experiment once they actually begin doing it. This is a good thing to do. Make careful notes about all the changes that you make.

Finally, write up your experiment, including:
- a heading, and the hypothesis that you tested
- a diagram of the apparatus that you used, and a full description of your method

(continued...)

(... continued)

- a neat and carefully headed table of results, including means if you decided to do repeats
- a conclusion, in which you say whether or not your results support your hypothesis
- a discussion, in which you use what you know about the energy content of different nutrients (look back

at pages **64** and **65**) to try to explain the pattern in your results
- an evaluation, in which you explain the main limitations that you feel might have affected the reliability of your data.

9.3 Anaerobic respiration does not involve oxygen.

It is possible to release energy from sugar without using oxygen. It is not such an efficient process as aerobic respiration and not much energy is released, but the process is used by some organisms. It is called **anaerobic respiration** ('an' means without).

> **Key definition**
>
> **anaerobic respiration**   the release of a relatively small amount of energy by the breakdown of food substances in the absence of oxygen

Yeast, a single-celled fungus, can respire anaerobically. It breaks down glucose to alcohol.

> glucose $\longrightarrow$ alcohol + carbon dioxide

> $C_6H_{12}O_6 \longrightarrow 2C_2H_5OH + 2CO_2$

As in aerobic respiration, carbon dioxide is made. Plants can also respire anaerobically like this, but only for short periods of time.

Some of the cells in your body, particularly muscle cells, can respire anaerobically for a short time. They make lactic acid instead of alcohol and no carbon dioxide is produced.

> glucose $\longrightarrow$ lactic acid

> $C_6H_{12}O_6 \longrightarrow 2C_3H_6O_3$

9.4 Yeast is used for baking and brewing.

Breaking down sugar to alcohol and carbon dioxide, which is the way in which yeast respires anaerobically, is called **fermentation**. Fermentation is used to make drinks such as beer and wine.

Brewing   To make beer, yeast is dissolved in a warm liquid containing the sugar maltose. The maltose comes from germinating barley seeds. The yeast respires anaerobically, breaking down the maltose and making alcohol and carbon dioxide. The carbon dioxide makes the beer fizzy.

Bread making   When making bread, flour is mixed with water to make a dough. Flour contains starch and some of this breaks down to the sugar maltose when the flour is moistened. Yeast is added to the dough and breaks down the sugar as it respires.

There is air in the dough, so the yeast respires aerobically at first, until the oxygen is used up. It makes carbon dioxide, and bubbles of this gas get caught in the dough, making it rise. The yeast is killed when the bread is cooked.

Table 9.1 A comparison of aerobic and anaerobic respiration

aerobic respiration	anaerobic respiration
uses oxygen	does not use oxygen
no alcohol or lactic acid made	alcohol (in yeast and plants) or lactic acid (in animals) is made
large amount of energy released from each molecule of glucose	much less energy released from each molecule of glucose
carbon dioxide made	carbon dioxide is made by yeast and plants, but not by animals

Activity 9.3
Investigating the production of carbon dioxide by anaerobic respiration

skills

C1 *Using techniques, apparatus and materials*
C2 *Observing, measuring and recording*
C3 *Interpreting and evaluating*
C4 *Planning*

1   Boil some water, to drive off any dissolved air.

2   Dissolve a small amount of sugar in the boiled water, and allow it to cool.

3   When it is cool, add yeast and stir with a glass rod.

4   Set up the apparatus as in the diagram. Add the liquid paraffin by trickling it gently down the side of the tube, using a pipette.

5   Set up an identical piece of apparatus, but use boiled yeast instead of living yeast.

6   Leave your apparatus in a warm place.

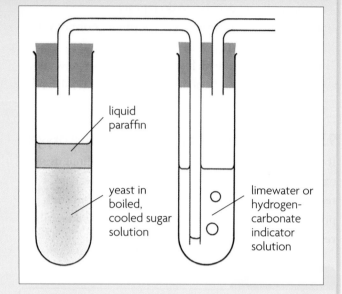

liquid paraffin

yeast in boiled, cooled sugar solution

limewater or hydrogen-carbonate indicator solution

7   Observe what happens to the limewater after half an hour.

Questions

1   Why is it important to boil the water?

2   Why must the sugar solution be cooled before adding the yeast?

3   What is the liquid paraffin for?

4   What happened to the limewater or hydrogencarbonate indicator solution in each of your pieces of apparatus? What does this show?

5   What new substance would you expect to find in the sugar solution containing living yeast at the end of the experiment?

6   Describe a method you could use to compare the rate of carbon dioxide production by yeast using different kinds of sugar. Remember to describe the variables you will change, those you will control and how, and how you will collect, record and analyse your results.

Questions

9.1   What is the purpose of respiration?

9.2   What is the energy released by respiration used for?

9.3   What is anaerobic respiration?

9.4   Name an organism which can respire anaerobically.

9.5   What is fermentation?

9.6   Why does bread not taste of alcohol after baking?

Gas exchange in humans

9.5 Gas exchange occurs at special surfaces.

If you look back at the aerobic respiration equation in section **9.2**, you will see that two substances are needed. They are glucose and oxygen. The way in which cells obtain glucose is described in Chapters **6** and **7**. Animals get sugar from carbohydrates which they eat. Plants make theirs by photosynthesis.

Oxygen is obtained in a different way. Animals and plants get their oxygen directly from their surroundings.

If you look again at the aerobic respiration equation you can see that carbon dioxide is made. This is a waste product and it must be removed from the organism. In organisms, there are special areas where the oxygen enters and carbon dioxide leaves. One gas is entering, and the other leaving, so these are surfaces for **gas exchange**. These surfaces have to be permeable. They have other characteristics which help the process to be quick and efficient.

1 They are thin to allow gases to diffuse across them quickly.
2 They are close to an efficient transport system to take gases to and from the exchange surface.

3 They are kept moist, to stop the cells dying.
4 They have a large surface area, so that a lot of gas can diffuse across at the same time.
5 They have a good supply of oxygen (often brought by breathing movements).

9.6 The structure of the breathing system.

Figure **9.1** shows the structures which are involved in gas exchange in a human. The most important are the two **lungs**. Each lung is filled with many tiny air spaces called air sacs or **alveoli**. It is here that oxygen diffuses into the blood. Because they are so full of spaces, lungs feel very light and spongy to touch. The lungs are supplied with air through the windpipe or **trachea**.

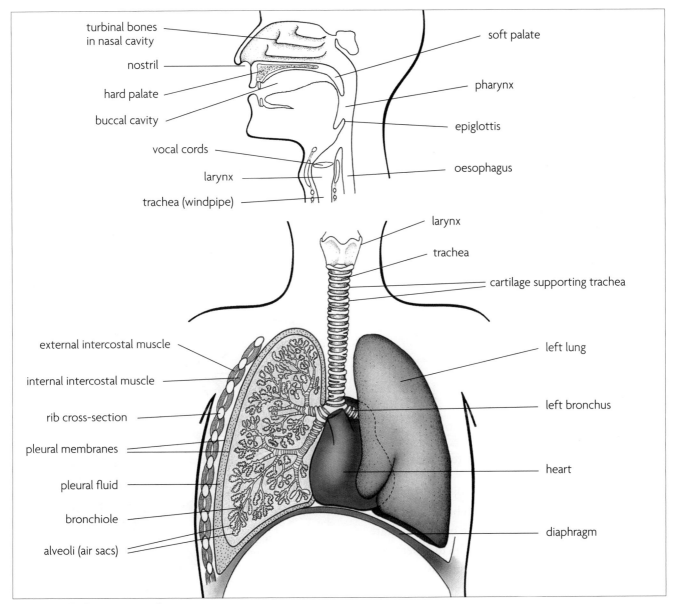

Figure 9.1 The human gas exchange system.

9.7 Air flows to the lungs.

The nose and mouth Air can enter the body through either the nose or mouth. The nose and mouth are separated by the **palate** (Figure **9.1**, page **117**), so you can breathe through your nose even when you are eating.

It is better to breathe through your nose, because the structure of the nose allows the air to become warm, moist and filtered before it gets to the lungs. Inside the nose are some thin bones called turbinal bones which are covered with a thin layer of cells. Some of these cells, called **goblet** cells, make a liquid containing water and mucus which evaporates into the air in the nose and moistens it (Figure **9.2**).

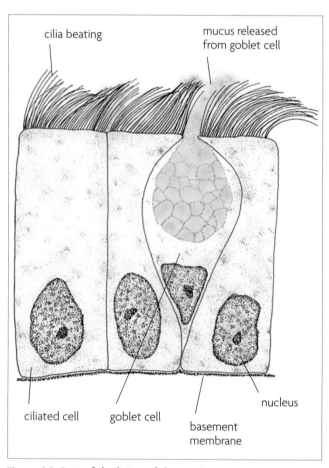

Figure 9.2 Part of the lining of the respiratory passages.

Other cells have very tiny hair-like projections called **cilia**. The cilia are always moving and bacteria or particles of dust get trapped in them and in the mucus. Cilia are found all along the trachea and bronchi, too. Here they waft the mucus, containing bacteria and dust, up to the back of the throat, so that it does not block up the lungs.

The trachea From the nose or mouth, the air then passes into the windpipe or **trachea**. At the top of the trachea is a piece of cartilage called the **epiglottis**. This closes the trachea and stops food going down the trachea when you swallow. This is a reflex action that happens automatically when a bolus of food touches the soft palate.

Just below the epiglottis is the voice box or **larynx**. This contains the **vocal cords**. The vocal cords can be tightened by muscles so that they make sounds when air passes over them. The trachea has rings of cartilage around it which keep it open.

The bronchi The trachea goes down through the neck and into the thorax. The thorax is the upper part of your body from the neck down to the bottom of the ribs and diaphragm. In the thorax the trachea divides into two. The two branches are called the right and left **bronchi** (singular: **bronchus**). One bronchus goes to each lung and then branches out into smaller tubes called **bronchioles**.

The alveoli At the end of each bronchiole are many tiny air sacs or alveoli (Figure **9.3**). This is where gas exchange takes place.

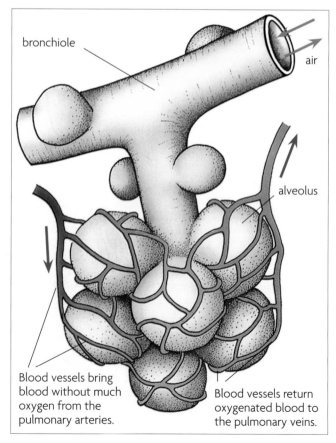

Figure 9.3 Alveoli.

9.8 Alveolar walls are the surface for gas exchange.

The walls of the alveoli are the gas exchange surface. Tiny capillaries are closely wrapped around the outside of the alveoli (Figure **9.3**). Oxygen diffuses across the walls of the alveoli into the blood (Figures **9.4** and **9.5**). Carbon dioxide diffuses the other way.

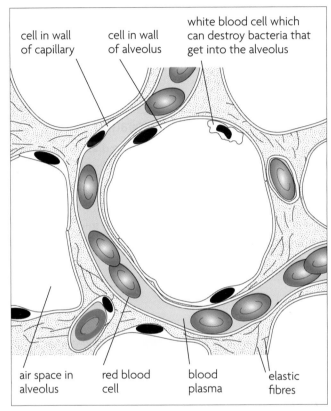

Figure **9.4** Section through part of the lung, magnified.

The walls of the alveoli have several features which make them an efficient gas exchange surface.

- They are very thin. They are only one cell thick. The capillary walls are also only one cell thick. An oxygen molecule only has to diffuse across this small thickness to get into the blood.
- They have an excellent transport system. Blood is constantly pumped to the lungs along the pulmonary artery. This branches into thousands of capillaries which take blood to all parts of the lungs. Carbon dioxide in the blood can diffuse out into the air spaces in the alveoli and oxygen can diffuse into the blood. The blood is then taken back to the heart in the pulmonary vein, ready to be pumped to the rest of the body.

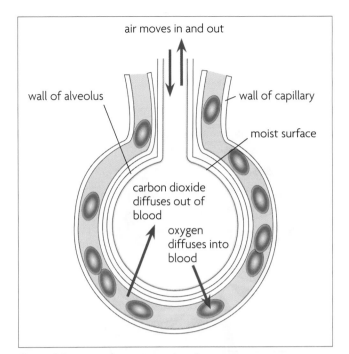

Figure **9.5** Gas exchange in an alveolus.

- They are moist. Special cells in the alveoli secrete a watery liquid. This covers the surface of the cells in the alveoli and prevents them from drying out.
- They have a large surface area. In fact, the surface area is enormous. The total surface area of all the alveoli in your lungs is over 100 m^2.
- They have a good supply of oxygen. Your breathing movements keep your lungs well supplied with oxygen.

Questions

9.7 What is the function of the cilia in the respiratory passages?

9.8 What is the larynx?

9.9 Where does gas exchange take place in a human?

9.10 How many cells does an oxygen molecule have to pass through, to get from an alveolus into the blood?

9.9 Muscles cause breathing movements.

To make air move in and out of the lungs, you must keep changing the volume of your thorax. First, you make it large so that air is sucked in. Then you make it smaller again so that air is squeezed out. This is called **breathing**.

There are two sets of muscles which help you to breathe. One set is in between the ribs. This set is called the intercostal muscles (Figure **9.6**). The other set is in the **diaphragm**. The diaphragm is a large sheet of muscle and elastic tissue which stretches across your body, underneath the lungs and heart.

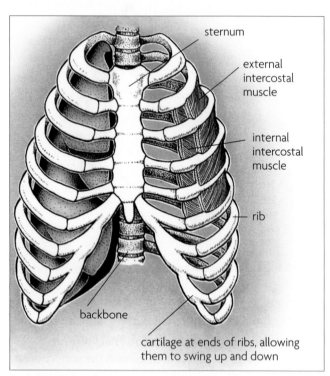

Figure 9.6 The rib cage.

Labels: sternum; external intercostal muscle; internal intercostal muscle; rib; backbone; cartilage at ends of ribs, allowing them to swing up and down

9.10 Breathing in is called inspiration.

When breathing in, the muscles of the diaphragm contract. This pulls the diaphragm downwards, which increases the volume in the thorax (Figure **9.7**). At the same time, the external intercostal muscles contract. This pulls the rib cage upwards and outwards (Figure **9.8**). This also increases the volume of the thorax.

As the volume of the thorax increases, the pressure inside it falls below atmospheric pressure. Extra space has been made and something must come in to fill it up. Air therefore rushes in along the trachea and bronchi into the lungs.

9.11 Breathing out is called expiration.

When breathing out, the muscles of the diaphragm relax. The diaphragm springs back up into its domed shape because it is made of elastic tissue. This decreases the volume in the thorax. The external intercostal muscles also relax. The rib cage drops down again into its normal position. This also decreases the volume of the thorax (Figure **9.7**).

Activity 9.4
Examining lungs

skills

C2 *Observing, measuring and recording*

Safety Wash your hands throughly after touching the lungs. If your teacher lets you blow down a tube into the lungs, the tube must be sterilised before you use it. Make sure you blow and don't suck!

Examine some ox lungs obtained from a butcher's shop or abattoir.

Questions

1 What colour are the lungs? Why are they this colour?

2 Push them gently with your finger. What do they feel like? Why do they feel like this?

3 Feel the smooth surface covering of the lungs. Look at Figure **9.1** (page **117**) and find its name. Why is it important for the lungs to have a very smooth surface?

4 Find the two tubes leading down to the lungs. Which one is the oesophagus? Follow it along, and notice that it goes right past the lungs. Where is it going to?

5 The other tube is the trachea. What does it feel like? Why does it feel like this?

6 What is the name of the wide part at the top of the trachea? What is its function?

7 If the lungs have not been badly cut, take a long glass tube (such as a burette tube) and push it down through the trachea. Hold the trachea tightly against it, and blow down it. What happens?

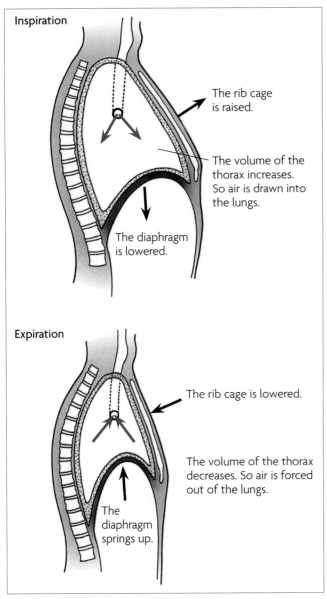

Inspiration

The rib cage is raised.

The volume of the thorax increases. So air is drawn into the lungs.

The diaphragm is lowered.

Expiration

The rib cage is lowered.

The volume of the thorax decreases. So air is forced out of the lungs.

The diaphragm springs up.

Figure 9.7 How the thorax changes shape during breathing.

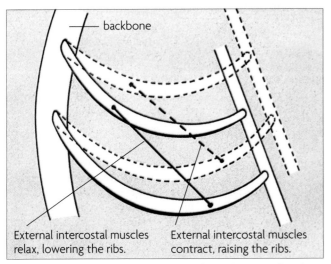

backbone

External intercostal muscles relax, lowering the ribs.

External intercostal muscles contract, raising the ribs.

Figure 9.8 How the external intercostal muscles raise the ribs.

As the volume of the thorax decreases, the pressure inside it increases. Air is squeezed out through the trachea into the nose and mouth, and on out of the body.

9.12 Internal intercostal muscles can force air out.

Usually, you breathe out by relaxing the external intercostal muscles and the muscles of the diaphragm, as explained in section **9.11**.

Sometimes, though, you breathe out more forcefully – when coughing, for example. Then the internal intercostal muscles contract strongly, making the rib cage drop down even further. The muscles of the abdomen wall also contract, helping to squeeze extra air out of the thorax.

Tables **9.2** and **9.3** (overleaf) compare the differences between respiration, gas exchange and breathing, and the composition of inspired and expired air.

Table 9.2 The differences between respiration, gas exchange and breathing

respiration	a series of chemical reactions which happen in all living cells, in which food is broken down to release energy, usually by combining it with oxygen
gas exchange	the exchange of gases across a respiratory surface; for example, oxygen is taken into the body, and carbon dioxide is removed from it; gas exchange also takes place during photosynthesis and respiration of plants
breathing	muscular movements which keep the respiratory surface supplied with oxygen

9.13 Exercise can create an oxygen debt.

All the cells in your body need oxygen for respiration and all of this oxygen is supplied by the lungs. The oxygen is carried by the blood to every part of the body.

Sometimes, cells may need a lot of oxygen very quickly. Imagine you are running in a race. The muscles in your legs are using up a lot of energy. The cells in the muscles will be combining oxygen with glucose as fast as they can, to release energy for muscle contraction.

A lot of oxygen is needed to work as hard as this. You breathe deeper and faster to get more oxygen into your blood. Your heart beats faster to get the oxygen to the

Activity 9.5
Comparing the carbon dioxide content of inspired air and expired air

C1 *Using techniques, apparatus and materials*

C2 *Observing, measuring and recording*

C3 *Interpreting and evaluating*

Safety   The rubber tubing must be sterilised before you use it. Don't blow or suck hard when doing this experiment, just breathe gently.

You can use either limewater or hydrogencarbonate indicator solution for this experiment. Limewater changes from clear to cloudy when carbon dioxide dissolves in it. Hydrogencarbonate indicator solution changes from red to yellow.

1   Set up the apparatus as in the diagram.

2   Breathe in and out gently through the rubber tubing.

Do not breathe too hard. Keep doing this until the liquid in one of the tubes changes colour.

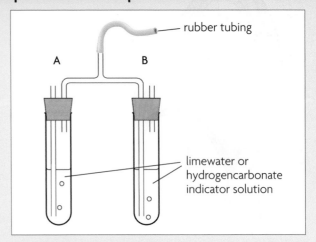

Questions

1   In which tube did bubbles appear when you breathed out? Explain why.

2   In which tube did bubbles appear when you breathed in? Explain why.

3   What happened to the liquid in tube **A**?

4   What happened to the liquid in tube **B**?

5   What do your results tell you about the relative amounts of carbon dioxide in inspired air and expired air?

Table 9.3   A comparison of inspired and expired air

	inspired air	expired air	reason for difference
oxygen	21%	16%	oxygen is absorbed across the gas exchange surface, then used by cells in respiration
carbon dioxide	0.04%	4%	carbon dioxide is made inside respiring cells, and diffuses out across the gas exchange surface
argon and other noble gases	1%	1%	
water content (humidity)	variable	always high	gas exchange surfaces are made of living cells, so must be kept moist; some of this moisture evaporates into the air
temperature	variable	always warm	air is warmed as it passes through the respiratory passages

Figure 9.9 These sprinters will pay back their oxygen debts after the race.

leg muscles as quickly as possible. Eventually a limit is reached. The heart and lungs cannot supply oxygen to the muscles any faster. But more energy is still needed for the race. How can that extra energy be found?

Extra energy can be produced by anaerobic respiration. Some glucose is broken down without combining it with oxygen.

glucose ⟶ lactic acid + energy

As explained in section **9.3**, this does not release very much energy, but a little extra might make all the difference.

When you stop running, you will have quite a lot of lactic acid in your muscles and your blood. This lactic acid must be broken down by combining it with oxygen in the liver. So, even though you do not need the energy any more, you go on breathing hard. You are taking in extra oxygen to break down the lactic acid.

While you were running, you built up an **oxygen debt**. You 'borrowed' some extra energy, without 'paying' for it with oxygen. Now, as the lactic acid is combined with oxygen, you are paying off the debt. Not until all the lactic acid has been used up, does your breathing rate and rate of heart beat return to normal (Figure **9.9**).

9.14 Changes in blood pH affect breathing rate.

The rate at which your breathing muscles work – and therefore your breathing rate – is controlled by the brain.

The brain constantly monitors the pH of the blood that flows through it. If there is a lot of carbon dioxide or lactic acid in the blood, this causes the pH to fall. When the brain senses this, it sends nerve impulses to the diaphragm and the intercostal muscles, stimulating them to contract harder and more often. The result is a faster breathing rate.

Gas exchange in flowering plants

9.15 Plants respire and photosynthesise.

Green plants photosynthesise. They make glucose by combining water and carbon dioxide.

$$\text{carbon dioxide} + \text{water} \xrightarrow[\text{chlorophyll}]{\text{sunlight}} \text{glucose} + \text{oxygen}$$

$$6CO_2 + 6H_2O \xrightarrow[\text{chlorophyll}]{\text{sunlight}} C_6H_{12}O_6 + 6O_2$$

This needs energy which comes from sunlight. The energy is trapped by chlorophyll. The glucose which is made contains some of this energy.

When the plant needs energy, it releases it from the glucose in the same way that an animal does – that is, by respiration.

$$\text{glucose} + \text{oxygen} \longrightarrow \text{carbon dioxide} + \text{water} + \text{energy}$$

$$C_6H_{12}O_6 + 6O_2 \longrightarrow 6CO_2 + 6H_2O + \text{energy}$$

At first sight, this reaction looks like photosynthesis going 'backwards'. In some ways it is. The photosynthesis reaction makes glucose and the respiration reaction breaks it down. However, the reactions are really very different. In photosynthesis, the energy that goes into the reaction is light energy which is trapped by the chlorophyll. In respiration, the energy which comes out is chemical energy.

9.16 Plants, like animals, need energy.

Plants do not need as much energy as animals. They are not so active, partly because they do not have to move to find their food. However, all living cells need some energy. Plant cells need energy for growth, reproduction, for transporting food material between cells and inside cells and many other reasons. They need energy all the time. So all living plant cells, like animal cells, are always respiring (Figure 9.10).

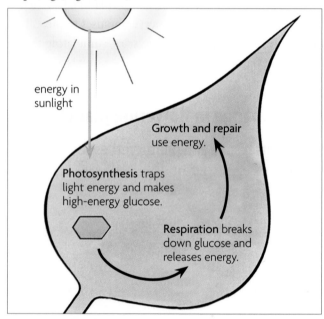

Figure 9.10 How the energy in sunlight is changed to useful energy in a plant.

9.17 The balance between photosynthesis and respiration.

Some plant cells, however, also photosynthesise. The cells in a leaf have chloroplasts and they use carbon dioxide and release oxygen during the daytime. At the same time, respiration is happening inside these cells (Figure 9.11).

In the daytime, photosynthesis is going on much faster than respiration. All of the carbon dioxide that the plant makes by respiration is used up by the chloroplasts in

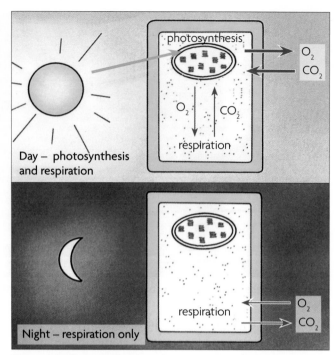

Figure 9.11 Photosynthesis and respiration in plants.

photosynthesis. Even this is not enough, and the plant takes in extra carbon dioxide from the air.

Some of the oxygen which is made by photosynthesis is used up for respiration. There is a lot left over, however, and this diffuses out of the cell.

At night, the chloroplasts stop photosynthesising. The cells, however, continue to respire. Oxygen is taken in, and carbon dioxide is released.

9.18 Plants get oxygen by diffusion.

Plants have a branching shape, so they have quite a large surface area in comparison to their volume. Therefore, diffusion alone can supply all their cells with as much oxygen as they need for respiration.

Leaves   In the daytime, leaves are photosynthesising. This supplies plenty of oxygen for respiration. At night, oxygen diffuses into the leaves through the stomata (Figure 6.7, page 53). It dissolves in the thin layer of moisture around the cells and diffuses in across their cell walls and membranes.

Roots   Roots get their oxygen from the air spaces in the soil. If the soil is waterlogged for very long, they become short of oxygen. Under these conditions the roots will respire anaerobically, producing alcohol (section 9.3). This may kill the plant.

Key ideas

◆ Respiration is a series of metabolic reactions that takes place in every living cell. The purpose of respiration is to release energy from glucose, so that the cell can make use of the energy.

◆ In aerobic respiration, the glucose is combined with oxygen, forming carbon dioxide and water.

> glucose + oxygen ⟶ carbon dioxide + water

> $C_6H_{12}O_6 + 6O_2 \longrightarrow 6CO_2 + 6H_2O$

◆ In anaerobic respiration, the glucose is broken down without being combined with oxygen. In plants and fungi, this produces alcohol and carbon dioxide.

> glucose ⟶ alcohol + carbon dioxide

> $C_6H_{12}O_6 \longrightarrow 2C_2H_5OH + 2CO_2$

In animals (including humans) it produces lactic acid.

> glucose ⟶ lactic acid

> $C_6H_{12}O_6 \longrightarrow 2C_3H_6O_3$

◆ Muscles respire aerobically when they are working so fast that they cannot be supplied with oxygen quickly enough. The lactic acid that is made is transported to the liver, and later is broken down by combining it with oxygen. This extra oxygen is breathed in after the exercise has stopped, and it is known as the oxygen debt.

◆ At gas exchange surfaces, oxygen diffuses into the body and carbon dioxide diffuses out. In a human, the gas exchange surface is the alveoli in the lungs.

◆ All gas exchange surfaces need to be thin, have a large surface area, be kept moist, and have a good supply of oxygen. In larger animals, a transport system is needed to carry away the carbon dioxide and bring oxygen.

◆ The air we breathe in travels down the trachea and bronchi, through the bronchioles and into the alveoli.

◆ Some of these tubes are lined with goblet cells which make mucus, and ciliated cells. The mucus traps dirt, bacteria and other particles and the cilia sweep the mucus up and away from the lungs.

◆ Air is drawn into the lungs by the contraction of the external intercostal muscles and the muscles in the diaphragm. These muscle contractions increase the volume of the thorax, which decreases the pressure. Air flows down the pressure gradient and into the lungs.

Revision questions

1 Which of these descriptions applies to aerobic respiration, which to anaerobic respiration and which to both?
  a lactic acid or alcohol made
  b energy released from glucose
  c carbon dioxide made
  d heat produced

2 a Explain the meaning of the term gas exchange surface.
  b List **three** features of gas exchange surfaces.
  c Explain how these features are shown by the gas exchange surface in human lungs.

3 Construct a table to compare the processes of photosynthesis and respiration in a plant leaf.

4 Copy and complete this table to summarise what happens during breathing.

	breathing in	breathing out
external intercostal muscles		
diaphragm muscles		
volume of thorax		
pressure in lungs		

10 Coordination and response

Response and coordination in animals

10.1 Nerves and hormones allow communication.

Changes in an organism's environment are called **stimuli** (singular: stimulus) and are sensed by specialised cells called **receptors**. The organism responds using **effectors**. Muscles are effectors, and may respond to a stimulus by contracting. Glands can also be effectors. For example, if you smell good food cooking, your salivary glands may respond by secreting saliva.

Animals need fast and efficient communication systems between their receptors and effectors. This is partly because most animals move in search of food. Many animals need to be able to respond very quickly to catch their food, or to avoid predators.

To make sure that the right effectors respond at the right time, there needs to be some kind of communication system between receptors and effectors. If you touch something hot, pain receptors on your fingertips send an impulse to your arm muscles to tell them to contract,

pulling your hand away from the hot surface. The way in which receptors pick up stimuli, and then pass information on to effectors, is called **coordination**.

Most animals have two methods of sending information from receptors to effectors. The fastest is by means of **nerves**. The receptors and nerves make up the animal's **nervous system**. A slower method, but still a very important one, is by means of chemicals called **hormones**. Hormones are part of the **endocrine system**.

The human nervous system

10.2 Neurones carry nerve impulses.

The human nervous system is made of special cells called neurones. Figure **10.1** illustrates a particular type of neurone called a **motor neurone**.

Neurones contain the same basic parts as any animal cell. Each has a nucleus, cytoplasm, and a cell membrane. However, their structure is specially adapted to be able to carry messages very quickly.

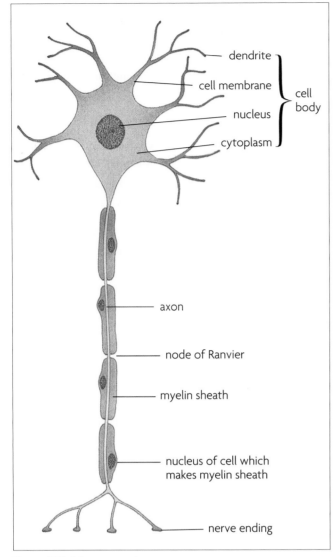

Figure 10.1 A human motor neurone.

The signals that neurones transmit are in the form of electrical impulses. Myelin insulates the nerve fibres, so that they can carry these impulses much faster. For example, a myelinated nerve fibre in a cat's body can carry impulses at up to 100 metres per second. A fibre without myelin can only carry impulses at about 5 metres per second.

10.4 Humans have a central nervous system.

All mammals (and many other animals) have a **central nervous system** (CNS) and a **peripheral nervous system**. The CNS is made up of the **brain** and **spinal cord** (Figure **10.2**). The peripheral nervous system is made up of nerves and receptors.

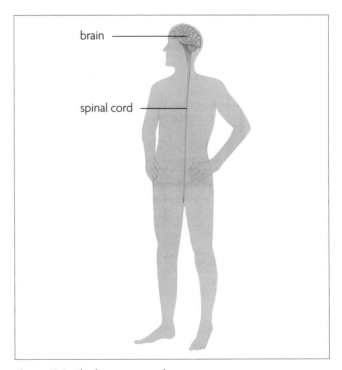

Figure 10.2 The human central nervous system.

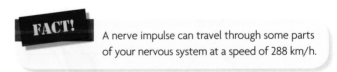

A nerve impulse can travel through some parts of your nervous system at a speed of 288 km/h.

To enable them to do this, they have long, thin fibres of cytoplasm stretching out from the cell body. They are called **nerve fibres**. The longest fibre in Figure **10.1** is called an **axon**. Axons can be more than a metre long. The shorter fibres are called **dendrons** or **dendrites**.

The dendrites pick up electrical signals from other neurones lying nearby. They pass the signal to the cell body, and then along the axon. The axon might then pass it on to another neurone.

10.3 Myelinated neurones carry impulses quickly.

Some of the nerve fibres of active animals like mammals are wrapped in a layer of fat and protein called **myelin**. Every now and then, there are narrow gaps in the myelin sheath.

Like the rest of the nervous system, the CNS is made up of neurones. Its role is to coordinate the messages travelling through the nervous system.

When a receptor detects a stimulus, it sends an electrical impulse to the brain or spinal cord. The brain or spinal cord receives the impulse, and sends an impulse on, along the appropriate nerve fibres, to the appropriate effector.

10.5 Reflex arcs allow rapid response.

Figures **10.3** and **10.4** show how these impulses are sent. If your hand touches a hot plate, an impulse is picked up by a sensory receptor in your finger. It travels to the spinal cord along the axon from the receptor cell. This cell is called a **sensory neurone**, because it is carrying an impulse from a sensory receptor (Figure **10.5**).

In the spinal cord, the neurone passes an impulse on to several other neurones. Only one is shown in Figure **10.4**. These neurones are called **relay neurones**, because they relay the impulse on to other neurones.

The relay neurones pass the impulse on to the brain. They also pass it on to an effector.

In this case, the effectors are the muscles in your arm. The impulse travels to the muscle along the axon of a motor neurone. The muscle then contracts, so that your hand is pulled away.

This sort of reaction is called a **reflex action**. You do not need to think about it. Your brain is made aware of it, but you only consciously realise what is happening after the message has been sent on to your muscles.

Reflex actions are very useful, because the message gets from the receptor to the effector as quickly as possible. You do not waste time in thinking about what to do. The pathway along which the nerve impulse passes – the sensory neurone, relay neurones and motor neurone – is called a **reflex arc**.

Figure **10.6** shows a person's reflex actions being tested – you may have had this test yourself. Another reflex action is described in section **10.9** on page **132**.

Reflex actions are examples of **involuntary actions**. They are not under conscious control. Many of our actions, however, are voluntary. They happen because we decide to carry them out. For example, reading this book is a **voluntary action**.

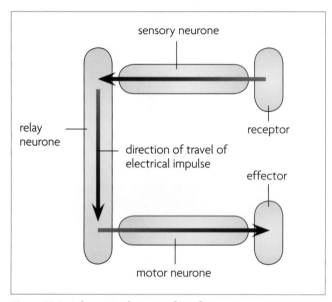

Figure 10.3 Schematic diagram of a reflex arc.

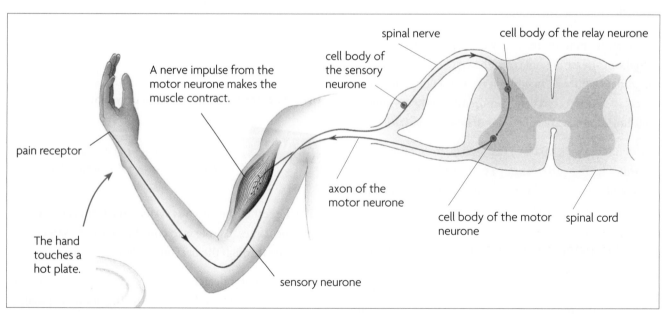

Figure 10.4 A reflex arc.

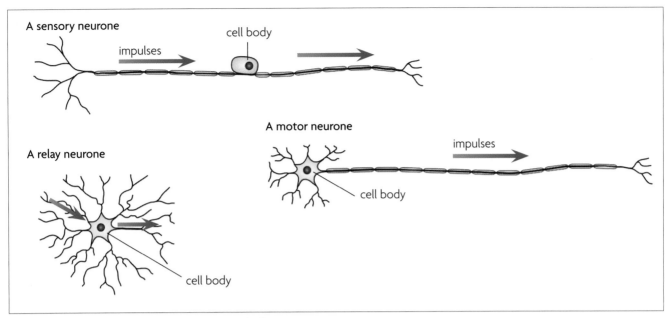

A sensory neurone

impulses

cell body

A motor neurone

A relay neurone

impulses

cell body

cell body

Figure 10.5 The structure of motor, sensory and relay neurones.

Figure 10.6 The knee jerk reflex is an example of a reflex action. A sharp tap just below the knee stimulates a receptor. This sends impulses along a sensory neurone into the spinal cord. The impulse then travels along a motor neurone to the thigh muscle, which quickly contracts and raises the lower leg.

FACT!

Your brain contains about 100 000 000 000 neurones. After the age of 18, you will lose about 1000 of these every day.
The animal with the heaviest brain is the sperm whale. Its brain weighs about 9.2 kg.

Questions

10.1 Give **two** examples of effectors.

10.2 What are the **two** main communication systems in an animal's body?

10.3 List **three** ways in which neurones are similar to other cells.

10.4 List **three** ways in which neurones are specialised to carry out their function of transmitting electrical impulses very quickly.

10.5 What is the function of the central nervous system?

10.6 Where are the cell bodies of each of these types of neurone found: **(a)** sensory neurone, **(b)** relay neurone, and **(c)** motor neurone?

10.7 What is the value of reflex actions?

10.8 Describe **two** reflex actions, other than the ones described in section 10.5.

Activity 10.1
To measure reaction time

skills
C2 Observing, measuring and recording
C3 Interpreting and evaluating

The time taken for a nerve impulse to travel from a receptor, through your CNS and back to an effector is very short. It can be measured, but only with special equipment. However, you can get a reasonable idea of the time it takes if you use a large number of people and work out an average time.

1   Get as many people as possible to stand in a circle, holding hands.

2   One person lets go of his or her neighbour with the left hand, and holds a stopwatch in it. When everyone is ready, this person simultaneously starts the stopwatch, and squeezes his or her neighbour's hand with the right hand.

3   As soon as each person's left hand is squeezed, he or she should squeeze his or her neighbour with the right hand. The message of squeezes goes all round the circle.

4   While the message is going round, the person with the stopwatch puts it into the right hand, and holds his or her neighbour's hand with the left hand. When the squeeze arrives, he or she should stop the watch.

5   Keep repeating this, until the message is going round as fast as possible. Record the time taken, and also the number of people in the circle.

6   Now try again, but this time make the message of squeezes go the other way around the circle.

Questions

1   Using the fastest time you obtained, work out the average time it took for one person to respond to the stimulus they received.

2   Did people respond faster as the experiment went on? Why might this happen?

3   Did the message go as quickly when you changed direction? Explain your answer.

4   If you have access to the Internet, find a site that allows you to measure your reaction time and try it out. Do you think the website gives you more reliable results than the 'circle' method? Compare the results you obtain, and discuss the advantages and disadvantages of each method.

Receptors

10.6 Receptors are often part of a sense organ.

The parts of an organism's body that detect stimuli, the receptors, may be specialised cells or just the endings of sensory neurones. In animals, the receptors are often part of a **sense organ** (Figure 10.7). Your eye, for example, is a sense organ, and the rod and cone cells in the retina are receptors. They are sensitive to light.

Key definition

sense organs   groups of receptor cells responding to specific stimuli: light, sound, touch, temperature and chemicals

eye (vision)

ear (hearing, balance)

nose (smell)

tongue (taste)

skin (touch, temperature, pain)

Figure 10.7 Sense organs.

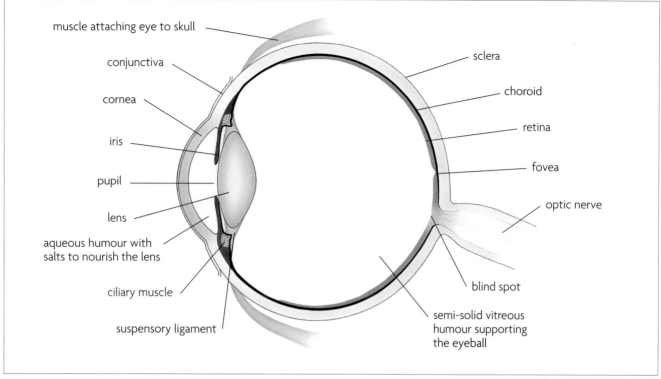

muscle attaching eye to skull
conjunctiva
cornea
iris
pupil
lens
aqueous humour with salts to nourish the lens
ciliary muscle
suspensory ligament

sclera
choroid
retina
fovea
optic nerve
blind spot
semi-solid vitreous humour supporting the eyeball

Figure 10.8 Section through a human eye (seen from above).

10.7 The eye is well protected.

The part of the eye that contains the receptor cells is the **retina** (Figure 10.8). This is the part which is actually sensitive to light. The rest of the eye simply helps to protect the retina, or to focus light onto it.

Each eye is set in a bony socket in the skull, called the **orbit**. Only the very front of the eye is not surrounded by bone (Figure 10.9).

The front of the eye is covered by a thin, transparent membrane called the **conjunctiva**, which helps to protect the parts behind it. The conjunctiva is always kept moist by a fluid made in the **tear glands**. This fluid contains an enzyme called **lysozyme**, which can kill bacteria.

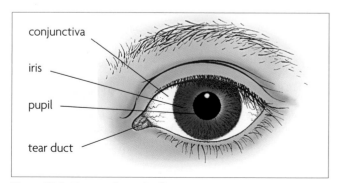

conjunctiva
iris
pupil
tear duct

Figure 10.9 The eye from the front.

The fluid is washed across your eye by your **eyelids** when you blink. The eyelids, eyebrows and eyelashes also help to stop dirt from landing on the surface of your eyes.

Even the part of the eye inside the orbit is protected. There is a very tough coat surrounding it called the **sclera**.

10.8 Cells in the retina are receptive to light.

The retina is at the back of the eye. When light falls on a receptor cell in the retina, the cell sends an electrical impulse along the optic nerve to the brain. The brain sorts out all the impulses from each receptor cell, and builds up an image.

The closer together the receptor cells are, the clearer the image the brain will get. The part of the retina where the receptor cells are packed most closely together is called the **fovea**. This is the part of the retina where light is focused when you look straight at an object.

There are no receptor cells where the optic nerve leaves the retina. This part is called the **blind spot**. If light falls on this place, no impulses will be sent to the brain. Try Activity 10.2.

Activity 10.2
Can you always see the image?

Hold this page about 45 cm from your face. Close the left eye, and look at the cross with your right eye. Gradually bring the page closer to you. What happens? Can you explain it?

Behind the retina is a black layer called the **choroid**. The choroid absorbs all the light after it has been through the retina, so it does not get scattered around the inside of the eye. The choroid is also rich in blood vessels which nourish the eye.

We have two kinds of receptor cells in the retina (Figure 10.10). **Rods** are sensitive to quite dim light, but they do not respond to colour. **Cones** are able to distinguish between the different colours of light, but they only function when the light is quite bright.

Rods therefore allow us to see in dim light but only in black and white, while cones give us colour vision.

The fovea contains almost entirely cones, packed tightly together. When we look directly at an object, we use our cones to produce a sharp image, in colour. Rods are found further out on the retina, and are less tightly packed. They show us a less detailed image.

10.9 The iris adjusts how much light enters the eye.

In front of the lens is a circular piece of tissue called the **iris**. The iris contains pigments, which absorb light and stop it getting through to the retina.

In the middle of the iris is a gap called the **pupil**. The size of the pupil can be adjusted. The wider the pupil is, the more light can get through to the retina. In strong light, the iris closes in, and makes the pupil small. This stops too much light getting in and damaging the retina.

To allow it to adjust the size of the pupil, the iris contains muscles. Circular muscles lie in circles around the pupil.

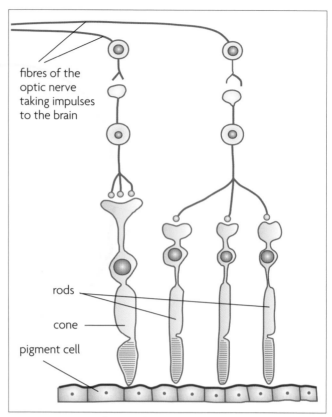

Figure 10.10 A small part of the retina, showing rods and a cone.

When they contract, they make the pupil constrict, or get smaller. Radial muscles run outwards from the edge of the pupil. When they contract, they make the pupil dilate, or get larger (Figure 10.11). This is called the **iris reflex** (or sometimes the **pupil reflex**).

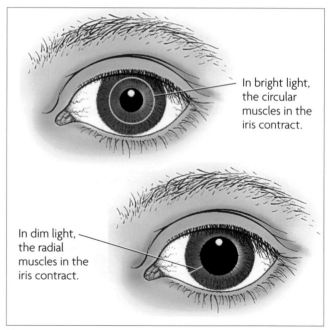

In bright light, the circular muscles in the iris contract.

In dim light, the radial muscles in the iris contract.

Figure 10.11 The iris reflex.

These responses of the iris are examples of a **reflex action**. Although the nerve impulses go into the brain, we do not need to think consciously about what to do. The response of the iris to light intensity (the stimulus) is fast and automatic. Like many reflex actions, this is very advantageous: it prevents damage to the retina that could be caused by very bright light falling onto it.

Activity 10.3
Looking at human eyes

skills

C2 *Observing, measuring and recording*

C3 *Interpreting and evaluating*

It is best to perform this experiment with a partner, although it is possible to use a mirror and look at your own eyes.

1 First identify all the following structures: eyebrows; eyelashes; eyelids; conjunctiva; pupil; iris; cornea; sclera; small blood vessels; openings to tear ducts.

Figure **10.9** will help you to do this.

2 Make a diagram of a front view of the eye and label each of these structures on it.

3 Use sections **10.7** to **10.9** to find out the functions of each structure you have labelled. Write down these functions, as briefly as you can, next to each label or beneath your diagram.

4 Ask your partner to close his or her eyes, and cover them with something dark to cut out as much light as possible. (Alternatively, you may be able to darken the whole room.) After about 3 or 4 minutes, quickly remove the cover (or switch on the lights) and look at your partner's eyes as they adapt to the light. What happens? What is the purpose of this change?

5 Read section **10.9**, and then explain how this change is brought about.

Activity 10.4
Dissecting a sheep's eye

skills

C2 *Observing, measuring and recording*

C3 *Interpreting and evaluating*

Safety: Take great care with the scalpel or scissors. Wear a lab coat or old shirt to keep your clothes clean. Wash your hands with soap and warm water after handling the eye.

1 Carefully examine the eye. Using forceps and a scalpel (ask to be shown how to use them correctly), remove as much of the white fat as you can. Be careful, though, not to damage the brownish-coloured muscles attached to the outside of the eye, or the white optic nerve which comes out at the back of it.

2 Draw the eye, and label: conjunctiva and cornea; iris; sclera; fat; eye muscles; optic nerve; pupil.

3 Using sharp scissors, make a small incision into the eye about half way between the front and the back. What comes out? What happens to the shape of the eye? So what is one of the functions of this substance?

4 Continue cutting around the eye until you have cut it completely in half.

5 First, look at the back half. The retina may have detached itself from here, and may have floated away in the fluid. The next layer is the black choroid. What is the function of the choroid?

6 Behind the choroid is the sclera. What is it like? What is its function?

7 Now investigate the front half of the eye. The lens will probably be floating loose. What normally holds the lens in position? What does the lens look like? If the lens is not too cloudy, put it over some writing and look through it. What does it do?

8 Try to find other structures at the front of the eye – for example, the iris. Identify and describe any structures you can find.

10.10 The cornea and lens focus light.

For the brain to see a clear image, there must be a clear image focused on the retina. Light rays must be bent, or refracted, so that they focus exactly onto the retina. The humours inside the eye are clear so that light can pass through them easily.

The cornea is responsible for most of the bending of the light. The lens makes fine adjustments.

Figure **10.12** shows how the cornea and lens focus light onto the retina. The image on the retina is upside down. The brain interprets this so that you see it the right way up.

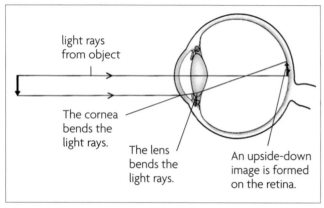

Figure 10.12 How an image is focused onto the retina.

10.11 The lens adjusts the focusing.

Not all light rays need bending the same amount to focus them onto the retina. Light rays coming from a nearby object are going away from one another, or diverging. They will need to be bent inwards quite strongly (Figure **10.13**).

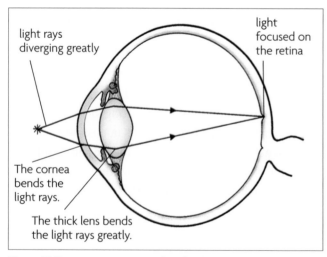

Figure 10.13 Focusing on a nearby object.

Light rays coming from an object in the distance will be almost parallel to one another. They will not need bending so much (Figure **10.14**).

The shape of the lens can be adjusted to bend light rays more, or less. The fatter it is, the more it will bend the light rays. The thinner it is, the less it will bend them. This adjustment in the shape of the lens, to focus light coming from different distances, is called **accommodation**.

Figure **10.15** shows how the shape of the lens is changed. It is held in position by a ring of **suspensory ligaments**. The tension on the suspensory ligaments, and thus the shape of the lens, is altered by means of the **ciliary muscle**. When this muscle contracts, the suspensory ligaments are loosened. When it relaxes, they are pulled tight. When the suspensory ligaments are tight, the lens is pulled thin. When they are loosened, the lens gets fatter.

Questions

10.9 What is a stimulus?

10.10 Name **two** parts of the body which contain receptors of chemical stimuli.

10.11 Which part of the eye contains cells which are sensitive to light?

10.12 Your brain can build up a very clear image when light is focused onto the fovea. Explain why it can do this.

10.13 If you look straight at an object when it is nearly dark, you may find it difficult to see it. It is easier to see if you look just to one side of it. Explain why this is.

10.14 What is the choroid, and what is its function?

10.15 List, in order, the parts of the eye through which light passes to reach the retina.

10.16 Name **two** parts of the eye which refract light rays.

10.17 What is meant by accommodation?

10.18 **a** What do the ciliary muscles do when you are focusing on a nearby object?
   b What effect does this have on **(i)** the suspensory ligaments, and **(ii)** the lens?

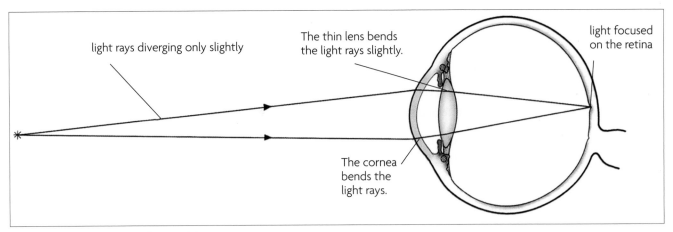

Figure 10.14 Focusing on a distant object.

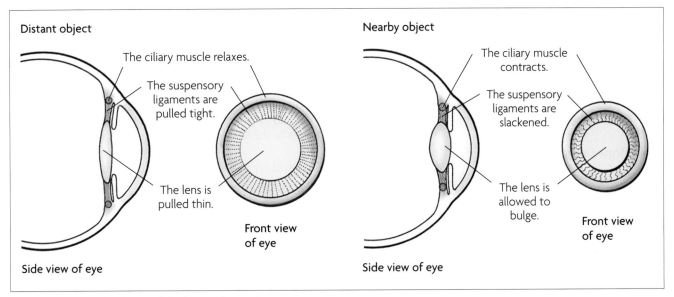

Figure 10.15 How the shape of the lens is changed.

Effectors

10.12 Muscles

Muscles are effectors – they carry out actions. We have already seen the roles of some muscles in reflex actions. For example, the thigh muscle contracts in the knee jerk reflex (Figure 10.6). The radial or circular muscles in the iris contract in the iris reflex (Figure 10.11).

We have several different kinds of muscles in the body.

Cardiac muscle This is only found in the heart. It makes up the walls of the atria and ventricles.

Smooth muscle This is found in organs such as the walls of the alimentary canal, the bladder and blood vessels. The muscle is also called involuntary muscle, because you do not have conscious control over it.

Striated muscle All of the muscles attached to your bones are striated muscle. They are sometimes called skeletal muscles, or voluntary muscles, as they are normally under conscious control.

FACT! Muscles make up about 40% of your body weight.

10.13 Each kind of muscle contracts in a different way.

Muscles cause movement by getting shorter, which is called **contraction**. They need energy to do this. They get their energy from respiration, and so muscles must have a good blood supply to bring nutrients (especially glucose) and oxygen to them.

All of the three types of muscle can contract, but they do it in slightly different ways. Cardiac muscle contracts and relaxes rhythmically all through your life. It never tires. It does not need conscious messages from your brain to make it contract – it will do it anyway. Nervous impulses from the brain can, however, alter its rate of contraction.

Smooth muscle, too, can contract of its own accord. For example, the muscles in the wall of the alimentary canal do this during peristalsis. In other places, however, smooth muscle needs to be stimulated by nerves, in the same way as striated muscle.

The contractions of smooth muscle are much slower than those of cardiac muscle. Smooth muscle contracts and relaxes slowly and rhythmically.

Striated muscle only contracts when electrical impulses are sent to it along nerves. Striated muscle can contract quickly, and very strongly. But it gets tired more quickly than smooth or cardiac muscle.

10.14 Bones meet at joints.

Wherever two bones meet each other, a **joint** is formed. **Synovial joints** are found where two bones need to move freely. The elbow joint and shoulder joint are examples of synovial joints.

Figure **10.16** shows the structure of a typical synovial joint. The two bones are held together by ligaments. Ligaments are very strong, but can stretch when the bones move.

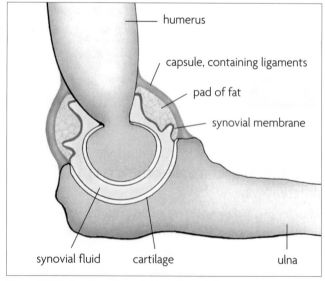

Figure 10.16 Section through the elbow joint.

10.15 Tendons and ligaments help joints to function.

Tendons are strong cords that attach your muscles to your bones (Figure **10.17**). If you hold your hand out palm upwards and tip it downwards, you will see the strong tendons in your wrist. You can also feel your Achilles tendon at the back of your ankle.

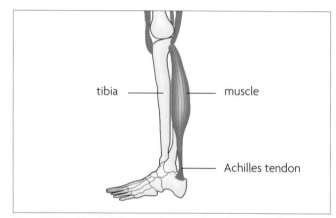

Figure 10.17 The Achilles tendon attaches the strong leg muscles to the heel bone.

As we will see in section **10.16**, when a muscle contracts it pulls on a bone and makes it move. The tendons transmit the force from the muscle to the bone. They do not stretch. If they did, your muscles could pull very hard and just stretch the tendon, rather than making the bones move.

Ligaments are also strong cords, but they are stretchy. They attach one bone to another at a joint. They have to be a bit stretchy, so that the bones can move.

10.16 Movements of the forearm.

Figure **10.18** shows the bones and two of the muscles in your arm. The arm can bend at the elbow, which is a hinge joint.

The **biceps muscle** is attached to the scapula at the top and the radius at the bottom. When it contracts, it pulls the radius and ulna up towards the scapula, so the arm bends. This is called flexing your arm, so the biceps is a **flexor muscle**.

But muscles can only pull, not push. The biceps cannot push your arm back down again. Another muscle is needed to pull it down. The **triceps muscle** does this.

When it contracts, the triceps straightens or extends your arm. It is called an **extensor muscle**.

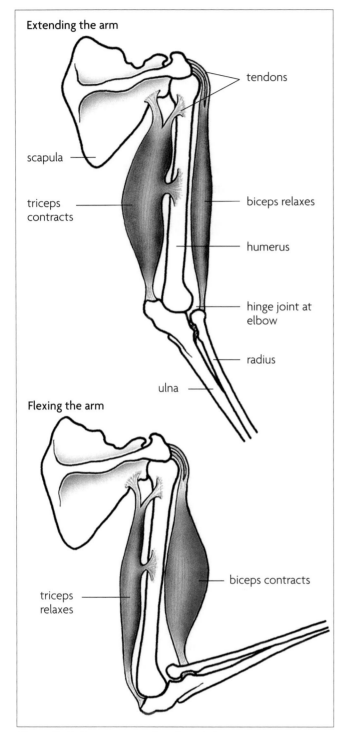

Extending the arm

scapula

triceps
contracts

tendons

biceps relaxes

humerus

hinge joint at
elbow

radius

ulna

Flexing the arm

triceps
relaxes

biceps contracts

Figure 10.18 Movement at the elbow joint.

The flexor and extensor muscles work together. When the biceps contracts, the triceps relaxes. When the triceps contracts, the biceps relaxes. The muscles are said to be **antagonistic muscles** because, in a way, they work against each other. There are many other examples of antagonistic muscles in your body.

The endocrine system

10.17 Endocrine glands make hormones.

So far in this chapter, we have seen how nerves can carry electrical impulses very quickly from one part of an animal's body to another. But animals also use chemicals to transmit information from one part of the body to another.

The chemicals are called **hormones**. Hormones are made in special glands called **endocrine glands**. Figure **10.19** (overleaf) shows the positions of the most important endocrine glands in the human body.

Endocrine glands have a good blood supply. They have blood capillaries running right through them. When the endocrine gland makes a hormone, it releases it directly into the blood.

Other sorts of gland do not do this. The salivary glands, for example, do not secrete saliva into the blood. Saliva is secreted into the salivary duct, which carries it into the mouth. Endocrine glands do not have ducts, so they are sometimes called **ductless glands**.

Key definition

hormone   a chemical substance produced by a gland, carried by the blood, which alters the activity of one or more specific target organs and is then destroyed by the liver

Once the hormone is in the blood, it is carried to all parts of the body, dissolved in the plasma. Although the blood is carrying many hormones, each affects only certain parts of the body. These are called its **target organs**.

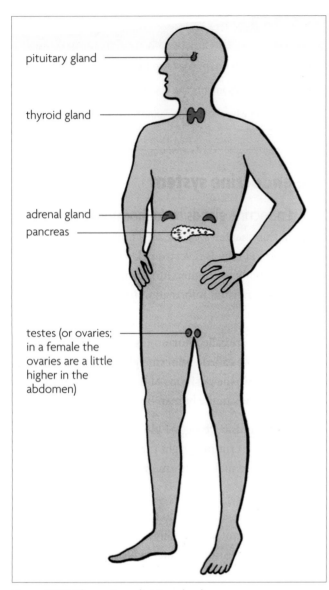

pituitary gland

thyroid gland

adrenal gland
pancreas

testes (or ovaries;
in a female the
ovaries are a little
higher in the
abdomen)

Figure 10.19 The main endocrine glands.

10.18 Adrenaline prepares the body for action.

There are two adrenal glands, one above each kidney. They make a hormone called **adrenaline**. When you are frightened, excited or keyed up, your brain sends impulses along a nerve to your adrenal glands. This makes them secrete adrenaline into the blood.

Adrenaline has several effects which are designed to help you to cope with danger. For example, it makes your heart beat faster, supplying oxygen to your brain and muscles more quickly. This gives them more energy for fighting or running away.

The blood vessels in your skin and digestive system contract so that they carry very little blood. This makes you go pale, and gives you 'butterflies in your stomach'.

As much blood as possible is needed for your brain and muscles in the emergency.

Adrenaline also causes the liver to release glucose into the blood. This provides extra glucose for the muscles, so that they can release energy from it (by respiration) and use the energy for contracting.

Table 10.1 A comparison of the nervous and endocrine systems of a mammal

nervous system	endocrine system
made up of neurones	made up of secretory cells
information transmitted in the form of electrical impulses	information transmitted in the form of chemicals called hormones
impulses transmitted along nerve fibres (axons and dendrons)	chemicals carried dissolved in the blood plasma
impulses travel very quickly	chemicals travel more slowly
effect of a nerve impulse usually only lasts for a very short time	effect of a hormone may last longer

10.19 Animal hormones are sometimes used in food production.

Farmers sometimes use hormones to make their animals grow faster, or to produce more of a particular product. One hormone used in this way is called bovine somatotropin, or **BST**.

BST is a hormone which is naturally produced by cattle. However, if cows are given extra BST, they make more milk. Some people think it would be a good idea to give cows BST, to get higher milk yields. You would need fewer cows to get the same amount of milk.

There are several arguments against it.

- Some people are worried about drinking milk from cows treated with BST. They think the BST might damage their health. In fact, this is very unlikely, because the hormone does not get into the milk in any significant quantity.
- It is difficult to see why we need BST. For example, the European Union already produces more milk than it needs, so milk quotas have to be imposed, to stop farmers from producing too much milk.

- There are concerns that the BST might harm the cows. Cows treated with BST make very large amounts of milk, far beyond the 'natural' levels which they produce. This makes them more likely to get infections of their udders, and may make them feel less comfortable.

Coordination and response in plants

10.20 Some plant responses result in growth.

Like animals, plants are able to respond to their environment, although usually with much slower responses than those of animals.

In general, plants respond to stimuli by changing their rate or direction of growth. They may grow either towards or away from a stimulus. Growth towards a stimulus is said to be a positive response, and growth away from a stimulus is a negative response.

These responses are called **tropisms**. A tropism is a growth response by a plant, in which the direction of the growth is affected by the direction of the stimulus.

Two important stimuli for plants are light and gravity. Shoots normally grow towards light. Roots do not usually respond to light, but a few grow away from it.

Shoots tend to grow away from the pull of gravity, while roots normally grow towards it.

> ### Key definitions
>
> **geotropism** a response in which a plant grows towards or away from gravity
>
> **phototropism** a response in which a plant grows towards or away from the direction from which light is coming

Activity 10.5
To find out how shoots respond to light

skills
C2 Observing, measuring and recording
C3 Interpreting and evaluating

1 Label three Petri dishes **A**, **B** and **C**. Line each with moist cotton wool or filter paper, and put about six peas or beans in each.

2 Leave all three dishes in a warm place for a day or two, until the seeds begin to germinate. Check that they do not dry out.

3 Now put dish **A** into a light-proof box with a slit in one side, so that the seedlings get light from one side only.

4 Put dish **B** onto a clinostat (see diagram) in a light place. The clinostat will slowly turn the seedlings around, so that they get light from all sides equally. If you do not have a clinostat, arrange to turn the dish by hand three or four times per day to achieve a similar effect.

5 Put dish **C** into a completely light-proof box.

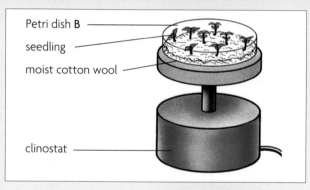

6 Leave all the dishes for a week, checking that they do not dry out.

7 Make labelled drawings of one seedling from each dish.

> ### Questions
>
> 1 How did the seedlings in **A** respond to light from one side? What is the name for this response?
>
> 2 Why was dish **B** put onto a clinostat, and not simply left in a light place?
>
> 3 Explain what happened to the seedlings in dish **C**.
>
> 4 What was the control in this experiment?

Activity 10.6
To find out how roots respond to gravity

skills

C2 *Observing, measuring and recording*

C3 *Interpreting and evaluating*

C4 *Planning*

You are going to design this investigation yourself. You can use similar techniques to those in Activity **10.5**.

This is the hypothesis you are going to test:

> Roots grow towards gravity.

When you have written your plan, get it checked by your teacher before you try to carry it out. Write it up in the usual way, including a discussion and evaluation.

Activity 10.7
To find out which part of a shoot is sensitive to light

skills

C1 *Using techniques, apparatus and materials*

C2 *Observing, measuring and recording*

C3 *Interpreting and evaluating*

Safety Take care with the scalpel blade.

1 Germinate several maize grains in three pots, labelled **A**, **B** and **C**. Space the seeds well out from each other. The seeds will grow shoots called coleoptiles.

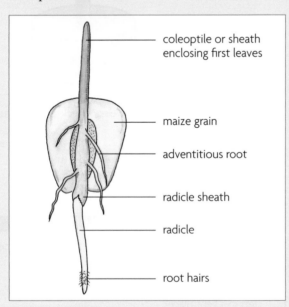

2 Cut the tips from each coleoptile in pot **A**.

3 Cover the tips of each coleoptile in pot **B** with foil.

Light-proof boxes, allowing light from one side only.

A	B	C
coleoptiles with their tips removed	coleoptiles with their tips covered	untreated coleoptiles

4 Measure the length of each coleoptile in each pot. Find the average length of the coleoptiles in each pot and record it.

5 Put pots **A**, **B** and **C** into light-proof boxes with light shining in from one side, as in the diagram. Leave them for one or two days.

6 Find the new average length of the coleoptiles in each pot and record it. Compare it with the original average length, to see whether the coleoptiles have grown or not.

7 Draw a results table and record your results fully.

Questions

1 Explain why some coleoptiles grew, and some did not.

2 Which coleoptiles grew towards the light, and which did not? Explain why.

10.21 These growth movements aid plant survival.

It is very important to the plant that its roots and shoots grow in appropriate directions. Shoots must grow upwards, away from gravity and towards the light, so that the leaves are held out into the sunlight. The more light they have, the better they can photosynthesise. Flowers, too, need to be held up in the air, where insects, birds or the wind can pollinate them.

Roots, though, need to grow downwards, into the soil in order to anchor the plant in the soil, and to absorb water and minerals from between the soil particles.

10.22 How a shoot responds to light.

In section **10.1** we saw that for an organism to respond to a stimulus, there must be a receptor to pick up the stimulus, an effector to respond to it, and some kind of communication system in between. In mammals, the receptor is often part of a sense organ, and the effector is a muscle or gland. Information is sent between them along nerves, or sometimes by means of hormones.

Plants, however, do not have complex sense organs, muscles or nervous systems. So how do they manage to respond to stimuli like light and gravity?

Figure **10.20** shows an experiment that can be done to find out which part of a shoot picks up the stimulus of light shining onto it. The sensitive region is the tip of the shoot. This is where the receptor is.

The part of the shoot which responds to the stimulus is the part just below the tip. This is the effector.

These two parts of the shoot must be communicating with one another somehow. They do it by means of chemicals called **plant hormones**.

10.23 Differences in auxin concentration cause growth movements.

One kind of plant hormone is called **auxin**. Auxin is being made all the time by the cells in the tip of a shoot. The auxin diffuses downwards from the tip, into the rest of the shoot.

Auxin makes the cells just behind the tip get longer. The more auxin there is, the faster they will grow. Without auxin, they will not grow (Figure **10.20**).

When light shines onto a shoot from all around, auxin is distributed evenly around the tip of the shoot. The cells all grow at about the same rate, so the shoot grows straight upwards. This is what normally happens in plants growing outside.

When, however, light shines onto a shoot from one side, the auxin at the tip concentrates on the shady side (Figure **10.21**, overleaf). This makes the cells on the shady side grow faster than the ones on the bright side, so the shoot bends towards the light.

If a potted *Coleus* plant is placed on its side in a dark room overnight, the shoot will bend upwards (Figure **10.22**, overleaf). Since there is no light, we can presume the result to be a response to gravity. (What other precaution should we take to be sure of this?)

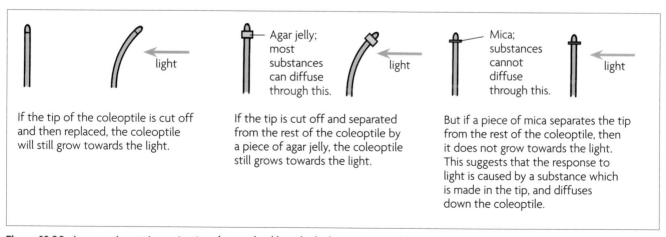

If the tip of the coleoptile is cut off and then replaced, the coleoptile will still grow towards the light.

If the tip is cut off and separated from the rest of the coleoptile by a piece of agar jelly, the coleoptile still grows towards the light.

But if a piece of mica separates the tip from the rest of the coleoptile, then it does not grow towards the light. This suggests that the response to light is caused by a substance which is made in the tip, and diffuses down the coleoptile.

Figure 10.20 An experiment investigating the method by which shoots respond to light.

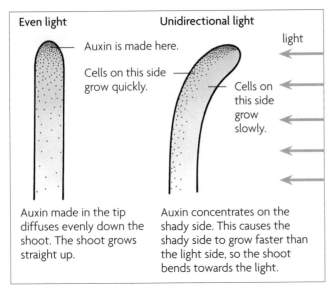

Even light | Unidirectional light

Auxin is made here.

Cells on this side grow quickly.

light

Cells on this side grow slowly.

Auxin made in the tip diffuses evenly down the shoot. The shoot grows straight up.

Auxin concentrates on the shady side. This causes the shady side to grow faster than the light side, so the shoot bends towards the light.

Figure 10.21 Auxin and phototropism.

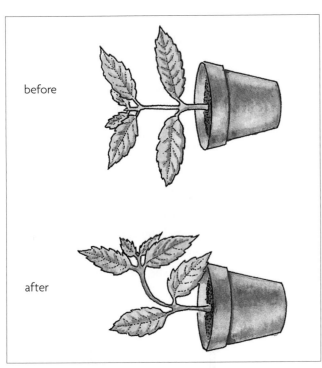

before

after

Figure 10.22 The response to gravity in a *Coleus* shoot.

With the stem in the horizontal position, auxin tends to collect on the lower side of the stem, causing faster growth there. Therefore, the stem curves upward.

In the same way, in the bean seedlings shown in Figure **10.23**, auxin has built up on the lower surface of the root. The effect here, however, is the opposite to that in the *Coleus* shoot. This amount of auxin slows down the growth on this side, and so the radicle bends downwards.

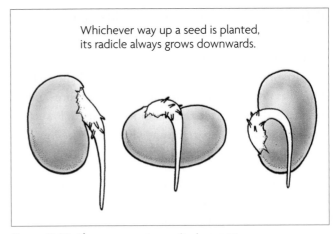

Whichever way up a seed is planted, its radicle always grows downwards.

Figure 10.23 The response to gravity in a root.

10.24 Plants become etiolated in the dark.

Seedlings grown in the dark are very pale, tall and thin. In darkness, auxin is also distributed evenly around the tip, and the shoot grows rapidly upwards. But chloroplasts do not develop properly in darkness. Therefore plants without light become yellow and spindly. They grow very tall and thin, and have smaller leaves, which are often further apart than in a normal plant. Plants like this are said to be **etiolated**.

If these plants reach the light, chlorophyll will develop, and the plants will begin to grow normally. If they do not reach light, they will die because they cannot photosynthesise.

Questions

10.22 What part of the shoot is sensitive to light?

10.23 What part of the shoot responds to light?

10.24 How do these parts communicate with each other? How is this like or unlike a similar system in a mammal?

10.25 How does the normal response of a shoot to light help the plant?

10.26 How does a root respond to gravity?

10.27 Describe **three** features of an etiolated plant.

Activity 10.8
To find out how auxin affects shoots

skills

C1 *Using techniques, apparatus and materials*

C2 *Observing, measuring and recording*

C3 *Interpreting and evaluating*

Safety Use a wooden splint or a cotton bud to smear substances onto the coleoptiles, and avoid getting IAA onto your fingers.

In this experiment, you will use a kind of auxin called indoleacetic acid, or IAA. When you put it onto a shoot, you need to mix it with lanolin, so that it will stick on.

1 Germinate some maize grains in three pots, labelled **A**, **B** and **C**.

2 Mix some IAA with a little warm lanolin. Gently smear the mixture down one side only of each coleoptile in pot **A**. Put the IAA on the same side of each coleoptile. Put a label in the pot to show which side of the coleoptiles the IAA was put on.

3 Do the same with the coleoptiles in pot **B**, but use pure lanolin with no IAA in it.

4 Leave pot **C** untreated.

5 Put all three pots onto clinostats (see Activity **10.5**, page **139**) in a light place, and leave them for a day.

Questions

1 What has happened to the coleoptiles in pots **A**, **B** and **C**? Explain why.

2 What was the reason for smearing the coleoptiles in pot **B** with lanolin?

3 Why were all the pots put onto clinostats?

10.25 Plant hormones are used in food production

Gardeners and horticulturists often use plant hormones to improve the look of their gardens, to increase yields from plants, or to speed up the rate at which they can produce new plants.

Many people use weedkillers in their gardens. Most weedkillers contain plant hormones. These hormones are often a type of auxin, usually a synthetic form (that is, it has been made in a factory and not extracted from plants). The weedkillers used to kill weeds in lawns are **selective weedkillers**. When they are sprayed onto the lawn, the weeds are affected by the auxin, but the grass is not. The weeds respond by growing very fast. Then the weeds die, leaving more space, nutrients and water for the grass to grow. Farmers use similar weedkillers to kill weeds growing in cereal crops such as wheat, millet, maize or sorghum.

Fruit growers often use plant hormones to help the fruits to grow larger, or to ripen well (Figure **10.24**). For example, many fruits produce the gas ethene when they are ripening. This encourages other fruits near them to ripen, as well. (It may seem odd, but there is no reason why a gas cannot be a hormone.) If tomatoes are picked while they are still green, they can be stored or transported without worrying about them going bad. When the suppliers want them to ripen, they can expose them to ethene.

Figure 10.24 This is an ethene (ethylene) generator in a banana store. Ethene is a plant hormone that makes the bananas ripen.

- All organisms are able to sense changes in their environment, called stimuli, and respond to them. The part of the body that senses the stimulus is a receptor, and the part that responds is an effector.

- The human nervous system contains specialised cells called neurones. The brain and spinal cord make up the central nervous system, which coordinates responses to stimuli.

- Reflex actions are fast, automatic responses to a stimulus. They involve a series of neurones making up a reflex arc. A sensory neurone takes the impulse to the CNS and a motor neurone takes it from the CNS to an effector.

- Receptors are generally found within sense organs.

- The receptors in the eye are rod and cone cells, found in the retina. Rods respond to dim light and cones to bright light. Cones give colour vision.

- The cornea and lens focus light rays onto the fovea, the part of the eye where cone cells are most densely packed.

- The shape of the lens is changed by the contraction or relaxation of the ciliary muscle. When focusing on a distant object, the muscle relaxes so that the suspensory ligaments are pulled taut and the lens is pulled into a thin shape. When focusing on a near object, the muscle contracts and the lens falls into its natural, more rounded shape.

- Muscles can pull when they contract, but they cannot push. A pair of muscles is therefore needed to pull in different directions – for example, at the elbow joint. They are antagonistic muscles.

- Hormones are chemicals made in endocrine glands and carried in the blood plasma. Adrenaline is secreted by the adrenal glands, and brings about changes that supply the muscles with extra glucose. This gives them energy for contraction for 'fight or flight'.

- Plants respond to some stimuli by growing towards or away from them. These responses are tropisms.

- Auxin is made in the tips of shoots. It collects on the shady side of the shoot, making this side grow faster so the shoot bends towards the light. Auxins are used as selective weedkillers.

Revision questions

1 Explain the difference between each of the following pairs of terms, giving examples whenever they make your answer clearer.

   a cornea, conjunctiva
   b choroid, sclera
   c receptor, effector
   d sensory neurone, motor neurone
   e negative geotropism, positive geotropism

2 If you walk from a brightly lit street into a dark room, your pupils will rapidly dilate.

   a What type of action is this?

   b Using each of the following words at least once, but not necessarily in this order, explain how this reaction is brought about.

      receptor   motor neurone

      sensory neurone   relay neurone   radial muscles

   c As well as the muscles in the iris, the eye also contains the ciliary muscle. What is its function?

11 Homeostasis and excretion

Homeostasis

11.1 Homeostasis keeps the internal environment constant.

The environment (surroundings) of a living organism is always changing. Think about your own environment. The temperature of the air around you changes. For example, it might be –10 °C outside on a cold day in winter, and 23 °C indoors. The amount of water in the air around you changes. On a rainy day, the air could hold a lot of water, while on a hot, dry, sunny day there could be very little water in the air.

The cells inside your body, however, do not have a changing environment. Your body keeps the environment inside your body almost the same, all the time. In the tissue fluid surrounding your cells, the temperature and amount of water are kept almost constant. So is the concentration of glucose. Keeping this internal environment constant is called **homeostasis**.

> **Key definition**
>
> **homeostasis**   the maintenance of a constant internal environment

Homeostasis is very important. It helps your cells to work as efficiently as possible. Keeping a constant temperature of around 37 °C helps enzymes to work at the optimum rate. Keeping a constant amount of water means that your cells are not damaged by absorbing or losing too much water by osmosis. Keeping a constant concentration of glucose means that there is always enough fuel for respiration.

In this chapter, you will see how homeostasis is carried out in humans. The nervous system and various endocrine glands are involved, and so are the kidneys.

The control of body temperature

11.2 Mammals and birds are homeothermic.

Some animals – including ourselves – are very good at controlling their body temperature. They can keep their temperature almost constant, even though the temperature of their environment changes. Animals that can do this are called **homeothermic** animals. They are also termed **endothermic** because they get their heat energy from within themselves ('endo' means within). Mammals and birds are homeothermic (Figure **11.1**). Animals that don't do this are called **poikilothermic** or **ectothermic**.

Being homeothermic has great advantages. If the internal body temperature can be kept at around 37 °C, then enzymes can always work very efficiently, no matter what the outside temperature is. Metabolism can keep going, even when it is cold outside. In cold weather, or at night, a homeothermic animal can be active when a poikilothermic animal is too cold to move.

But there is a price to pay. The energy to keep warm has to come from somewhere. Homeothermic animals get their heat energy from food, by respiration. Because of this, homeothermic animals have to eat far more food than poikilothermic ones.

11.3 Skin has two layers.

One of the most important organs involved in temperature regulation in mammals is the skin. Figure **11.2** shows a section through human skin.

Human skin is made up of two layers. The top layer is called the **epidermis**, and the lower layer is the **dermis**.

11.4 The epidermis protects the deeper layers.

All the cells in the epidermis have been made in the layer of cells at the base of it, called the **Malpighian layer**. These cells are always dividing by a type of cell division called **mitosis** (section **14.3**). The new cells that are made gradually move towards the surface of the skin.

Outside temperature 0 °C

At 0 °C, a poikilothermic animal's metabolic rate slows down, because its body temperature is also 0 °C. The animal is inactive.

Outside temperature 20 °C

At 20 °C, a poikilothermic animal's body temperature is 20 °C. Its metabolic rate speeds up, and it becomes active.

At 0 °C, a homeothermic animal remains active. Its cells produce heat by breaking down food through respiration. Its body temperature stays high enough to keep its metabolism going.

At 20 °C, a homeothermic animal is no more active than at 0 °C, because its body temperature does not change. It may even be less active, to avoid overheating.

Figure 11.1 Poikilothermic and homeothermic animals.

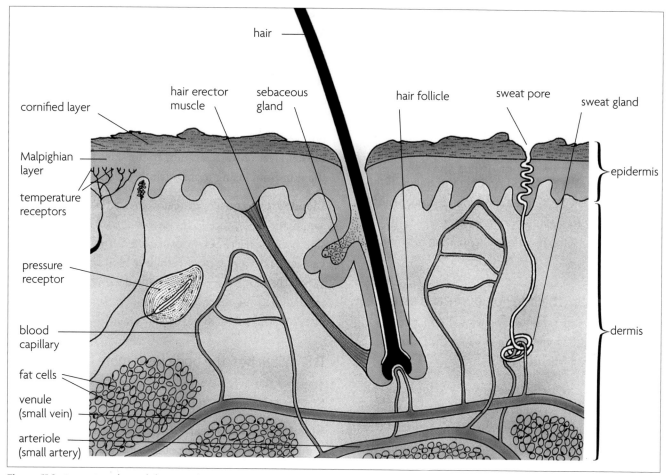

Figure 11.2 A section through human skin.

As they go, they die, and fill up with a protein called **keratin**. The top layer of the skin is made up of these dead cells. It is called the **cornified layer**.

The cornified layer protects the softer, living cells underneath, because it is hard and waterproof. It is always being worn away, and replaced by cells from beneath. On the parts of the body which get most wear – for example, the soles of the feet – it grows thicker.

Some of the cells in the epidermis contain a dark brown pigment, called **melanin**. Melanin absorbs the harmful ultraviolet rays in sunlight, which would damage the living cells in the deeper layers of the skin.

Here and there, the epidermis is folded inwards, forming a **hair follicle**. A hair grows from each one. Hairs are made of keratin.

Each hair follicle has a **sebaceous gland** opening from the side of it. These glands make an oily liquid called sebum. Sebum keeps the hair and skin soft and supple.

11.5 The dermis has many functions.

Most of the dermis is made of connective tissue. This tissue contains elastic fibres and collagen fibres. As a person gets older, the fibres lose their elasticity, so the skin becomes loose and wrinkled.

The dermis also contains **sweat glands**. These secrete a liquid called sweat. Sweat is mostly water, with small amounts of salts and urea dissolved in it. It travels up the sweat ducts, and out onto the surface of the skin through the sweat pores. As we will see, sweat helps in temperature regulation.

The dermis contains blood vessels and nerve endings. These nerve endings are sensitive to touch, pain, pressure and temperature, so they help to keep you aware of changes in your environment.

Underneath the dermis is a layer of fat, called **adipose tissue**. This is made up of cells which contain large drops of oil. This layer helps to insulate your body, and also acts as an energy reserve.

11.6 The hypothalamus coordinates temperature control.

A part of the brain called the **hypothalamus** is at the centre of the control mechanism that keeps internal temperature constant. The hypothalamus coordinates the activities of the parts of the body that can bring about temperature changes.

The hypothalamus acts like a thermostat. It can sense the temperature of the blood running through it. If this is above or below 37 °C, then the hypothalamus sends electrical impulses, along nerves, to the parts of the body which have the function of regulating your body temperature.

11.7 When cold, the body produces and saves heat.

If your body temperature drops below 37 °C, nerve impulses from the hypothalamus cause the following things to happen (Figure **11.3**).

Muscles work   Muscles in some parts of the body contract and relax very quickly. This produces heat. It is called shivering. The heat generated in the muscles warms the blood as it flows through them. The blood distributes this heat all over the body.

Metabolism may increase   The speed of chemical reactions such as respiration may increase. This also releases more heat.

Hair stands up   The erector muscles in the skin contract, pulling the hairs up on end. In humans, this does not do anything very useful – it just produces 'goose pimples'. In a hairy animal though, like a cat, it traps a thicker layer of warm air next to the skin. This prevents the skin from losing more warmth. It acts as an insulator.

Blood system conserves heat   The arterioles that supply the blood capillaries near to the surface of the skin become narrower, or constricted. This is called **vasoconstriction**. Only a very little blood can flow in them. The blood flows through the deep-lying capillaries instead. Because these are deep under the skin, the blood does not lose so much heat to the air.

11.8 When hot, the body loses more heat.

Hair lies flat   The erector muscles in the skin relax, so that the hairs lie flat on the skin.

Blood system loses heat   The arterioles supplying the capillaries near the surface of the skin get wider – they become dilated. This is called **vasodilation**. More blood

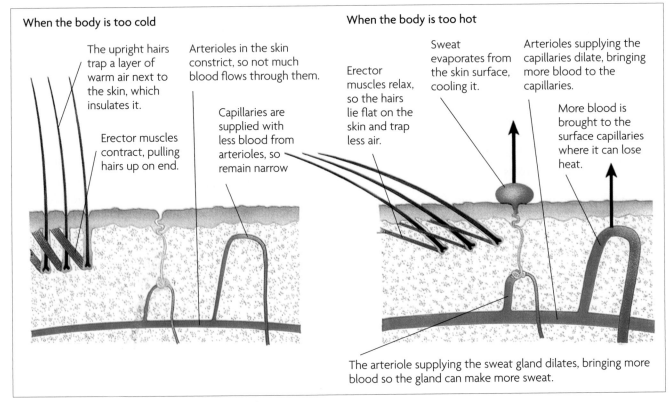

Figure **11.3** How skin helps with temperature regulation.

therefore flows through them. Because a lot of blood is so near the surface of the skin, heat is readily lost from the blood into the air.

Sweat   The sweat glands secrete sweat. The sweat lies on the surface of the hot skin. The water in it then evaporates, taking heat from the skin with it, thus cooling the body.

 The highest body temperature a person has ever had and survived is 46.5 °C.

11.9   Temperature regulation involves negative feedback.

Figure **11.4** summarises the way in which the hypothalamus, skin and muscles work together to keep your body temperature constant.

We have seen that, when the temperature of your blood rises above the norm, the hypothalamus senses this. It responds by sending nerve impulses to your skin that bring about actions to help cool the blood. When the cooler blood reaches the hypothalamus, it responds by sending nerve impulses to your skin that bring about actions to help reduce the rate at which heat is lost from the blood. At the same time, the rate of heat production in the muscles is increased.

So, all the time, the hypothalamus is monitoring small changes in the temperature of your blood. As soon as this rises above normal, actions take place that help to reduce the temperature. Then, as soon as the hypothalamus senses the lowered temperature, it stops these actions taking place and starts off another set of actions that help to raise the blood temperature.

This process is called **negative feedback**. The term 'feedback' refers to the fact that, when the hypothalamus has made your skin take action to increase heat loss, information about the effects of these actions is 'fed back' to it, as it senses the drop in the blood temperature. It is called 'negative' because the information that the blood has cooled down **stops** the hypothalamus making your skin do these things.

Questions

11.1   Outline **two** advantages and **one** disadvantage of maintaining a constant internal body temperature.

11.2   Give **two** functions of the fat layer beneath the skin.

11.3   Explain how sweating helps to cool the body.

11.4   Explain what vasodilation is, and how it helps to cool the body.

11.5   Name the organ which coordinates temperature regulation.

11.6   Explain what is meant by negative feedback.

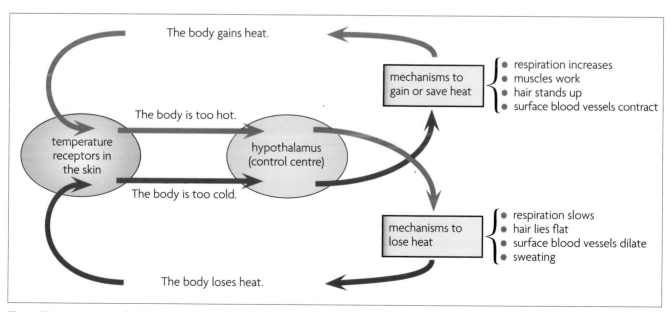

Figure 11.4   Maintaining body temperature in a steady state.

Activity 11.1
Experiment to investigate the effect of size on rate of cooling

skills
C2 *Observing, measuring and recording*
C3 *Interpreting and evaluating*

Temperature regulation is an important part of homeostasis. We lose heat from our bodies to the air around us. Cells produce more heat to prevent the body temperature from dropping.

In this investigation, you will use containers of hot water to represent a human body. The experiment will test this hypothesis:

> A large body cools more slowly than a small one.

1   Take two test tubes or other containers, identical except that one is large and one is small. You will also need two thermometers.

2   Read through what you are going to do. Draw a results chart in which you can write your results as you go along. Remember to put the units in your table headings.

3   Now collect some hot water. Pour water into each of your containers until they are almost full. Immediately take the temperature of each one and record your results for time 0.

4   Take readings every 2 minutes for at least 14 minutes.

5   Draw a line graph to display your results.

Questions

1   a   State **two** variables that are kept constant in this experiment.

    b   Why is it important to keep these variables constant?

2   a   Calculate the number of °C by which the large container cooled during your experiment.
    b   Calculate the number of °C by which the small container cooled during your experiment.

3   Do your results support the hypothesis that you were testing? Explain your answer.

Activity 11.2
Investigating the effect of evaporation on the rate of cooling

skills
C2 *Observing, measuring and recording*
C3 *Interpreting and evaluating*
C4 *Planning*

Sweating helps to cool the body. You are going to plan and carry out an experiment to test this hypothesis:

> Evaporation of water from the surface of a hot object causes it to cool faster.

You can use a technique similar to the one you used in Activity **11.1**. You will need to use two or three test tubes all the same size. You will also need to use some water-absorbent material, which you can wrap round one or more of the tubes.

1   What are you going to vary in your experiment? How will you do this?

2   Make a list of the things that you will keep the same in your experiment.

3   What will you measure? How will you measure it, and how often?

4   How will you display your results?

5   Predict the results you will obtain, if the hypothesis is correct.

Now get your plan checked by your teacher, before you carry it out.

Control of blood glucose content

11.10 Insulin and glucagon regulate blood sugar levels.

The control of the concentration of glucose in the blood is a very important part of homeostasis. Cells need a steady supply of glucose to allow them to respire; without this, they cannot release the energy they need. Brain cells are especially dependent on glucose for respiration, and die quite quickly if they are deprived of it.

On the other hand, too much glucose in the blood is not good either, as it can cause water to move out of cells and into the blood by osmosis. This leaves the cells with too little water for them to carry out their normal metabolic processes.

The control of blood glucose concentration is carried out by the **pancreas** and the **liver** (Figure **11.5**).

The pancreas is two glands in one. Most of it is an ordinary gland with a duct. It makes pancreatic juice, which flows along the pancreatic duct into the duodenum (section **7.28**).

Scattered through the pancreas, however, are groups of cells called **islets of Langerhans**. These cells do not make pancreatic juice. They make two hormones called **insulin** and **glucagon**. These hormones help the liver to control the amount of glucose in the blood. Insulin has the effect of lowering blood sugar, and glucagon does the opposite.

If you eat a meal which provides a lot of glucose, the concentration in the blood goes up. The islets of Langerhans detect this, and secrete insulin into the blood. When insulin reaches the liver, it causes the liver to absorb glucose from the blood. Some is used for respiration, but some is stored in the liver as the insoluble polysaccharide, **glycogen**.

If the blood sugar concentration falls too low, the pancreas secretes glucagon. This causes liver cells to break down glycogen to glucose, and release it into the blood.

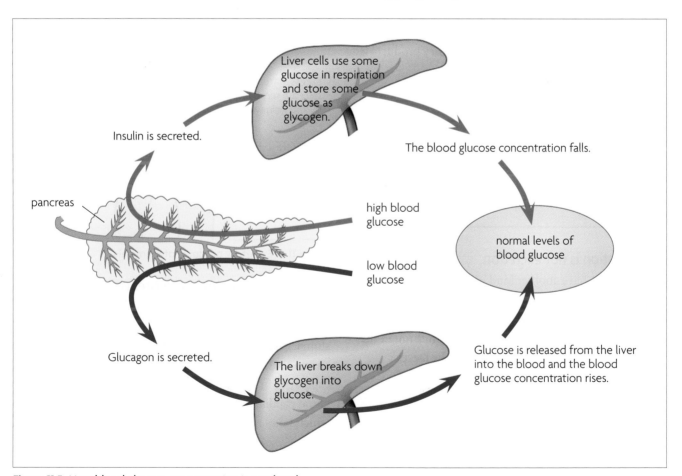

Figure 11.5 How blood glucose concentration is regulated.

Excretion

11.11 Waste products of metabolism are excreted.

All living cells have a great many metabolic reactions going on inside them. The reactions of respiration (section **9.2**), for example, provide energy for the cell. These reactions often produce other substances as well, which the cells do not need. If allowed to remain in the cells, these substances may become poisonous or toxic.

Respiration, for example, produces not only energy, but also water and carbon dioxide. Animal cells need the energy, and may be able to make use of the water. They do not, however, need the carbon dioxide. The carbon dioxide is a waste product.

During daylight hours, plant cells can use the carbon dioxide that they produce in respiration for photosynthesis, so it is not a waste product for them at that time. However, at night, when they cannot photosynthesise but continue to respire, carbon dioxide *is* a waste product.

A waste product like carbon dioxide, which is made in a cell as a result of a metabolic reaction, is called an **excretory product.** The removal of excretory products is called **excretion**.

> **Key definition**
>
> **excretion**   the removal from organisms of toxic materials, the waste products of metabolism (chemical reactions in cells including respiration) and substances in excess of requirements

11.12 Egestion is not excretion.

Many animals have another kind of waste material to get rid of. Almost always, some of the food that an animal eats cannot be digested. Humans, for example, cannot digest cellulose in our food – it goes straight through the alimentary canal, and out of the anus in the faeces.

This cellulose is not an excretory product. It has never been involved in any metabolic reaction in the person's cells. It has not even been inside a cell – it has simply passed, unchanged, through the digestive system. So getting rid of undigested cellulose in faeces is not excretion. It is called **egestion**.

11.13 Excretory products of animals.

The carbon dioxide from respiration is excreted from the lungs, gills or other gas exchange surface (Figure **11.6**). Animals also produce **nitrogenous waste**. This is formed from excess proteins and amino acids. Animals are not able to store these in their bodies, so any that are surplus to requirements are broken down to form a nitrogen-containing excretory product. In mammals, this substance is mainly **urea.**

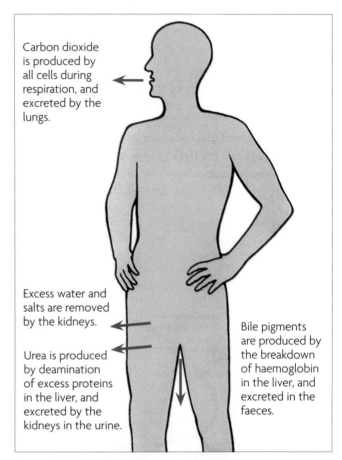

Carbon dioxide is produced by all cells during respiration, and excreted by the lungs.

Excess water and salts are removed by the kidneys.

Urea is produced by deamination of excess proteins in the liver, and excreted by the kidneys in the urine.

Bile pigments are produced by the breakdown of haemoglobin in the liver, and excreted in the faeces.

Figure 11.6 Excretory products of mammals.

11.14 In mammals, excess proteins are converted to urea.

When you eat proteins, digestive enzymes in your stomach, duodenum and ileum break them down into amino acids. The amino acids are absorbed into the blood capillaries in the villi in your ileum (section **7.31**). The blood capillaries all join up to the hepatic portal vein, which takes the absorbed food to the liver.

The liver allows some of the amino acids to carry on, in the blood, to other parts of your body. But if you have eaten more than you need, then some of them must be got rid of.

It would be very wasteful to excrete the extra amino acids just as they are. They contain energy which, if it is not needed straight away, might be needed later.

So enzymes in the liver split up each amino acid molecule (Figure **11.7**). The part containing the energy is kept, turned into carbohydrate and stored. The rest, which is the part that contains nitrogen, is turned into **urea**. This process is called **deamination**.

The urea dissolves in the blood plasma, and is taken to the kidneys to be excreted. A small amount is also excreted in sweat.

The liver has many other functions, as well as deamination. One of the more important ones is storage. Table **11.1** lists some of the functions.

Table 11.1 Some functions of the liver

1	Converts excess amino acids into urea and carbohydrates, in a process called deamination.
2	Controls the amount of glucose in the blood, with the aid of the hormones insulin and glucagon.
3	Stores carbohydrate as the polysaccharide glycogen.
4	Makes bile.
5	Breaks down old red blood cells, storing the iron and excreting the remains of the haemoglobin as bile pigments.
6	Breaks down harmful substances such as alcohol.
7	Stores vitamins A, B, D, E and K.
8	Stores potassium.
9	Makes cholesterol, which is needed to make and repair cell membranes.

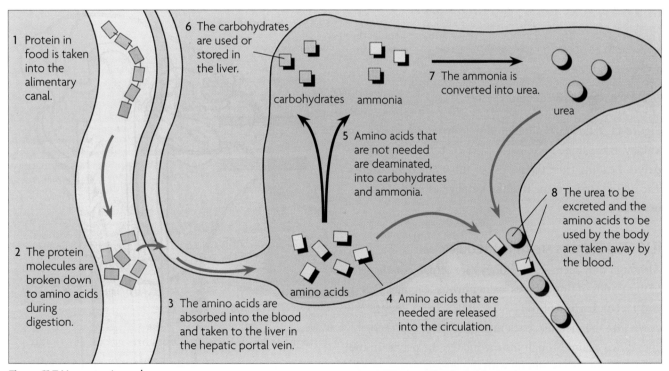

Figure **11.7** How urea is made.

Questions

11.7 What are the main excretory products of animals?

11.8 What processes produce these products?

11.9 What happens to the excess protein you eat?

The human excretory system

11.15 The kidneys are part of the excretory system.

Figure **11.8** illustrates the position of the two kidneys in the human body. They are at the back of the abdomen, behind the intestines.

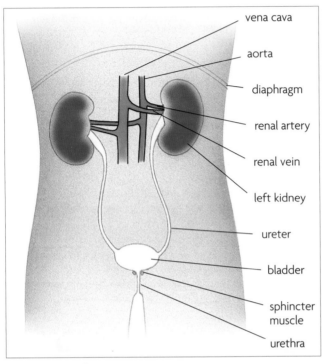

Figure 11.8 The human excretory system.

Figure **11.9** illustrates a longitudinal section through a kidney. It has three main parts – the **cortex**, **medulla** and **pelvis**. Leading from the pelvis is a tube, called the **ureter**. The ureter carries urine that the kidney has made to the **bladder**.

11.16 The kidneys are full of tubules.

Although they seem solid, kidneys are actually made up of thousands of tiny tubules, or **nephrons** (Figures **11.9** and **11.10**). Each nephron begins in the cortex, loops down into the medulla, back into the cortex, and then goes down again through the medulla to the pelvis. In the pelvis, the nephrons join up with the ureter.

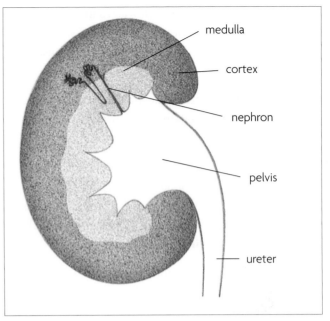

Figure 11.9 A longitudinal section through a kidney showing the position of the nephron (which is drawn much larger than its relative size).

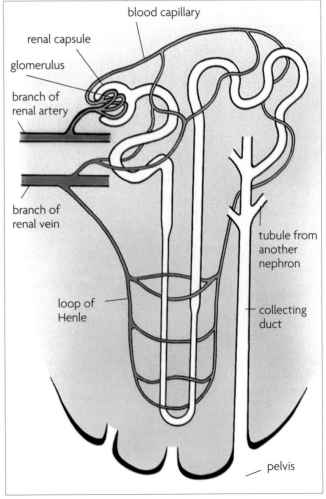

Figure 11.10 A nephron.

11.17 Urine is made by filtration and selective reabsorption.

As blood passes through the kidneys, it is filtered. This removes most of the urea from it, and also excess water and salts. As this liquid moves through the kidneys, any glucose in it is reabsorbed back into the blood. Most of the water is also reabsorbed along with some of the salts.

The final liquid produced by the kidneys is a solution of urea and salts in water. It is called **urine**, and it flows out of the kidneys, along the ureters and into the bladder. It is stored in the bladder for a while, before being released from the body through the **urethra**.

11.18 Filtration happens in renal capsules.

Blood is brought to the **renal capsule** in a branch of the renal artery. Small molecules, including water and most of the things dissolved in it, are squeezed out of the blood into the renal capsule.

There are thousands of renal capsules in the cortex of each kidney. Each one is shaped like a cup. It has a tangle of blood capillaries, called a **glomerulus**, in the middle. The blood vessel bringing blood to each glomerulus is quite wide, but the one taking blood away is narrow. This means that the blood in the glomerulus cannot get away easily. Quite a high pressure builds up, squeezing the blood in the glomerulus against the capillary walls.

These walls have small holes in them. So do the walls of the renal capsules. Any molecules small enough to go through these holes will be squeezed through, into the space in the renal capsule (Figure **11.11**).

Only small molecules can go through. These include **water**, **salt**, **glucose** and **urea**. Most protein molecules are too big, so they stay in the blood, along with the blood cells.

11.19 Useful substances are reabsorbed.

The fluid in the renal capsule is a solution of glucose, salts and urea, dissolved in water. Some of the substances in this fluid are needed by the body. All of the glucose, some of the water and some of the salts need to be kept in the blood.

Wrapped around each kidney tubule are blood capillaries. Useful substances from the fluid in the kidney tubule are reabsorbed, and pass back into the blood in these capillaries.

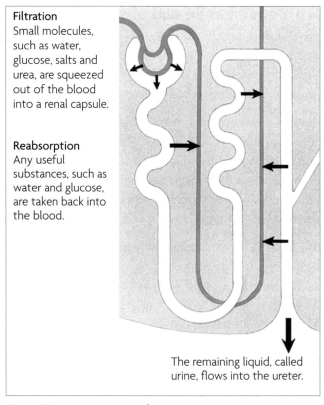

Filtration
Small molecules, such as water, glucose, salts and urea, are squeezed out of the blood into a renal capsule.

Reabsorption
Any useful substances, such as water and glucose, are taken back into the blood.

The remaining liquid, called urine, flows into the ureter.

Figure 11.11 How urine is made.

The remaining fluid continues on its way along the tubule. By the time it gets to the collecting duct, it is mostly water, with urea and salts dissolved in it. It is called urine.

The kidneys are extremely efficient at reabsorbing water. Over 99% of the water entering the tubules is reabsorbed. In humans, the two kidneys filter about $170\,dm^3$ of water per day, yet only about $1.5\,dm^3$ of urine are produced in the same period.

11.20 The bladder stores urine.

The urine from all the nephrons in the kidneys flows into the ureters. The ureters take it to the bladder.

The bladder stores urine. It has stretchy walls, so that it can hold quite large quantities.

Leading out of the bladder is a tube called the urethra. There is a sphincter muscle at the top of the urethra, which is usually tightly closed. When the bladder is full, the sphincter muscle opens, so that the urine flows along the urethra and out of the body.

Adult mammals can consciously control this sphincter muscle. In young mammals, it opens automatically when the bladder gets full.

11.21 Dialysis does the work of damaged kidneys.

Sometimes, a person's kidneys stop working properly. This might be because of an infection. Complete failure of the kidneys allows urea and other waste products to build up in the blood, and will cause death if not treated.

The best treatment is a kidney transplant, but this is not easy to arrange, because the 'tissue type' of the donor and the recipient must be a close match, or the recipient's body will reject the transplanted kidney. The donated kidney usually comes from a healthy person who has died suddenly – for example, in a car accident.

The usual treatment for a person with kidney failure is to have several sessions a week using a **dialysis unit** (Figure **11.12**), sometimes called a kidney machine. The person's blood flows through the machine and back into their body. Inside the machine, the blood is separated from a special fluid by a partially permeable membrane (like Visking tubing). This fluid contains water, glucose, salts and other substances that should be present in the blood.

As the patient's blood passes through the tubes, the substances in the fluid diffuse through the membrane, down their concentration gradients. For example, there is no urea in the dialysis fluid, so urea diffuses out of the patient's blood and into the fluid. The amount of other substances in the blood can be regulated by controlling their concentrations in the dialysis fluid. Proteins in the blood remain there, as their molecules are too big to pass through the membrane.

Patients need to be treated on a dialysis unit two or three times a week, and the treatment lasts for several hours.

11.22 The immune system can reject transplants.

Most people who have to use a dialysis machine would prefer to have to a kidney transplant. The person receiving the transplant is the recipient, and the person from whose body the organ was taken is the donor. Many people carry donor cards with them all the time, stating that they are happy for their organs to be used in a transplant operation. Organs for transplants must be removed quickly from a body and kept cold, so that they

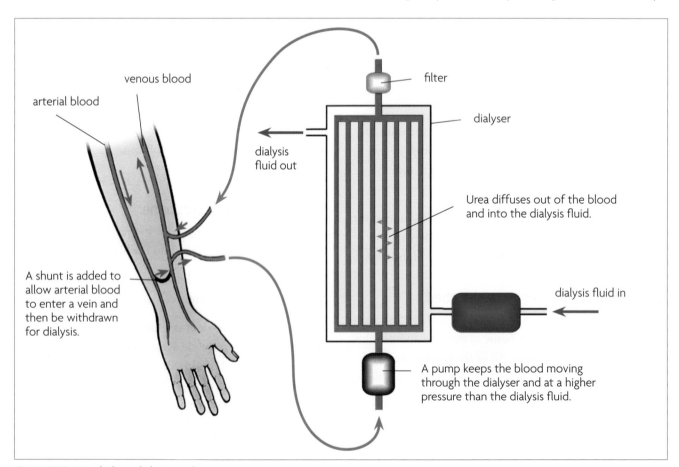

Figure 11.12 How kidney dialysis works.

do not deteriorate. Sometimes, however, the donor may be alive. A person may donate a kidney to a brother or sister who needs one urgently. You can manage perfectly well with just one kidney.

Surgeons now have very few problems with transplant operations – they can almost always make an excellent job of removing the old organ and replacing it with a better one. The big problem comes afterwards. The recipient's immune system recognises the donor organ as being 'foreign', and attacks it. This is called **rejection**.

The recipient is given drugs called immunosuppressants which stop the white blood cells working efficiently, to decrease the chances of rejection. The trouble with immunosuppressants is that they stop the immune system from doing its normal job, and so the person is more likely to suffer from all sorts of infectious diseases. The drugs have to be taken for the rest of the recipient's life.

The chances of rejection are reduced if the donor is a close relative of the recipient. Closely related people are more likely to have antigens (section **8.15**) on their cells which are similar to each other, so the recipient's immune system is less likely to react to the donated organ as if it were 'foreign'. If there is not a relative who can donate an organ, then a search may be made world-wide, looking for a potential donor with similar antigens to the recipient.

Questions

11.10 What is a nephron (kidney tubule)?

11.11 Which blood vessels bring blood to the kidneys?

11.12 What is a glomerulus?

11.13 How is a high blood pressure built up in a glomerulus?

11.14 Why is this high blood pressure needed?

11.15 Name **two** substances found in the blood which you would **not** find in the fluid inside a Bowman's capsule.

11.16 List **three** substances which are reabsorbed from the nephron into the blood.

11.17 What is urine?

Key ideas

◆ Homeostasis is the maintenance of a constant internal environment.

◆ Homeostasis is achieved using negative feedback.

◆ Organisms that can control their internal body temperature are called homeotherms. Mammals and birds are homeotherms. All other animals are poikilotherms, meaning that they have only limited ways of controlling their temperature.

◆ The control of body temperature in humans involves the hypothalamus, the skin and muscles. When the body becomes too hot, sweating and vasodilation increase the rate of heat loss from the skin. When the body becomes too cold, shivering increases heat production, and vasoconstriction reduces the rate of heat loss from the skin.

◆ The pancreas, working in conjunction with the liver, controls blood glucose concentration. When this rises too high, the pancreas secretes insulin which causes the liver to remove glucose from the blood and convert it to glycogen. When blood glucose concentration falls too low, the pancreas secretes glucagon which causes the liver to convert glycogen to glucose.

◆ Excretion is the removal from the body of waste products of metabolism. The main excretory products of mammals are carbon dioxide, urea, salts and excess water.

◆ Mammals excrete carbon dioxide from the lungs and urea from their kidneys.

◆ Urea is produced in the liver from excess amino acids. It is transported in solution in blood plasma to the kidneys, where it is excreted in urine.

◆ Urine is made in the nephrons of each kidney. First, blood is filtered. Then any substances to be retained in the blood are reabsorbed. The fluid that is left in the nephron flows into the ureters and then to the bladder, before leaving the body as urine through the urethra.

1 Explain the differences between each of the following pairs of terms.
  a poikilothermic, homeothermic
  b dilate, constrict
  c epidermis, dermis
  d sebaceous gland, sweat gland
  e excretion, egestion
  f urine, urea
  g ureter, urethra

2 Explain why homeothermic animals eat more food in proportion to their weight than poikilothermic animals.

3 An investigation was performed to find out how the rate of a chemical reaction is affected by temperature. The results are shown on the graph.

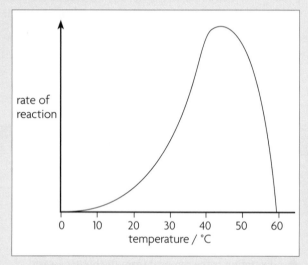

  a Is this reaction catalysed by an enzyme? Give reasons for your answer.
  b What is the optimum temperature for this reaction?
  c Suggest one reaction which might give results like this.
  d Use your answers to (b) and (c) to explain the advantages to an organism of being homeothermic rather than poikilothermic.

4 In an investigation, a number of newly hatched birds were kept at a temperature of 20 °C. Each day, their body temperature and the amount of oxygen they used were measured. The average temperature and oxygen consumption for each day were plotted on a graph.
  a What is (i) the body temperature, and (ii) the oxygen consumption of a three-day-old bird, when the air temperature is 20 °C?

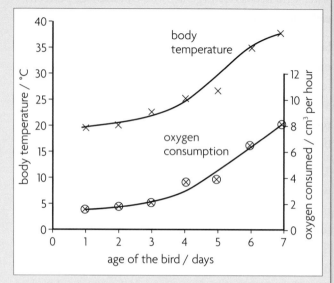

  b Adult birds are homeothermic, keeping their body temperature at around 38 °C. Using information in the graph, suggest the age at which young birds become able to maintain this body temperature.
  c Using the information provided by the graph, and your own knowledge of how homeothermic animals maintain their body temperature, explain why one-day-old birds consume less oxygen than seven-day-old birds.
  d What would you expect to happen to (i) the body temperature, and (ii) the oxygen consumption of a one-day-old bird, if the air temperature was raised to 25 °C? Explain your answer.

5 Explain how the principle of negative feedback operates in controlling each of the following.
  a the internal body temperature
  b the glucose concentration of the blood

6 a What is the main nitrogenous waste product excreted by human kidneys?
  b Where is this waste product formed?
  c Briefly describe how this waste product is formed.
  d Which blood vessels deliver this waste product to the kidneys?
  e Name two substances found in the blood plasma which are not found in the urine of a healthy person.
  f For each of the substances named in (e), explain how the structure and function of the kidney ensure that they are not lost in the urine.

12 Drugs

In this chapter, you will find out:

- ◆ the meaning of the term drug
- ◆ how antibiotics save lives
- ◆ the problems associated with the misuse of heroin
- ◆ how alcohol affects the body
- ◆ the effects of smoking on health.

Uses of drugs

12.1 What is a drug?

People have always used drugs. Long ago, people discovered that some plants could help to cure diseases or to heal wounds. They also used substances obtained from plants and animals to change their perception of the world around them, inducing hallucinations and feelings of contentment or excitement. Today, many of the drugs we use still come from plants.

> **Key definition**
>
> **drug**   any substance taken into the body that modifies or affects chemical reactions in the body

Without drugs, many people would live much shorter lives, or suffer greater pain. Drugs used in medical care, or to relieve mild pain, are very helpful to us. However, some people misuse drugs, so that they cause harm to themselves and to others around them.

Medicinal drugs

12.2 Antibiotics kill bacteria in the body.

Sometimes, a person's body needs help in its fight against a bacterial infection. Until 1944, there was little help that could be given. People died from diseases which we now think quite harmless, such as infected cuts.

Then a discovery was made which has had a tremendous effect on our ability to treat diseases. **Antibiotics** were discovered.

Antibiotics are substances which kill bacteria, but do not harm other living cells. Most of them are made by fungi. It is thought that the fungi make antibiotics to kill bacteria living near them – bacteria and fungi are both decomposers, so they might compete for food. We use the chemical warfare system of the fungus to wage our own war against bacteria.

The first antibiotic to be discovered was **penicillin**. It is made by the fungus *Penicillium*, which you might sometimes see growing on decaying fruit. The way in which penicillin is made is described on page 47. Penicillin kills bacteria by stopping them making their cell walls. Since the introduction of penicillin, many more antibiotics have been found (Figure **12.1**, overleaf).

We have to go on trying to find more and more antibiotics, because bacteria evolve to become resistant to them, as described in section **14.29**. The more we use antibiotics, the more selection pressure we put on bacteria to evolve resistance. People did not realise this when antibiotics were first discovered, and used them for all sorts of diseases where they did not help at all, such as diseases caused by viruses. Now doctors are much more careful about the amounts of antibiotics which they prescribe. We should only use antibiotics when they are

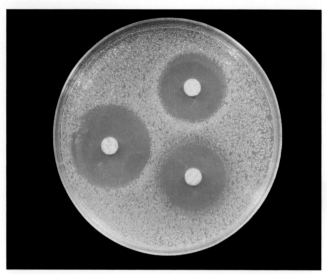

Figure 12.1 This Petri dish contains agar jelly on which the bacteria that cause typhoid fever are growing. The three white circles are little discs of filter paper soaked in different antibiotics. You can see how the bacteria are unable to grow close to the discs, showing that these antibiotics are effective against the bacteria.

really needed – then there is more chance that they will work when we need them to.

Many antibiotics kill bacteria by damaging their cell walls. Viruses do not have cell walls, so they are unharmed by antibiotics.

Misuse of drugs

12.3 Heroin is an addictive depressant.

Opium poppies produce a substance called opium, which contains a number of different chemicals. Some of these, especially morphine and codeine, are used in medicine for the relief of pain. Opium is also the raw material from which **heroin** is produced, which is also used in medicine, but can be **addictive**. An addictive drug is one which causes a person to become dependent on it – they are not able to stop taking it without suffering severe psychological and physical symptoms.

Heroin is a powerful **depressant**. This means that it slows down many functions of the brain. It reduces pain, and slows down breathing. It also slows down the functions of the hypothalamus. When a person takes heroin, it produces a feeling of euphoria – that is, they feel intensely happy. However, in many people it can rapidly become addictive. They feel so ill when they do not take it that

they will do anything to obtain more. As their bodies become more tolerant of the drug, they need to take more and more of it in order to obtain any feelings of pleasure.

Not everyone who takes heroin becomes addicted to it, but many do. Addiction can develop very rapidly, so that a person who has taken it for only one or two weeks may find that they cannot give it up.

A person who has become addicted to heroin may lose any ability to be a part of normal society. He or she may think only of how they will get their next dose. They may not be able to hold down a job, and therefore become unable to earn money, so many heroin addicts turn to crime in order to obtain money to buy their drug. They are not able to help and support their family.

Some people take heroin by injecting it into their veins. This can be dangerous as the needles used for injection are often not sterile, and pathogens such as the hepatitis virus can be introduced into the body. The sharing of needles by heroin addicts has been a major method by which HIV has spread from one person to another.

It is possible for a heroin addict to win the battle against his or her addiction, but it needs a great deal of will-power and much help from others. The withdrawal symptoms that an addict suffers after a few hours without the drug can be extremely unpleasant, and even life threatening.

12.4 Alcohol is also a powerful depressant.

Alcohol is a very commonly used drug in many different countries. People often drink alcoholic drinks because they enjoy the effect that alcohol has on the brain. Alcohol can make people feel more relaxed and release their inhibitions, making it easier for them to enjoy themselves and to mix and interact with other people.

Alcohol is quickly absorbed through the wall of the stomach, and carried all over the body in the blood. It is eventually broken down by the liver, but this takes quite a long time.

Drinking fairly small quantities of alcohol is not dangerous, but alcohol does have many effects on the body which can be very dangerous if care is not taken.

- **Alcohol lengthens reaction time.**

  Alcohol is a depressant, which means that even small amounts of alcohol slow down the actions of parts of the brain, so alcohol lengthens the time you take to respond to a stimulus. This can mean the difference between life and death – often someone else's death – if the affected person is driving a car. A very high proportion of road accidents involve people who have recently drunk alcohol – either drivers or pedestrians (Figure **12.2**). Most countries in which drinking alcohol is allowed have legal limits on blood alcohol level when you drive. However, we now know that even very small quantities of alcohol increase the risk of an accident, so the only safe rule is not to drink alcohol at all if you drive.

Figure 12.2 Many road accidents would not happen if no-one drank alcohol before driving.

- **Alcohol can increase aggression in some people.**

  Different people react differently to alcohol. In some people, it increases their feelings of aggression, and releases their inhibitions so that they are more likely to be violent or commit other crimes. They may be violent towards members of their family. Research has shown that at least 50% of violence in the home in many countries is related to drunkenness, and that alcohol has played a part in the criminal behaviour of around 60% of people in prison in western countries.

- **Large intakes of alcohol can kill.**

  Every year, people die as a direct result of drinking a lot of alcohol over a short period of time. Alcohol is a poison. Large intakes of alcohol can result in unconsciousness, coma and even death. Sometimes, death is caused by a person vomiting when unconscious, and then suffocating because their airways are blocked by vomit.

12.5 Alcoholism is a dangerous disease.

Alcoholism is a disease in which a person cannot manage without alcohol. The cause of the disease is not fully understood. Although it is obvious that you cannot become an alcoholic if you never drink alcohol, many people regularly drink large quantities of alcohol, but do not become alcoholics. Probably, there are many factors which decide whether or not a person becomes alcoholic. They may include a person's genes, their personality, and the amount of stress in their lives.

An alcoholic needs to drink quite large quantities of alcohol regularly. This causes many parts of the body to be damaged, because alcohol is poisonous to cells. The **liver** is often damaged, because it is the liver which has the job of breaking down drugs such as alcohol in the body. One form of liver disease resulting from alcohol damage is **cirrhosis**, where fibres grow in the liver (Figure **12.3**). This can be fatal.

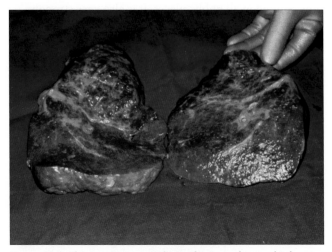

Figure 12.3 This was a person's liver. She was a heavy drinker, and you can see that there are fibres and dark areas in her liver. This is cirrhosis.

Excessive alcohol drinking also damages the **brain**. Over a long period of time, it can cause loss of memory and confusion. One way in which the damage is done is that alcohol in the body fluids draws water out of cells by osmosis. When this happens to brain cells, they shrink, and may be irreversibly damaged. This osmotic effect is made worse because alcohol inhibits the release of a hormone which stops the kidneys from allowing too much water to leave the body in the urine. So drinking alcohol causes a lot of dilute urine to be produced, resulting in low levels of water in the blood.

The effects of smoking on health

12.6 Tobacco smoke contains irritants and carcinogens.

Everyone knows that smoking damages your health, but still people do it. It is especially worrying that so many young people smoke cigarettes (Figure 12.4).

Figure 12.5 shows the main components of tobacco smoke. There are, in fact, many more substances in tobacco smoke, and researchers are still finding out more about them, and the damage that each of them can do to the smoker's health.

One public health concern is that these dangers exist for both smokers and non-smokers. The possible damage is just as real for non-smokers who are in a smokers' environment. They breathe in smoke from burning cigarettes, and from smoke exhaled by smokers. This is termed passive smoking. In many countries, smoking is now banned in all public places. It is also very strongly recommended that parents do not smoke anywhere near their children.

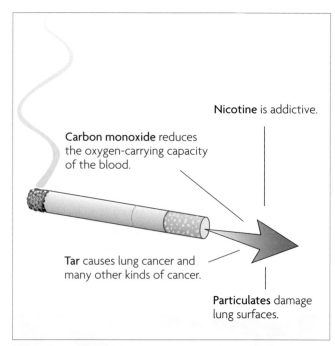

Figure 12.5 Some of the substances in tobacco smoke.

12.7 Nicotine is addictive.

Nicotine affects the brain. It is a stimulant, which means it makes a person feel more alert. Nicotine is an addictive drug. This is why smokers often find it extremely difficult to give up.

Nicotine damages the circulatory system, making blood vessels get narrower. This can increase blood pressure, leading to **hypertension**. Smokers have a much greater chance of developing heart disease than non-smokers.

Tar contains many different chemicals, some of which are **carcinogens** – that is, they can cause cancer. The chemicals can affect the behaviour of some of the cells in the respiratory passages and the lungs, causing them to divide uncontrollably. The cells divide over and over again, forming a lump or **tumour**. If this tumour is **malignant**, this is **cancer**. Cells may break away from the first tumour and spread to other parts of the body, where new tumours will grow. Almost everyone who gets lung cancer is a smoker, or has lived or worked in an environment where they have been breathing in other people's cigarette smoke. Smoking cigarettes increases the risk of developing many different kinds of cancer. All forms of cancer are more common in smokers than in non-smokers.

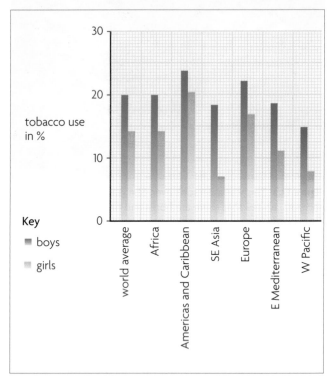

Figure 12.4 Cigarette smoking by 13- to 14-year-old students in different parts of the world.

Carbon monoxide is a poisonous gas which affects the blood. The carbon monoxide diffuses from the lungs into the blood, and combines with haemoglobin inside the red blood cells. This means that less oxygen can be carried. The body cells are therefore deprived of oxygen. This is not good for anyone, but it is especially harmful for a baby growing in its mother's uterus. When the mother smokes, the baby gets all the harmful chemicals in its blood. The carbon monoxide can prevent it from growing properly.

Smoke particles are little particles of carbon and other materials that are present in cigarette smoke. They get trapped inside the lungs. White blood cells try to remove them, and secrete chemicals that are intended to get rid of these invading particles. Unfortunately, the chemicals secreted by the white blood cells can do serious damage to the lungs themselves. Often, this causes the delicate walls of the alveoli to break down (Figure 12.6). There is therefore less surface area across which gas exchange can take place. The person is said to have **emphysema**.

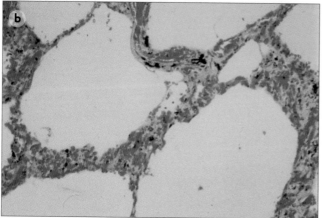

They find it difficult to get enough oxygen into their blood. A person with emphysema may not be able to do anything at all active, and eventually they may not even have the energy to walk.

Several of the chemicals in cigarette smoke harm the cells lining the respiratory passages. You may remember that these cells clean the air as it passes through, stopping bacteria and dust particles from getting down to the lungs (section 9.7). Figure 12.7 shows how smoking affects this cleaning mechanism.

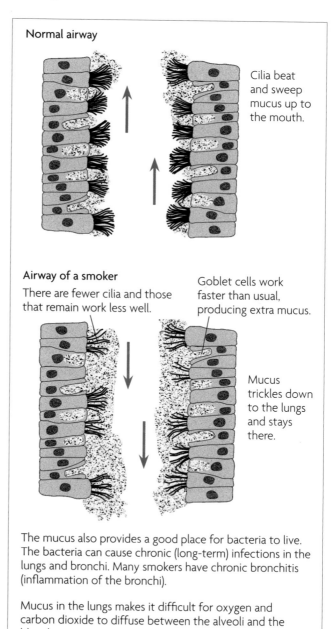

Normal airway

Cilia beat and sweep mucus up to the mouth.

Airway of a smoker
There are fewer cilia and those that remain work less well.

Goblet cells work faster than usual, producing extra mucus.

Mucus trickles down to the lungs and stays there.

The mucus also provides a good place for bacteria to live. The bacteria can cause chronic (long-term) infections in the lungs and bronchi. Many smokers have chronic bronchitis (inflammation of the bronchi).

Mucus in the lungs makes it difficult for oxygen and carbon dioxide to diffuse between the alveoli and the blood.

Figure 12.7 How smoking damages the cells lining the bronchi and bronchioles.

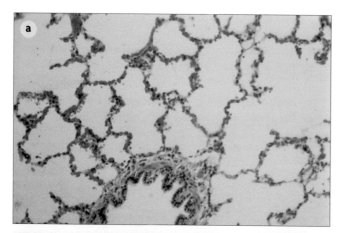

Figure 12.6 a Healthy lung tissue with many small air spaces.
b Lung tissue with emphysema – air spaces are fewer, larger and have thicker walls between (×60).

◆ A drug is a substance that affects chemical reactions in the body.

◆ Many drugs are used in medicine. For example, antibiotics are used to kill bacteria that are causing disease in the body.

◆ Heroin is a depressant that is often addictive. Use of heroin often leads to crime and misery for the user's family. People who inject heroin run a high risk of infection with HIV.

◆ Alcohol is also a depressant. Drinking alcohol lengthens reaction time, reduces self-control and may cause agression, causing serious problems for friends and family. Some people become addicted to alcohol. Over time, the liver is damaged by excessive alcohol intake.

◆ Tobacco smoke contains many different substances that harm health. Nicotine is an addictive stimulant, and its intake increases the risk of developing heart disease. Tar causes lung and other cancers. Carbon monoxide reduces the ability of red blood cells to transport oxygen. Smoke particles irritate the lungs and can contribute to the development of emphysema.

Revision questions

1 Explain the difference between each of the following pairs of terms.
   a stimulant, depressant
   b carbon monoxide, carbon dioxide
   c cirrhosis, emphysema
   d tar, nicotine

2 Suggest explanations for each of the following statements.
   a Antibiotics cannot be used to treat influenza.
   b People who smoke cigarettes usually find it very difficult to give up.
   c Heroin users have a high risk of getting HIV/AIDS.
   d Passive smoking can cause lung cancer.

13 Reproduction

Asexual reproduction

13.1 Asexual reproduction produces genetically identical offspring.

Reproduction is one of the fundamental characteristics of all living things. Each kind of organism has its own particular method of reproducing, but all of these methods fit into one of two categories – asexual reproduction or sexual reproduction.

In reproduction, each new organism obtains a set of **chromosomes** from its parent or parents. Chromosomes are long threads of DNA found in the nucleus of a cell, and they contain sets of instructions known as **genes**. As you will find out in Chapter **14**, these genes vary slightly from one another in different individuals.

Asexual reproduction involves just one parent. Some of the parent organism's cells divide by a kind of cell division called mitosis (section **14.3**). This cell division produces new cells that contain exactly the same genes as the parent cell, and so they are said to be genetically identical. They grow into new organisms, which are all genetically identical to each other and to their single parent.

> **Key definition**
>
> **asexual reproduction**   the process resulting in the production of genetically identical offspring from one parent

13.2 Bacteria reproduce by binary fission.

The cells of bacteria, unlike the cells of animals and plants, do not have a nucleus containing chromosomes. They therefore cannot divide by mitosis. Instead, they use a process called binary fission – which just means 'splitting into two' (Figures **13.1** and **13.2**, overleaf).

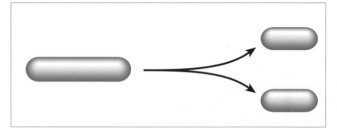

Figure 13.1 Binary fission in bacteria.

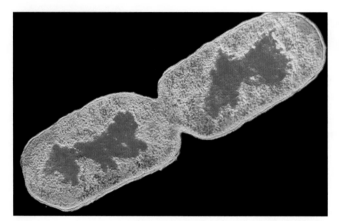

Figure 13.2 This is a bacterium called *Escherichia coli*, dividing into two (×48 000). The red part is the DNA.

The DNA of a bacterium is just a single, circular molecule. When the bacterium is about to divide, an exact copy is made of this DNA. The cell then divides into two, with one DNA molecule in each half. Two new cells have therefore been made, each genetically identical to the original cell.

When they are in suitable conditions, at the right temperature and with plenty of nutrients and water, bacteria can reproduce very quickly. A new cell may be able to divide after only 20 minutes. So, if you start with one cell, there will be two after 20 minutes, four after 40 minutes, eight after one hour and so on. After five hours, 32 768 genetically identical bacteria will have been formed from just one original parent cell. You might like to try to calculate how many there will be after 24 hours!

13.3 Fungi produce spores asexually.

Fungi are living organisms that are neither animals, plants or bacteria. Like plants, their cells have cell walls, but these walls often contain a substance called chitin, not cellulose. Like animals, they cannot photosynthesise, and they feed on organic food materials. Figures **13.3** and **13.4** show a fungus called *Mucor*, or bread mould, which often grows and feeds on bread.

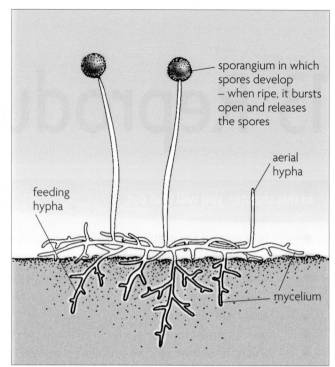

Figure 13.3 *Mucor.*

sporangium in which spores develop – when ripe, it bursts open and releases the spores

aerial hypha

feeding hypha

mycelium

Figure 13.4 This photograph shows a mature sporangium of *Mucor*, with the tiny, spherical spores (coloured blue) ready to be dipsersed.

You may be able to grow some *Mucor* if you leave a piece of damp bread lying around for a few days – put a loose cover over it to stop it drying out. The fungus looks like a furry growth on the bread. This 'fur' is called a **mycelium**, and it is made up of many threads called **hyphae** (singular: **hypha**). Each hypha is just one cell thick. Some of the hyphae grow through the bread, where they secrete enzymes which digest the starch, protein and fat in the bread to produce glucose, amino acids, fatty acids and glycerol. These substances then diffuse into the hypha.

Other hyphae grow upwards, to form aerial hyphae. A swelling called a **sporangium** forms at the tip of each aerial hypha, and inside it cells divide asexually to form many genetically identical **spores**. Each spore is surrounded by a hard, protective coat that stops it from drying out. The spores are very small and light, and can be blown away on air currents, or carried away stuck to the feet of a housefly. Some of them may land on another suitable piece of food, where they will germinate and grow into a new mycelium.

13.4 Potatoes reproduce by producing stem tubers.

Many plants are able to reproduce asexually, and gardeners and farmers make use of this. Asexual reproduction can quickly and efficiently produce many new plants, all genetically identical to one another. This is advantageous to the grower if the original plant had exactly the characteristics that are wanted, such as large and attractive flowers, or good flavour, or high yield.

Potatoes, for example, reproduce using **stem tubers** (Figure 13.5). Some of the plant's stems grow normally, above ground, producing leaves, which photosynthesise. Other stems grow under the soil. Swellings called tubers form on them. Sucrose is transported from the leaves into these underground stem tubers, where it is converted into starch and stored. The tubers grow larger and larger. Each plant can produce many stem tubers.

The tubers are harvested, to be used as food. Some of them, however, are saved to produce next year's crop. These tubers are planted underground, where they grow shoots and roots to form a new plant. Because each

Tubers form on stems that grow on or under the ground.

bud ('eye') from which new shoots will grow next year

Next year, each tuber grows into a new plant.

Figure 13.5 Tuber formation in potatoes.

potato plant produces many tubers, one plant can give rise to many new ones. To get more plants, tubers can be cut into several pieces. As long as each piece has a bud on it, it can grow into a complete new plant.

Questions

13.1 Explain why offspring produced by asexual reproduction are genetically identical to each other.

13.2 What is binary fission?

13.3 Suggest why *Mucor* produces its spores in a sporangium held up above the surface of the bread on which it is growing.

13.4 Explain why a gardener might choose to propagate a plant asexually.

13.5 What is a stem tuber?

Sexual reproduction

13.5 Sexual reproduction involves fertilisation.

In sexual reproduction, the parent organism produces sex cells called **gametes**. Eggs and sperm are examples of gametes. Two of these gametes join and their nuclei fuse together. This is called **fertilisation**. The new cell which is formed by fertilisation is called a **zygote**. The zygote divides again and again, and eventually grows into a new organism.

13.6 Gametes have half the normal number of chromosomes.

Gametes are different from ordinary cells, because they contain only half as many chromosomes as usual. This is so that when two of them fuse together, the zygote they form will have the correct number of chromosomes.

Humans, for example, have 46 chromosomes in each of their body cells. But human egg and sperm cells only have 23 chromosomes each. When an egg and sperm fuse together at fertilisation, the zygote which is formed will therefore have 46 chromosomes, the normal number (Figure **13.6**).

The 46 chromosomes in an ordinary human cell are of 23 different kinds. There are two of each kind. This is because there are two sets of chromosomes in the cell. One set came from the father, and one set from the mother. A cell which has the full number of chromosomes, with two complete sets, is called a **diploid cell**.

An egg or sperm, though, only has 23 chromosomes – a single set. It is called a **haploid cell**. Gametes are always haploid. When two gametes fuse together, they form a diploid zygote.

13.7 Sexual reproduction produces genetically different offspring.

Gametes are made by ordinary body cells dividing. For example, human sperm are made when cells in a testis divide.

Because gametes need to have only half as many chromosomes as their parent cell, division by mitosis will not do. When gametes are being made, cells divide in a different way, called **meiosis**. This process is described in Chapter **14**.

> **Key definition**
>
> **sexual reproduction**   the process involving the fusion of haploid nuclei to form a diploid zygote and the production of genetically dissimilar offspring

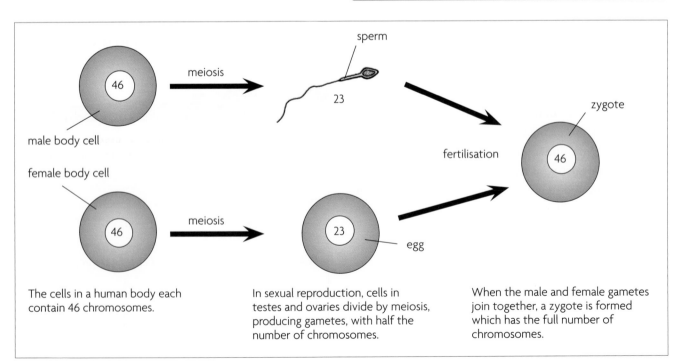

The cells in a human body each contain 46 chromosomes.

In sexual reproduction, cells in testes and ovaries divide by meiosis, producing gametes, with half the number of chromosomes.

When the male and female gametes join together, a zygote is formed which has the full number of chromosomes.

Figure 13.6 Sexual reproduction.

In flowering plants and animals, meiosis only happens when gametes are being made. Meiosis produces new cells with only half as many chromosomes as the parent cell.

13.8 Male gametes move – female ones stay still.

In many organisms, there are two different kinds of gamete. One kind is quite large, and does not move much. This is called the female gamete. In humans, the female gamete is the **egg**.

The other sort of gamete is smaller, and usually moves actively in search of the female gamete. This is called the male gamete. In humans, the male gamete is the **sperm**. In flowering plants, the male gamete is found inside the pollen grain. It does not move by itself, but is carried to the female gamete by a pollen tube (Figure **13.33**, page **186**).

Often, one organism can only produce one kind of gamete. Its sex is either male or female, depending on what kind of gamete it makes. All mammals, for example, are either male or female.

Sometimes, though, an organism can produce both sorts of gamete. Earthworms and slugs, for example, can produce both eggs and sperm. An organism which produces both male and female gametes is a **hermaphrodite**. Many flowering plants are also hermaphrodite.

Questions

13.6 What is a gamete?

13.7 What is a zygote?

13.8 Why do gametes contain only half the normal number of chromosomes?

13.9 What is meant by a diploid cell?

13.10 Name one part of your body where you have diploid cells.

13.11 What is meant by a haploid cell?

13.12 Give one example of a haploid cell.

13.13 When do cells divide by meiosis?

13.14 What is the purpose of meiosis?

13.15 What does hermaphrodite mean?

13.16 Give one example of a hermaphrodite organism.

Sexual reproduction in humans

13.9 The female reproductive organs.

Figure **13.7** shows the reproductive organs of a woman. The female gametes, called eggs, are made in the two **ovaries**. Leading away from the ovaries are the **oviducts**, sometimes called **Fallopian tubes**. They do not connect directly to the ovaries, but have a funnel-shaped opening just a short distance away.

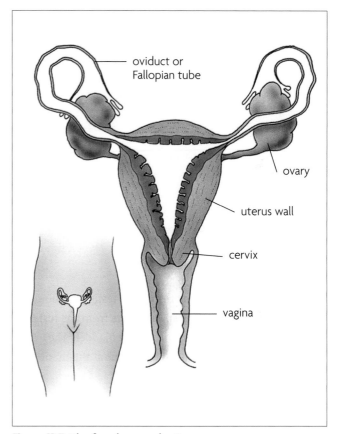

oviduct or Fallopian tube

ovary

uterus wall

cervix

vagina

Figure 13.7 The female reproductive organs.

The two oviducts lead to the womb or **uterus**. This has very thick walls, made of muscle. It is quite small – only about the size of a clenched fist – but it can stretch a great deal when a woman is pregnant.

At the base of the uterus is a narrow opening, guarded by muscles. This is the neck of the uterus, or **cervix**. It leads to the **vagina**, which opens to the outside.

The opening from the bladder, called the **urethra**, runs in front of the vagina, while the **rectum** is just behind it. The three tubes open quite separately to the outside.

13.10 The male reproductive organs.

Figure **13.8** shows the reproductive organs of a man. The male gametes, called spermatozoa or **sperm**, are made in two **testes**. These are outside the body, in two sacs of skin called the **scrotum**.

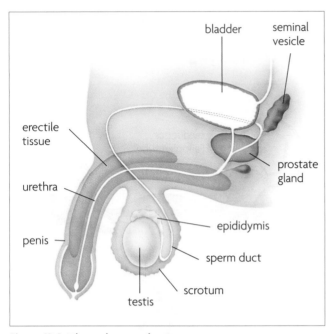

Figure 13.8 The male reproductive organs.

The sperm are carried away from each testis in a tube called the **sperm duct**. The sperm ducts from the testes join up with the urethra just below the bladder. The urethra continues downwards and opens at the tip of the **penis**. The urethra can carry both urine and sperm at different times.

Where the sperm ducts join the urethra, there is a gland called the **prostate gland**. This makes a fluid which the sperm swim in. Just behind the prostate gland are the **seminal vesicles**, which also secrete fluid.

13.11 Ovaries make eggs.

Eggs begin to be formed inside a girl's ovaries before she is born. At birth, she will already have thousands of partly developed eggs inside her ovaries.

When she reaches puberty (section **13.23**), some of these follicles will begin to develop. Usually, only one develops at a time. When it is mature (Figure **13.9**), an egg bursts out of the ovary and into the funnel at the end of the oviduct. This is called **ovulation**. In humans, it happens once a month.

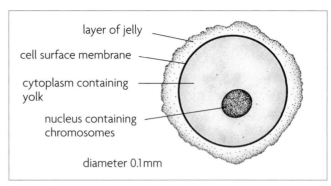

Figure 13.9 A human egg.

13.12 Testes make sperm.

Figure **13.10** shows a section through a testis. It contains thousands of very narrow, coiled tubes or **tubules**. These are where the sperm are made. Sperm develop from cells in the walls of the tubules, which divide by meiosis. Sperm are made continually from puberty onwards. Figure **13.11** shows the structure of a sperm.

Sperm production is very sensitive to heat. If they get too hot, the cells in the tubules will not develop into sperm. This is why the testes are outside the body, where they are cooler than they would be inside.

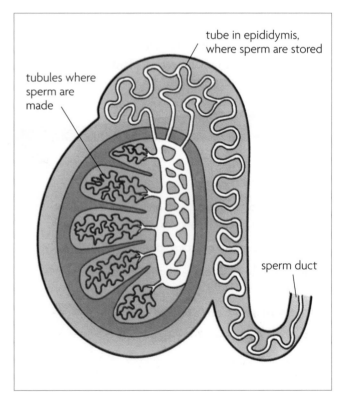

Figure 13.10 Section through a testis.

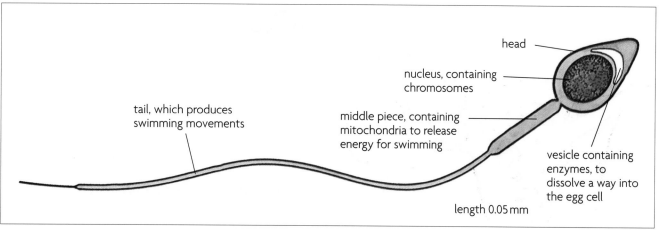

head

nucleus, containing
chromosomes

tail, which produces
swimming movements

middle piece, containing
mitochondria to release
energy for swimming

vesicle containing
enzymes, to
dissolve a way into
the egg cell

length 0.05 mm

Figure 13.11 A human sperm.

13.13 Mating introduces sperm into the vagina.

After ovulation, the egg is caught in the funnel of
the oviduct. The funnel is lined with cilia which beat
rhythmically, wafting the egg into the entrance of the
oviduct.

Very slowly, the egg travels towards the uterus. Cilia
lining the oviduct help to sweep it along. Muscles in the
wall of the oviduct also help to move it, by **peristalsis**.
(Figure **7.18** on page **78** shows peristalsis in the
alimentary canal.)

If the egg is not fertilised by a sperm within 8–24 hours
after ovulation, it will die. By this time, it has only
travelled a short way along the oviduct. So a sperm must
reach an egg while it is quite near the top of the oviduct if
fertilisation is to be successful.

When the man is sexually excited, blood is pumped into
spaces inside the penis, so that it becomes erect. To bring
the sperm as close as possible to the egg, the man's penis
is placed inside the vagina of the woman. This is called
sexual intercourse.

Sperm are pushed out of the penis into the vagina.
This happens when muscles in the walls of the tubes
containing the sperm contract rhythmically. The wave of
contraction begins in the testes, travels along the sperm
ducts, and into the penis. The sperm are squeezed along,
out of the man's urethra and into the woman's vagina. This
is called **ejaculation**.

The fluid containing the sperm is called **semen**.
Ejaculation deposits the semen at the top of the vagina,
near the cervix.

13.14 Fertilisation happens in the oviduct.

The sperm are still quite a long way from the egg. They
swim, using their tails, up through the cervix, through
the uterus, and into the oviduct (Figures **13.12** and **13.13**,
overleaf).

Figure 13.12 This sperm cell is swimming over the surfaces of
the ciliated cells in the oviduct.

Sperm can only swim at a rate of about 4 mm per minute,
so it takes quite a while for them to get as far as the oviducts.
Many will never get there at all. But one ejaculation
deposits about a million sperm in the vagina, so there is a
good chance that some of them will reach the egg.

One sperm enters the egg. Only the head of the sperm
goes in; the tail is left outside. The nucleus of the sperm
fuses with the nucleus of the egg. This is **fertilisation**
(Figure **13.14**, overleaf).

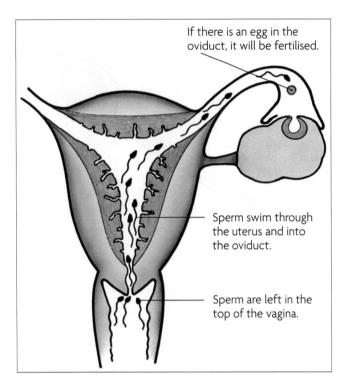

Figure 13.13 How sperm get to the egg (sperm and egg are drawn to different scales).

If there is an egg in the oviduct, it will be fertilised.

Sperm swim through the uterus and into the oviduct.

Sperm are left in the top of the vagina.

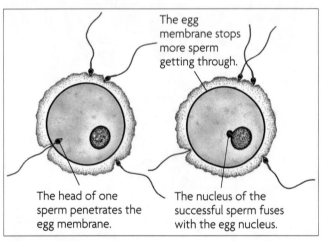

The egg membrane stops more sperm getting through.

The head of one sperm penetrates the egg membrane.

The nucleus of the successful sperm fuses with the egg nucleus.

Figure 13.14 Fertilisation.

As soon as the successful sperm enters the egg, the egg membrane becomes impenetrable, so that no other sperm can get in. The unsuccessful sperm will all die.

13.15 The zygote implants in the uterus wall.

When the sperm nucleus and the egg nucleus have fused together, they form a zygote. The zygote continues to move slowly down the oviduct. As it goes, it divides by mitosis. After several hours, it has formed a ball of cells. This is called an **embryo**. The embryo obtains food from the yolk of the egg.

It takes several hours for the embryo to reach the uterus, and by this time it is a ball of 16 or 32 cells. The uterus has a thin, spongy lining, and the embryo sinks into it. This is called **implantation** (Figure 13.15).

13.16 The embryo's life-support system is its placenta.

The cells in the embryo, now buried in the soft wall of the uterus, continue to divide. As the embryo grows, a **placenta** also grows, which connects it to the wall of the uterus (Figure 13.16). The placenta is soft and dark red, and has finger-like projections called villi. The villi fit closely into the uterus wall.

After eleven weeks, the embryo has developed into a **fetus**. The placenta is joined to the fetus by the **umbilical cord**. Inside the cord are two arteries and a vein. The arteries take blood from the fetus into the placenta, and the vein returns the blood to the fetus.

In the placenta are capillaries filled with the fetus's blood (Figure 13.17). In the wall of the uterus are large spaces filled with the mother's blood. The fetus's and mother's blood do not mix. They are separated by the wall of the placenta. But they are brought very close together, because the wall of the placenta is very thin.

Oxygen and food materials in the mother's blood diffuse across the placenta into the fetus's blood, and are then

Questions

13.17 What is the name for the narrow opening between the uterus and the vagina?

13.18 Where is the prostate gland, and what is its function?

13.19 Explain how ovulation happens.

13.20 Where are sperm made?

13.21 How does an egg travel along the oviduct?

13.22 Where does fertilisation take place?

13.23 Construct a table to compare the size, structure and ability to move of a sperm and an egg.

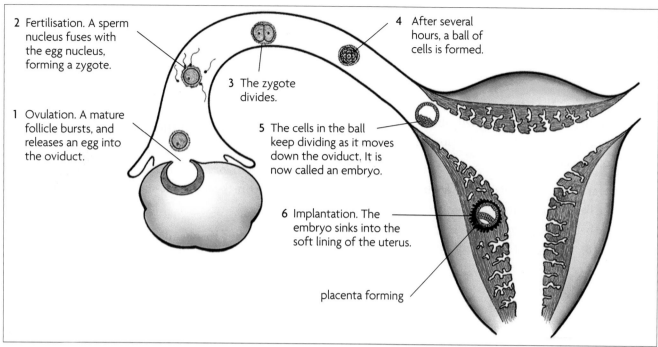

2 Fertilisation. A sperm nucleus fuses with the egg nucleus, forming a zygote.

3 The zygote divides.

4 After several hours, a ball of cells is formed.

1 Ovulation. A mature follicle bursts, and releases an egg into the oviduct.

5 The cells in the ball keep dividing as it moves down the oviduct. It is now called an embryo.

6 Implantation. The embryo sinks into the soft lining of the uterus.

placenta forming

Figure 13.15 Stages leading to implantation.

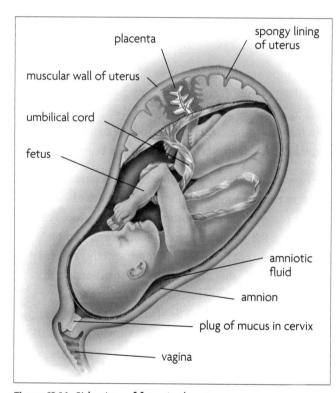

placenta

spongy lining of uterus

muscular wall of uterus

umbilical cord

fetus

amniotic fluid

amnion

plug of mucus in cervix

vagina

Figure 13.16 Side view of fetus in the uterus.

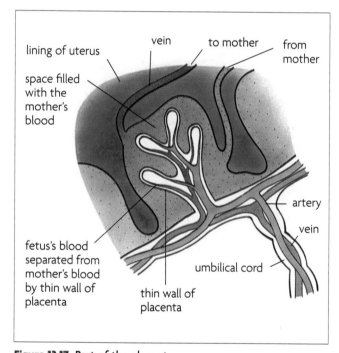

lining of uterus

vein

to mother

from mother

space filled with the mother's blood

fetus's blood separated from mother's blood by thin wall of placenta

thin wall of placenta

artery

vein

umbilical cord

Figure 13.17 Part of the placenta.

carried along the umbilical cord to the fetus. Carbon dioxide and waste materials diffuse the other way, and are carried away in the mother's blood. As the fetus grows, the placenta grows too. By the time the baby is born, the placenta will be a flat disc, about 12 cm in diameter, and 3 cm thick.

13.17 An amnion protects the fetus.

The fetus is surrounded by a strong membrane, called the **amnion**. Inside the amnion is a liquid called **amniotic fluid**. This fluid helps to support the embryo, and to protect it.

13.18 Muscular contractions cause birth.

A few weeks before birth, the fetus usually turns over in the uterus, so that it is lying head downwards. Its head lies just over the opening of the cervix.

Birth begins when the strong muscles in the wall of the uterus start to contract. This is called **labour**. To begin with, the contractions of the muscles slowly stretch the opening of the cervix.

After several hours, the cervix is wide enough for the head of the baby to pass through. Now, the muscles start to push the baby down through the cervix and the vagina (Figure **13.18**). This part of the birth happens quite quickly.

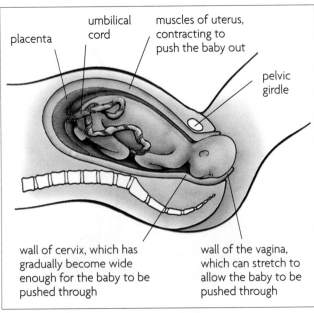

Figure 13.18 Birth.

The baby is still attached to the uterus by the umbilical cord and the placenta. Now that it is in the open air, the baby can breathe for itself, so the placenta is no longer needed. The placenta falls away from the wall of the uterus, and passes out through the vagina. It is called the **afterbirth**.

The umbilical cord is cut, and clamped just above the point where it joins the baby. This is completely painless, because there are no nerves in the cord. The stump of the cord forms the baby's navel.

The contractions of the muscles of the uterus are painful. They feel rather like cramp. However, there is now no need for any mother to suffer really bad pain. She can help herself a lot by preparing her body with exercises before labour begins, by breathing in a special way during labour, and she can also be given pain-killing drugs if she needs them.

13.19 Pregnant women should care for their health.

When a woman is pregnant, she should take extra care of her health, both for her own benefit and that of her baby. This is sometimes called **ante-natal care**, meaning 'before birth'.

She should ensure that her diet contains plenty of calcium, to help to form the growing fetus's bones. She also needs extra iron, because her body will produce a lot of extra blood to help to carry oxygen and nutrients to the placenta, and her growing baby is also forming blood. Iron is needed to make the haemoglobin in the red blood cells. She may also need a little extra carbohydrate, because she needs extra energy to help to move her heavier body around, and extra protein, to help to form her growing fetus's new cells.

She should continue to take exercise. Most people consider that steady, gentle exercise is best, such as swimming or walking. She may also be given special exercises to do which will help her to stay fit during pregnancy, and also allow her to take an active part when she is giving birth.

We have seen that many useful substances cross the placenta from the mother's blood to the fetus's blood. Unfortunately, harmful substances can cross, too. For example, if the mother smokes, nicotine and carbon monoxide can enter the baby's blood, and this can cause the baby to grow more slowly and be born smaller than if the mother was a non-smoker. A woman should never smoke during pregnancy. She also needs to take care not to drink too much alcohol, or to take any drug without advice from her doctor.

The mother also needs to avoid some illnesses. Rubella is caused by a virus, producing a rash and a fever. If the rubella virus crosses the placenta, it can cause serious harm to the fetus, who may be born deaf or with other disabilities. In many countries, teenage girls are offered vaccination against rubella. Another disease which is very dangerous to a fetus is AIDS. The virus that causes AIDS, called HIV, can cross the placenta so that a baby may be born with AIDS.

13.20 Mammals care for their young.

Although it has been developing for nine months, a human baby is very helpless when it is born. Usually both parents help to care for it.

During pregnancy, the glands in the mother's breasts will have become larger. Soon after the birth of the baby, they begin to make milk. This is called **lactation** (Figure **13.19**). Lactation happens in all mammals, but not in other animals.

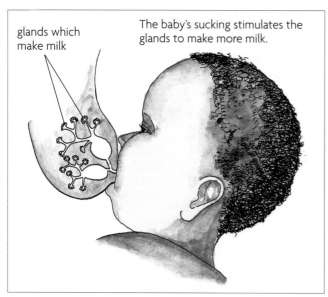

glands which make milk

The baby's sucking stimulates the glands to make more milk.

Figure 13.19 Lactation.

Milk contains all the nutrients that the baby needs. It also contains antibodies (section **8.15**) which will help the baby to resist infection.

As well as being fed, the baby needs to be kept warm. Because it is so small, a baby has a large surface area in relation to its volume, so it loses heat very quickly.

It is extremely important that a young baby is cared for emotionally, as well as physically. Babies need a lot of close contact with their parents.

Most mammals care for their young by feeding them and keeping them warm. In humans, parental care also involves teaching the baby and young child how to look after itself, and how to live in society. This continues into its 'teens' – a much longer time than for any other animal.

13.21 Breast milk has advantages over formula milk.

Most people consider that feeding a baby on breast milk is much better than bottle-feeding. Formula milk is bought as powder that is mixed with boiled (sterilised) water. The baby then sucks this milk from a bottle.

This can make life easier for the mother, because she can hand over the feeding of her baby to someone else. This can also help the father to bond with the baby, if he is involved in feeding it.

However, formula milk is much more expensive than breast milk, which is free! And, unless the equipment used for making up the formula milk is kept clean, it is easy for bacteria to get into the milk and make the baby ill.

Another advantage of breast milk is that it contains antibodies from the mother, which help the baby to fight off infectious diseases. Breast-feeding also helps a close relationship to develop between the mother and her baby, which is beneficial to both of them.

The composition of breast milk changes as the baby grows, so that the nutrients it contains are exactly right for the different stages of its development.

FACT! The most children one woman has ever had is 69. A Russian woman who lived between 1707 and 1782 had sixteen pairs of twins, seven sets of triplets and four sets of quadruplets, all born between 1725 and 1765.

Questions

13.24 What is implantation?

13.25 What is a fetus?

13.26 How is the fetus connected to the placenta?

13.27 List **two** substances which pass from the mother's blood into the fetus's blood.

13.28 Describe what happens to each of the following during the birth of a baby: **(a)** muscles in the uterus wall, **(b)** the cervix, **(c)** the placenta.

13.29 Construct a comparison table to show the advantages and disadvantages of breast-feeding and bottle-feeding.

13.22 The menstrual cycle lasts about 28 days.

Usually, one egg is released into the oviduct every month in an adult woman. Before the egg is released, the lining of the uterus becomes thick and spongy, to prepare itself for a fertilised egg. It is full of tiny blood vessels, ready to supply the embryo with food and oxygen if it should arrive.

If the egg is not fertilised, it is dead by the time it reaches the uterus. It does not sink into the spongy wall, but continues onwards, down through the vagina. As the spongy lining is not needed now, it gradually disintegrates. It, too, is slowly lost through the vagina. This is called **menstruation**, or a period. It usually lasts for about five days. After menstruation, the lining of the uterus builds up again, so that it will be ready to receive the next egg, if it is fertilised.

Figure **13.20** shows what happens during the human **menstrual cycle**.

13.23 Sexual maturity is reached at puberty.

The time when a person approaches sexual maturity is called adolescence. Sperm production begins in a boy, and ovulation in a girl.

During adolescence, the secondary sexual characteristics develop. In boys, these include growth of facial and pubic hair, breaking of the voice, and muscular development. In girls, pubic hair begins to grow, the breasts develop, and the pelvic girdle becomes broader.

These changes are brought about by hormones. The male hormone is **testosterone**. It is produced in the testes. The female hormone is **oestrogen**. It is produced in the ovaries.

The point at which sexual maturity is reached is called **puberty**. This is often several years earlier for girls than for boys. At puberty, a person is still not completely adult, because emotional development is not complete.

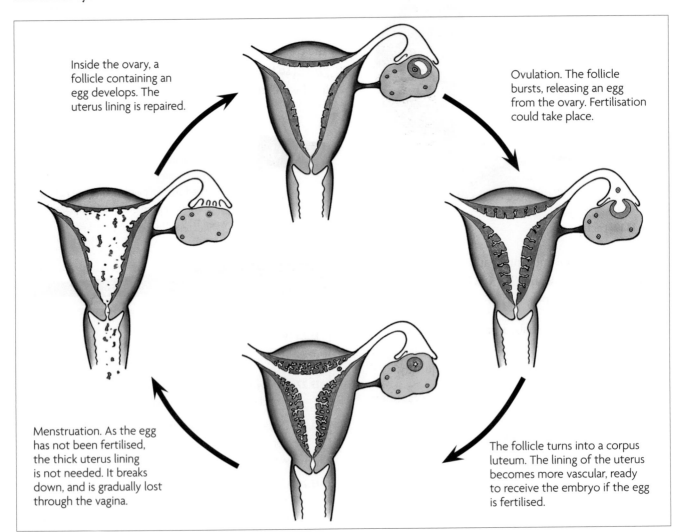

Inside the ovary, a follicle containing an egg develops. The uterus lining is repaired.

Ovulation. The follicle bursts, releasing an egg from the ovary. Fertilisation could take place.

Menstruation. As the egg has not been fertilised, the thick uterus lining is not needed. It breaks down, and is gradually lost through the vagina.

The follicle turns into a corpus luteum. The lining of the uterus becomes more vascular, ready to receive the embryo if the egg is fertilised.

Figure 13.20 The menstrual cycle.

13.24 Female sex hormones control the menstrual cycle.

Oestrogen is not the only female sex hormone. The ovaries also produce a hormone called **progesterone** during certain stages of the menstrual cycle, and during pregnancy. The secretion of these hormones is controlled by two other hormones secreted by the pituitary gland in the head, called **LH** and **FSH** (Figure 13.21.)

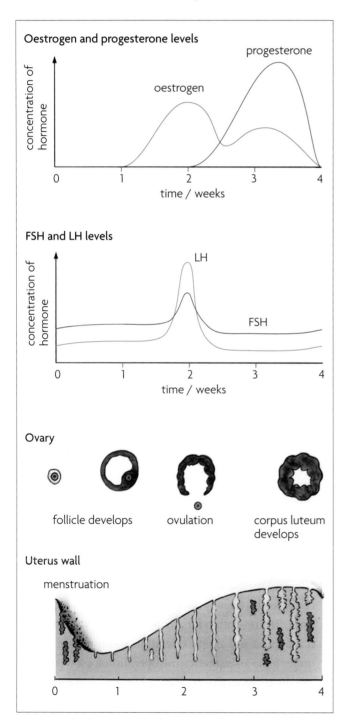

Ovary

follicle develops   ovulation   corpus luteum develops

Uterus wall

menstruation

Figure 13.21 Hormones and the menstrual cycle.

Whereas male mammals make sperm all the time, females only produce eggs at certain times. We have seen that, in humans, ovulation happens once a month. Ovulation is part of the menstrual cycle.

First, a follicle develops inside an ovary. The developing follicle secretes oestrogen, and the concentration of oestrogen in the blood steadily increases. The oestrogen makes the lining of the uterus grow thick and spongy. Throughout this time, the pituitary gland secretes LH and FSH. These two hormones stimulate the follicle to keep on secreting oestrogen.

When the follicle is fully developed, there is a surge in the production of LH and also – to a lesser extent – FSH. This causes ovulation to take place.

The now empty follicle stops secreting oestrogen. It becomes a **corpus luteum**. The corpus luteum starts to secrete another hormone – progesterone. Levels of FSH and LH fall.

Progesterone keeps the uterus lining thick, spongy, and well supplied with blood, in case the egg is fertilised. If it is not fertilised, then the corpus luteum gradually disappears. Progesterone is not secreted any more, and so the lining of the uterus breaks down. Menstruation happens. A new follicle starts to develop in the ovary, and the cycle begins again.

But if the egg is fertilised, the corpus luteum does not degenerate so quickly. It carries on secreting progesterone until the embryo sinks into the uterus wall, and a placenta develops. Then the placenta secretes progesterone, and carries on secreting it all through the pregnancy. The progesterone maintains the uterus lining, so that menstruation does not happen during the pregnancy.

Questions

13.30 Why does the uterus wall become thick and spongy before ovulation?

13.31 What happens if the egg is not fertilised?

13.32 What is meant by (a) adolescence, and (b) puberty?

13.33 What is testosterone?

13.34 List **two** effects of testosterone.

13.25 Birth control can prevent unwanted pregnancies.

Birth control can help couples to have no more children than they want. Birth control is important in keeping family sizes small, and in limiting the increase in the human population. Careful and responsible use of birth control methods means no unwanted children are born.

- **Natural methods** involve the couple avoiding sexual intercourse completely (abstinence) or ensuring that they do not have sexual intercourse when the woman has an egg in her oviducts. This is a risky method, and only works for women who have very regular and predictable menstrual cycles. However, it is useful for couples who do not wish to use other birth control methods for religious or other reasons. Figure **13.22** shows how a woman can work out the 'safe period', when an egg is least likely to be in her oviducts. She needs to keep a careful record of her body temperature. You can see from the graph that temperature rises slightly around the time of ovulation.

  If the woman has a regular cycle, then she can use this to predict when ovulation will take place. She should then avoid sexual intercourse on the three or four days either side of this date. However, few women have completely regular cycles, and this makes this method of birth control unreliable.

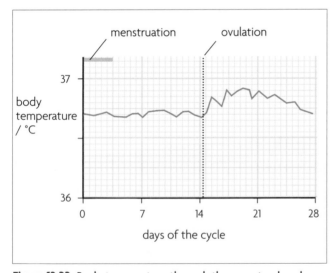

Figure 13.22 Body temperature through the menstrual cycle.

- **Chemical methods** can involve the use of chemicals called **spermicides**, which kill sperm. They are best used in combination with another method. For example, spermicides may be inserted into the vagina with a diaphragm, or cap (see below). Another type of chemical birth control is the use of sex hormones to disrupt the menstrual cycle. A woman can take the **contraceptive pill**, which stops eggs being produced in her ovaries. The pill contains progesterone and oestrogen. She may have to take a pill each day, or she may be given a long-lasting injection of contraceptive hormones.

- **Mechanical methods** work by putting a barrier between the eggs and sperm. The most widely used mechanical method of birth control is a **condom** – a thin sheath that is placed over the man's erect penis and that stops any sperm getting into the woman's vagina. This also has the advantage that it stops any pathogens passing between the couple, so it is good protection against the transmission of diseases such as gonorrhoea or HIV/AIDS (pages **180–181**). Women can use a female version of a condom, called a **femidom**, which is placed inside the vagina and works in a similar way.

  An alternative method for a woman is to use a **diaphragm**, sometimes called a cap. This is a circular, slightly domed piece of rubber which is inserted into the vagina and which covers the cervix, stopping sperm getting past it and into the uterus. To make absolutely sure that none can squeeze past, it is a good idea to use a spermicide cream as well. Yet another method is a device that is placed inside the uterus (and therefore has to be fitted by a doctor) called an **IUD** (standing for intra-uterine device). This interferes with the ability of sperm to find and fertilise an egg, and also prevents the implantation and development of any egg that does get fertilised.

- **Surgical methods** tend to be most suitable for couples who already have as many children as they want. The operation for a man is called a **vasectomy**. It is a quick and simple operation, usually done under local anaesthetic. The operation for a woman usually involves a short stay in hospital, and a general anaesthetic. Figure **13.23** shows what the operations entail.

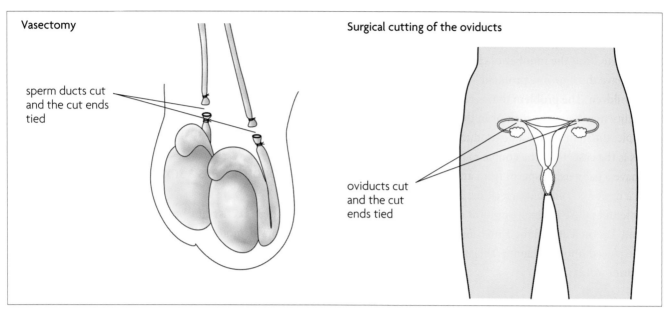

Vasectomy

sperm ducts cut
and the cut ends
tied

Surgical cutting of the oviducts

oviducts cut
and the cut
ends tied

Figure 13.23 Surgical methods of birth control.

The various methods of birth control, and their advantages
and disadvantages, are summarised in Table **13.1**.

Table 13.1 Some methods of birth control

method	how it works	advantages and disadvantages
condom (mechanical)	The condom is placed over the erect penis. It traps semen as it is released, stopping it from entering the vagina.	This is a very safe method of contraception if used correctly, but care must be taken that no semen is allowed to escape before it is put on or after it is removed. It can also help to prevent the transfer of infection, such as gonorrhoea and HIV, from one partner to another.
diaphragm, or cap (mechanical)	The diaphragm is a circular sheet of rubber, which is placed over the cervix, at the top of the vagina. Spermicidal (sperm-killing) cream is first applied round its edges. Sperm deposited in the vagina cannot get past the diaphragm into the uterus.	This is an effective method, if used and fitted correctly. Fitting must be done by a doctor, but after that a woman can put her own diaphragm in and take it out as needed.
the pill or oral contraceptive (chemical)	The pill contains the female sex hormones oestrogen and progesterone. One pill is taken every day. The hormones are like those that are made when a woman is pregnant, and stop egg production.	This is a very effective method, so long as the pills are taken at the right time. However, some women do experience unpleasant side-effects, and it is important that women on the pill have regular check-ups with their doctor.
sterilisation (surgical)	In a man, the sperm ducts are cut or tied, stopping sperm from travelling from the testes to the penis. In a woman, the oviducts are cut or tied, stopping eggs from travelling down the oviducts.	An extremely sure method of contraception, with no side-effects. However, the tubes often cannot be re-opened if the person later decides that they do want to have children, so it is not a method for young people.
spermicides (chemical)	Spermicidal cream in the vagina kills sperm.	This is quite easy to use. It is only effective, however, if used in combination with another method, such as the diaphragm.
natural	The woman keeps a careful record of her menstrual cycle over several months, so that she can predict roughly when an egg is likely to be present in her oviducts. She must avoid sexual intercourse for several days around this time.	This is a very unsafe method, because it is never possible to be 100% certain when ovulation is going to happen. Nevertheless, it is used by many people who do not want to use one of the other contraceptive methods.

13.26 Hormones can be used to increase fertility.

Whereas many couples want to use birth control methods to limit the number of children that they have, others have the opposite problem – they are not able to have children. The problem that is causing the couple's infertility may be in the man or in the woman. For example, the man may not be producing healthy sperm. If this is the case, then the couple may decide that they will have a baby using sperm from another man. Sperm from a donor is collected in a clinic, and can be stored at a low temperature for many months or even years. The woman can then attend the clinic, and some of the sperm can be placed into her vagina. This is called **artificial insemination**.

This may be a real help to a couple, as it allows them to have a child that they could not otherwise have. However, they need to think very carefully about this before they go ahead, and make sure that they are both happy with the idea. The man has to be able to accept that the child they have is not biologically his. Problems can also be caused when the child grows up and wants to know who his or her biological father is. It can be very difficult for a young person not to know this, so some people think that the identity of the sperm donor should be given to the child. Others, however, think this may cause more problems than it solves, because one sperm donor could end up being the father of many children. Indeed, fewer people would be likely to become sperm donors in many countries if this information was freely available.

Another way in which an infertile couple can be helped is using **fertility drugs**. This method is used when the woman is not producing enough eggs. She is given hormones, including FSH, that cause her to produce eggs. Sometimes, these are simply allowed to be released into the oviducts in the normal way. Sometimes, they are removed from her ovaries just before they are due to be released, and placed in a warm liquid in a Petri dish. Some of her partner's sperm are then added, and fertilisation takes place in the dish. Two or three of the resulting zygotes are then placed into her uterus, where they develop in the usual way.

This method is quite expensive, and some people think that it should not be freely available to anyone who wants it. Others think that the inability to have children can be so devastating to a couple that they should receive the treatment free of charge. The treatment is not always successful, and may have to be repeated many times before a woman becomes pregnant. Another problem is that, while usually only one of the embryos develops, sometimes two or three do, so that the couple might have twins or triplets when they really only wanted one child.

13.27 Gonorrhoea is a sexually transmissable disease.

Gonorrhoea is caused by bacteria that can be passed from one person to another during sexual intercourse.

The bacterium that causes gonorrhoea is a small, round cell (Figure **13.24**). It can only survive in moist places, such as the tissues lining the tubes in the reproductive systems of a man or a woman. If gonorrhoea bacteria are living in a woman's vagina or a man's urethra, the infection can be passed from one to the other during sexual intercourse.

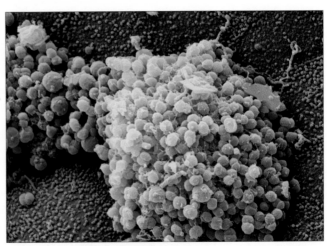

Figure 13.24 Each green sphere is a tiny *Neisseria* bacterium, which causes gonorrhoea (× 1700).

The first symptoms usually occur between two and seven days after infection. In a man, the bacteria reproduce inside the urethra, and this produces an unpleasant discharge and pain when urinating. In a woman, the bacteria reproduce mostly in the cervix, although they can also do so in the vagina. As in men, this produces a discharge, but many women do not notice this and they do not suffer pain as men do. Consequently, while most men with gonorrhoea know that they have it, many women are quite unaware that they have the infection.

Gonorrhoea can be treated with antibiotics (section **12.2**) such as penicillin, and this is almost always successful. However, it is much better to make sure that you do not get the disease in the first place! This can be achieved by:

- not being sexually active
- having only one sexual partner – if neither has the disease, then they cannot pass it on to each other
- ensuring that the man uses a condom, as the bacteria cannot pass through this from one person to the other
- tracing, warning and treating all possible sexual contacts of a person who is diagnosed with gonorrhoea to make sure that it does not spread any further.

13.28 HIV can be transmitted during sexual intercourse.

The disease **AIDS**, or acquired immune deficiency syndrome, is caused by **HIV**. HIV stands for human immunodeficiency virus. Figure **13.25** shows this virus.

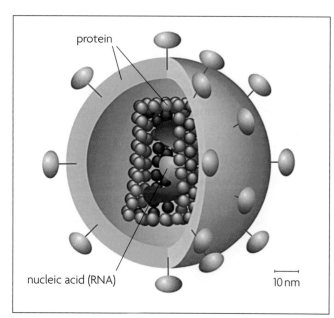

protein

nucleic acid (RNA)

10 nm

Figure 13.25 The human immunodeficiency virus, HIV. A nanometre (nm) is 1×10^{-9} m, so this virus is very, very small.

HIV infects lymphocytes, and in particular a type called T cells. Over a long period of time, HIV slowly destroys T cells. Several years after infection with the virus, the numbers of certain kinds of T cells are so low that they are unable to fight against other pathogens effectively. Because HIV attacks the very cells which would normally kill viruses – the T cells – it is very difficult for someone's own immune system to protect them against HIV.

About ten years after initial infection with HIV, a person is likely to develop symptoms of AIDS unless they are given effective treatment. They become very vulnerable to other infections, such as pneumonia. They may develop cancer, because one function of the immune system is to destroy body cells which may be beginning to produce cancers. Brain cells are also quite often damaged by HIV. A person with AIDS usually dies of a collection of several illnesses.

There is still no cure for AIDS, though drugs can greatly increase the life expectancy of a person infected with HIV. Researchers are always trying to develop new drugs, which will kill the virus without damaging the person's own cells. As yet, no vaccine has been produced either, despite large amounts of money being spent on research.

13.29 HIV is transmitted in body fluids.

The virus that causes AIDS cannot live outside the human body. In fact, it is an especially fragile virus – much less tough than the cold virus, for example. You can only become infected with HIV through direct contact of your body fluids with those of someone with the virus. This can be in one of the following ways.

Through sexual intercourse   HIV can live in the fluid inside the vagina, rectum and urethra. During sexual intercourse, fluids from one partner come into contact with fluids of the other. It is very easy for the virus to be passed on in this way.

The more sexual partners a person has, the higher the chance of them becoming infected with HIV. In some parts of the world, where it is common practice for men to have many different sexual partners, extremely high percentages of people have developed AIDS. This is so in some parts of Africa and Asia, and also amongst some homosexual communities in parts of Europe and the USA.

The best way of avoiding AIDS is never to have more than one sexual partner. If everyone did that, then AIDS would immediately stop spreading. Using condoms is a good way of lowering the chances of the viruses passing from one person to another during sexual intercourse – though it does not rule it out.

Through blood contact Many cases of AIDS have been caused by HIV being transferred from one person's blood to another. In the 1970s and 1980s, when AIDS first appeared, and before anyone knew what was causing it, blood containing HIV was used in transfusions. People being given the transfusions were infected with HIV, and later developed AIDS. Now all blood used in transfusions in most countries is screened for HIV before it is used.

Blood can also be transferred from one person to another if they share hypodermic needles. This most commonly happens in people who inject drugs, such as heroin. Many drug users have died from AIDS. It is essential that any hypodermic needle used for injection is sterile.

People who have to deal with accidents, such as police and paramedics, must always be on the guard against HIV if there is blood around. They often wear protective clothing, just in case a bleeding accident victim is infected with HIV.

However, in general, there is no danger of anyone becoming infected with HIV from contact with someone with AIDS. You can quite safely talk to the person, shake hands with them, drink from cups which they have used and so on. In fact, there is far more danger to the person who has AIDS from such contacts, because they are so vulnerable to any bacterium or virus which they might catch from you.

Sexual reproduction in flowering plants

13.30 Flowers are for sexual reproduction.

Many flowering plants can reproduce in more than one way. Often, they can reproduce asexually (section **13.1**) and also sexually, by means of flowers.

The function of a flower is to make gametes, and to ensure that fertilisation will take place. Figure **13.26** illustrates the structure of an insect-pollinated flower. Figure **13.27** shows flowers of *Eucryphia* which makes both male and female gametes, so it is a hermaphrodite flower. Most, but not all, flowers are hermaphrodite.

On the outside of the flower are the **sepals**. The sepals protect the flower while it is a bud. Sepals are normally green.

Just inside the sepals are the **petals**. These are often brightly coloured. The petals attract insects to the flower. The petals of some flowers have lines running from top to bottom. These lines are called **guide-lines**, because they guide insects to the base of the petal. Here, there is a gland called a **nectary**. The nectary makes a sugary liquid called nectar, which insects feed on.

Inside the petals are the **stamens**. These are the male parts of the flower. Each stamen is made up of a long **filament**,

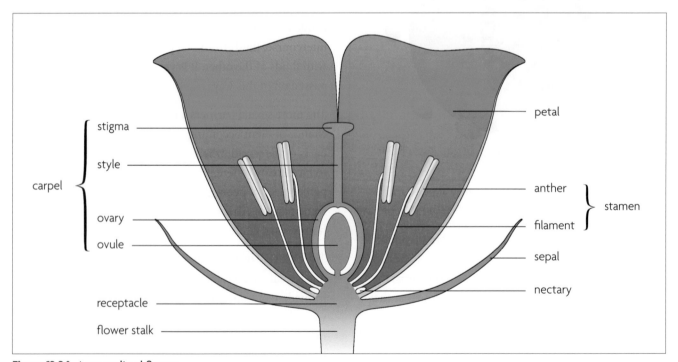

Figure 13.26 A generalised flower.

Figure 13.27 *Eucryphia* flowers.

One of the differences is the arrangement of the ovules in the ovary. Figure **13.28** shows one arrangement.

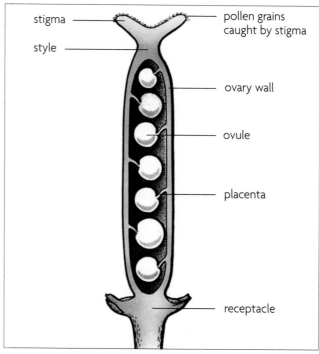

Figure 13.28 Section through the female part of a flower.

with an **anther** at the top. The anthers contain **pollen grains**, which contain the male gametes.

The female part of the flower is in the centre. It consists of one or more **carpels**. A carpel contains an **ovary**. Inside the ovary are many **ovules**, which contain the female gametes. At the top of the ovary is the **style**, with a **stigma** at the tip. The function of the stigma is to catch pollen grains. The female parts of different kinds of flower vary.

Activity 13.1
Investigating the structure of a flower

skills
C2 *Observing, measuring and recording*
C3 *Interpreting and evaluating*

Safety Take care with the sharp knife blade.

During this investigation, make large, labelled drawings of the structures that you observe.

1 Take an open, fresh-looking flower. Can you suggest two ways in which the flower advertises itself to insects?

2 Gently remove the sepals from the outside of the flower. Look at the sepals on a flower bud, near the top of the stem. What is the function of the sepals?

3 Now remove the petals from your flower. Make a labelled drawing of one of them, to show the markings. What is the function of these markings?

4 Find the stamens. If you have a young flower, there will be pollen on the anthers at the top of the stamens. Dust some onto a microscope slide, and look at it under a microscope. Draw a few pollen grains.

5 Now remove the stamens. What do you think is the function of the filaments?

6 Using a hand lens, try to find the nectaries at the bottom of the flower. What is their function?

7 The carpel is now all that is left of the flower. Find an ovary, style and stigma. Look at the stigma under a binocular microscope or a lens. What is its function, and how is it adapted to perform it?

8 Using a sharp blade, make a clean cut lengthways through the ovary, style and stigma. You have made a longitudinal section. Find the ovules inside the ovary. How big are they? What colour are they? About how many are there?

13.31 Pollen grains contain male gametes.

The male gametes are inside the pollen grains, which are made in the anthers.

Figure **13.29a** illustrates a young anther, as it looks before the flower bud opens. You can see in Figure **13.29b** that the anther has four spaces or pollen sacs inside it. Some of the cells around the edge of the pollen sacs divide by meiosis to make pollen grains. When the flower bud opens, the anthers split open (Figure **13.29c**). Now the pollen is on the outside of the anther.

The pollen looks like a fine powder. It is often yellow. Under the microscope, you can see the shape of individual grains (Figure **13.30**). Pollen grains from different kinds of flowers have different shapes. Each grain is surrounded by a hard coat, so that it can survive in difficult conditions if necessary. The coat protects the male gametes that are inside the grains, as the pollen is carried from one flower to another.

13.32 Each ovule contains a female gamete.

The female gametes are inside the ovules, in the ovary. They have been made by meiosis. Each ovule contains just one gamete.

13.33 Pollen must be carried from anther to stigma.

For fertilisation to take place, the male gametes must travel to the female gametes. The first stage of this journey

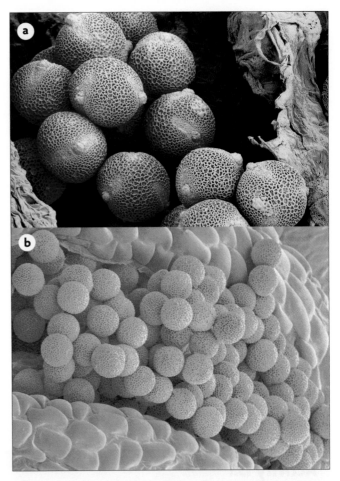

Figure 13.30 a These pollen grains are from a flower of a melon plant. The electron micrograph image has been magnified about 600 times. **b** This photo of an anther of a rape plant was also taken using a scanning electron microscope. You can see the pollen grains bursting out of a split in the anther.

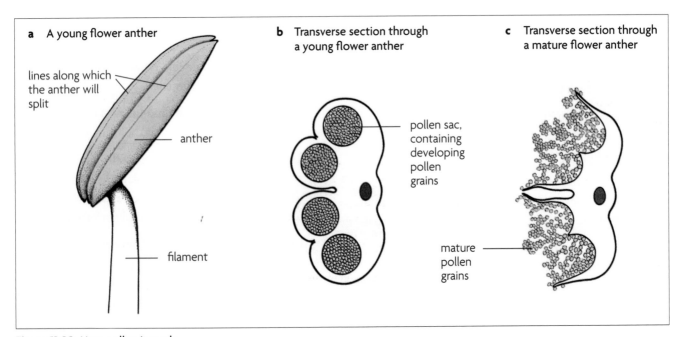

a A young flower anther

lines along which the anther will split

anther

filament

b Transverse section through a young flower anther

pollen sac, containing developing pollen grains

c Transverse section through a mature flower anther

mature pollen grains

Figure 13.29 How pollen is made.

is for pollen to be taken from the anther where it was made, to a stigma. This is called **pollination**.

Pollination is often carried out by insects (Figure **13.31**). Insects such as honey bees come to the flowers, attracted by their colour and strong sweet scent. The bee follows the guide-lines to the nectaries, brushing past the anthers as it goes. Some of the pollen sticks to its body.

Figure 13.31 The bee has come to the flower to collect nectar. Pollen gets stuck to its body, and the bee will then carry this to the next flower it visits.

The bee then goes to another flower, looking for more nectar. Some of the pollen it picked up at the first flower sticks onto the stigma of the second flower when the bee brushes past it. The stigma is sticky, and many pollen grains get stuck on it. If the second flower is from the same species of plant as the first, pollination has taken place.

Key definition

pollination   the transfer of pollen grains from the male part of the plant (anther of stamen) to the female part of the plant (stigma)

13.34 Flowers can be self- or cross-pollinated.

Sometimes, pollen is carried to the stigma of the same flower, or to another flower on the same plant. This is called **self-pollination**.

If pollen is taken to a flower on a different plant of the same species, this is called **cross-pollination**. If pollen lands on the stigma of a different species of plant, it usually dies.

Activity 13.2
Pollination

skills
C2 *Observing, measuring and recording*
C3 *Interpreting and evaluating*
C4 *Planning*

You are going to design and carry out an investigation to test this hypothesis:

> Bees visit yellow flowers more often than flowers of other colours.

You will need to carry out this investigation outdoors. It will be much easier to control variables if you make artificial flowers rather than using real ones. You can make them using coloured plastic to make 'petals', surrounding a central area where you can put a little pot of sugar solution. You will need to do your experiment on a sunny day, when there are plenty of bees flying.

Remember to think about controlling variables. Think carefully about exactly how you will count the bee visits, how you will record them and how you will display your results.

Write a simple conclusion from your results, and then discuss the results in the light of what you know about pollination. (You might also be interested in finding out about how bees see colour.) Evaluate your experiment, and suggest improvements you could make.

13.35 Some flowers are wind-pollinated.

In some plants, it is the wind which carries the pollen between flowers. Figure 13.32 shows a grass flower, which is an example of a wind-pollinated flower.

Table 13.2 compares insect-pollinated and wind-pollinated flowers.

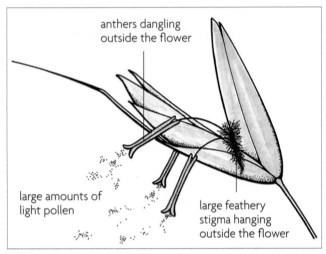

Figure 13.32 An example of a wind-pollinated flower.

13.36 Pollen tubes take male gametes to ovules.

After pollination, the male gamete inside the pollen grain on the stigma still has not reached the female gamete. The female gamete is inside the ovule, and the ovule is inside the ovary.

If it has landed on the right kind of stigma, the pollen grain begins to grow a tube. You can try growing some pollen tubes, in Activity 13.3. The pollen tube grows

down through the style and the ovary, towards the ovule (Figure 13.33). It secretes enzymes to digest a pathway through the style.

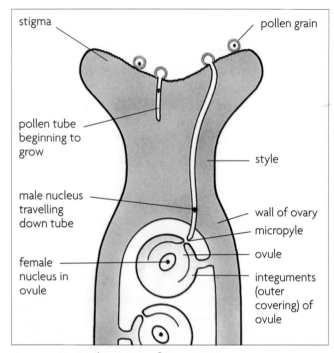

Figure 13.33 Fertilisation in a flower.

The ovule is surrounded by several layers of cells called the **integuments**. At one end, there is a small hole in the integuments, called the **micropyle**. The pollen tube grows through the micropyle, into the ovule.

The male gamete travels along the pollen tube, and into the ovule. It fuses with the female gamete. Fertilisation has now taken place.

Table 13.2 A comparison between insect-pollinated and wind-pollinated flowers

insect-pollinated	wind-pollinated
large, conspicuous petals, often with guide-lines	small, inconspicuous petals, or no petals at all
often strongly scented	no scent
often have nectaries at the base of petals	no nectaries
anthers inside flower, where insect has to brush past them to reach nectar	anthers dangling outside the flower, where they catch the wind
stigma inside flower, where insect has to brush past it to reach nectar	stigmas large and feathery and dangling outside the flower, where pollen in the air may land on it
sticky or spiky pollen grains, which stick to insects	smooth, light pollen, which can be blown in the wind
quite large quantities of pollen made, because some will be eaten or will be delivered to the wrong kind of flower	very large quantities of pollen made, because most will be blown away and lost

One pollen grain can only fertilise one ovule. If there are many ovules in the ovary, then many pollen grains will be needed to fertilise them all.

13.37 Fertilised ovules become seeds.

Once the ovules have been fertilised, many of the parts of the flower are not needed any more. The sepals, petals and stamens have all done their job. They wither, and fall off.

Inside the ovary, the ovules start to grow. Each ovule now contains a **zygote**, which was formed at fertilisation. The zygote divides by mitosis to form an **embryo plant**.

The ovule is now called a **seed**. The integuments of the ovule become hard and dry, to form the **testa** of the seed. Water is withdrawn from the seed, so that it becomes dormant.

The embryo consists of a **radicle**, which will grow into a root, and a **plumule**, which will grow into a shoot (Figure **13.34**).

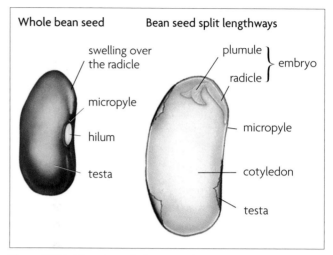

Figure 13.34 Structure of a bean seed.

Activity 13.3
Growing pollen tubes

skills

C1 *Using techniques, apparatus and materials*

C2 *Observing, measuring and recording*

C3 *Interpreting and evaluating*

When a stigma is ripe, it secretes a fluid which stimulates pollen grains on it to grow tubes. The fluid contains sugar. In this investigation, you can try germinating different kinds of pollen grains in different concentrations of sugar solution.

It is best if the class is divided into groups. Each group should use sugar solution of just one concentration.

1 Collect four cavity slides. Using your finger, make a neat ring of petroleum jelly around the outer edge of each cavity.

2 Stick a label on each slide. Write your initials on it, and the concentration of sugar solution your group is using.

3 Fill the cavity in each slide with sugar solution.

4 Choose one flower of each kind which has pollen on its anthers. Dust pollen from one flower onto the solution on one of your slides. Gently lower a cover slip over it, without squashing the petroleum jelly ring. Write the name of the flower on the label.

5 Repeat step **4** with the other three flowers.

6 Place each slide in a warm incubator, and leave for at least an hour.

7 Set up a microscope. Examine each of your slides under the microscope. Look carefully for pollen tubes. Record your results in a table, and collect results from groups using other concentrations of sugar solution.

Questions

1 Why was a ring of petroleum jelly put around the cavity in each slide?

2 In which solution did each of the four types of pollen germinate best?

3 Can you suggest why pollen dies if it lands on an unripe stigma, or a stigma of the wrong sort of flower?

4 Why do pollen grains grow tubes?

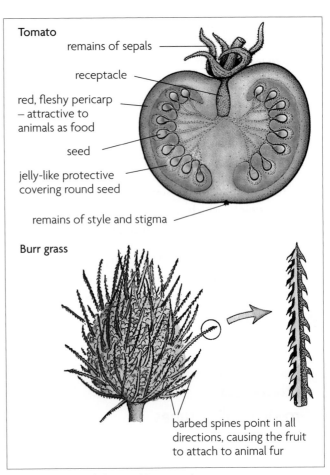

Figure 13.35 Seed dispersal by animals.

The seed also contains food for the embryo. In a bean seed, the food is stored in two cream-coloured **cotyledons**. These contain starch and protein. The cotyledons also contain enzymes. Surrounding the cotyledons is a tough, protective covering called the testa. The testa stops the embryo from being damaged and it prevents bacteria and fungi from entering the seed. The testa has a tiny hole in it called the **micropyle**. When a seed has been separated from the plant, near the micropyle there is a scar, the **hilum,** where the seed was joined to the pod (ovary).

The ovary also grows. It is now called a **fruit**. The wall of the fruit is called a **pericarp**.

13.38 Fruits protect and disperse seeds.

The function of the fruit is to protect the seeds inside it until they are ripe, and then to help disperse the seeds. **Dispersal** of seeds is important, because it prevents too many plants growing close together. If this happens, they compete for light, water and nutrients, so that none of them can grow properly. Dispersal also allows the plant to colonise new areas.

Fruits and seeds are dispersed by animals and water and through the air. Some examples of dispersal mechanisms are shown in Figures **13.35** and **13.36**.

13.39 Fruits are ovaries after fertilisation.

Plants have an enormous variety of fruits, all adapted to disperse their seeds as effectively as possible. It is important to remember that, in biology, the word 'fruit' has a very particular meaning. Most people use the word to mean sweet fruits eaten as snacks or dessert. The fruits of tomato, pepper and beans are commonly called vegetables.

The biological definition of a fruit is an ovary after fertilisation, containing seeds. Cherries, plums and oranges are true fruits, but so also are cucumbers. You can tell a fruit because it contains one or more seeds, and it has two scars – one where it was attached to the plant, and one where the style and stigma were attached to it.

 FACT!

The plant with the largest seed is the Coco de Mer, which grows on the Seychelles islands in the Indian Ocean. The seeds (coconuts) weigh up to 18 kg. The plants with the smallest seeds are orchids. Some kinds have 1 235 000 seeds to the gram.

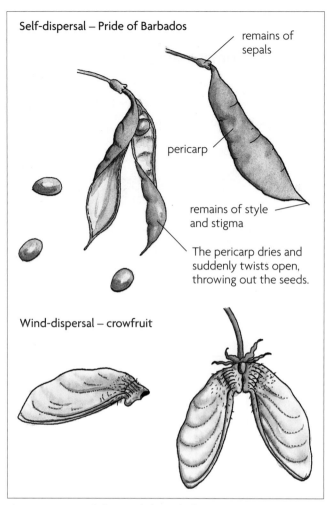

Self-dispersal – Pride of Barbados

remains of sepals

pericarp

remains of style and stigma

The pericarp dries and suddenly twists open, throwing out the seeds.

Wind-dispersal – crowfruit

Figure 13.36 Seed dispersal through the air.

Sometimes, it is not easy to tell a fruit from a seed. A seed, though, only has one scar, called the hilum, where it was joined to the fruit.

13.40 Uptake of water begins seed germination.

A seed contains hardly any water. When it was formed on the plant, the water in it was drawn out, so that it became dehydrated. Without water, almost no metabolic reactions

Questions

13.43 Give **two** functions of a fruit.

13.44 List **two** different ways in which seeds may be dispersed, giving **one** example for each.

13.45 Give **two** differences between fruits and seeds.

13.46 Which of the following are fruits, and which are not? (a) orange, (b) tomato, (c) potato, (d) cabbage, (e) bean pod, (f) cucumber.

can go on inside it. The seed is inactive or **dormant**. This is very useful, because it means that the seed can survive harsh conditions, such as cold or drought, which would kill a growing plant.

A seed must be in certain conditions before it will begin to germinate. You can find out what they are if you do Activity **13.4**.

When a seed germinates, it first takes up water through the micropyle. As the water goes into the cotyledons, they swell. Eventually, they burst the testa (Figure **13.37**).

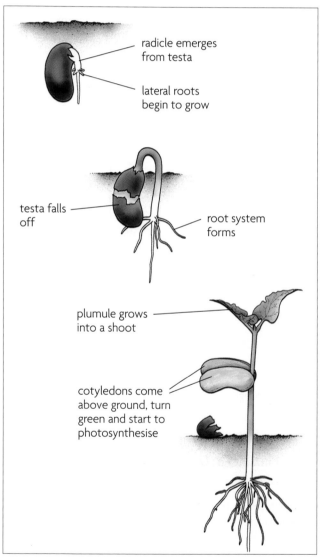

radicle emerges from testa

lateral roots begin to grow

testa falls off

root system forms

plumule grows into a shoot

cotyledons come above ground, turn green and start to photosynthesise

Figure 13.37 Stages in germination of one type of bean seed.

Once there is sufficient water, the enzymes in the cotyledons become active. **Amylase** begins to break down the stored starch molecules to maltose. **Proteases** break down the protein molecules to amino acids.

Activity 13.4
To find the conditions necessary for the germination of tomato seeds

skills

C2 *Observing, measuring and recording*

C3 *Interpreting and evaluating*

C4 *Planning*

Safety Pyrogallol is very caustic. Your teacher will handle it for you. You should not use it yourself.

1 Set up five tubes as shown in the diagram. Pyrogallol absorbs oxygen.

5 Copy the results table and fill it in to show what conditions the seeds in each tube have. The first line has been done for you.

	tube				
	A	B	C	D	E
water	✓	✓	✓	✓	
warmth					
oxygen					
light					
Did seeds germinate?					

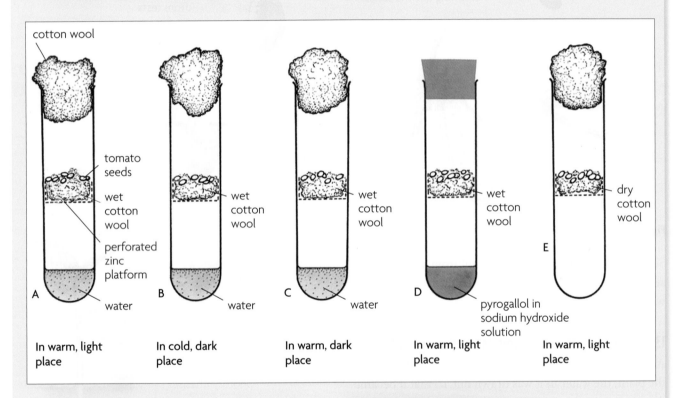

cotton wool

tomato seeds

wet cotton wool

perforated zinc platform

A   water

In warm, light place

B   water

In cold, dark place

wet cotton wool

C   water

In warm, dark place

wet cotton wool

wet cotton wool

D   pyrogallol in sodium hydroxide solution

In warm, light place

E

dry cotton wool

In warm, light place

2 Put tubes **A**, **D** and **E** in a warm place in the laboratory, in the light.

3 Put tube **B** in a refrigerator.

4 Put tube **C** in a warm, dark cupboard.

Questions

1 What three conditions do tomato seeds need for germination?

2 Read sections **13.40** and **13.41**, and then explain why each of these conditions is needed for successful germination.

Maltose and amino acids are soluble, so they dissolve in the water. They diffuse to the embryo plant, which uses these foods for growth. The way in which the embryo plant grows is shown in Figure 13.37 (page 189).

13.41 During germination, enzymes digest food stores.

When a seed first begins to germinate, it increases in mass. This is because it absorbs water from the soil.

As soon as it begins to grow, it starts to use its food stores. The stored protein is broken down to amino acids, which are used to make new protein molecules for cell membranes and cytoplasm. The stored starch is broken down to maltose and then to glucose. Some of the glucose will be made into cellulose, to make cell walls for the new cells.

All this requires energy. The seed, like all living organisms, gets its energy by breaking down glucose, in respiration. Quite a lot of the glucose from the stored starch will be used up in respiration, so the seed loses mass.

After a few days, the plumule of the seed grows above the surface of the ground. The first leaves open out and begin to photosynthesise. The plant can now make its own food faster than it is using it up. Its mass begins to increase.

Sexual and asexual reproduction

13.42 Sexual reproduction produces variation.

This chapter has described some methods of asexual reproduction and sexual reproduction. Now that you know something about them, we can look in more detail at some of the important differences between them.

In asexual reproduction, some of the parent's cells divide by mitosis. This makes new cells that are genetically identical to the parent cell. They are **clones**. Asexual reproduction does not produce variation.

But in sexual reproduction, some of the parent's cells divide by meiosis. The new cells that are made are called gametes, and they have only half as many chromosomes as the parent cell. When two sets of chromosomes in the two gametes combine at fertilisation, a new combination of genes is produced. So sexual reproduction produces offspring that are genetically different from their parents.

Questions

13.47 What do the cotyledons of a bean seed contain?

13.48 What does dormant mean?

13.49 What is the advantage of seed dormancy?

13.50 What activates the enzymes in the cotyledons of a germinating seed?

13.51 What do the enzymes do?

Activity 13.5
To find the effect of storage time on the germination rate of seeds

skills
- C2 *Observing, measuring and recording*
- C3 *Interpreting and evaluating*
- C4 *Planning*

Many seeds are able to survive in a dormant state for years. However, the percentage of seeds that germinate does tend to fall as the seeds age.

You are going to design and carry out an investigation to test this hypothesis:

> The older a batch of seeds is, the lower the percentage of the seeds that will germinate.

Remember to think about variables – what you will change, what you will keep the same and what you will measure. Think also about how you will record and display your results.

When you have written your plan, check it with your teacher before carrying out your experiment.

Analyse your results. Do they support or disprove your hypothesis? Discuss the main sources of error in your experiment, and suggest how you could improve it if you were able to do it again.

13.43 Sexual and asexual reproduction each have their advantages.

Is it useful or not to have genetic variation among offspring? This depends on the circumstances.

Sometimes, it is a good thing not to have any variation. If a plant, for example, is growing well in a particular place, then it must be well adapted to its environment. If its offspring all inherit the same genes, then they will be equally well adapted and are likely to grow well. This is especially true if there is plenty of space for them in that area. However, if it is getting crowded, then it may not be a good thing for the parent to produce new offspring that grow all around it.

Another advantage of asexual reproduction is that a single organism can reproduce on its own. It does not need to wait to be pollinated, or to find a mate. This can be good if there are not many of those organisms around – perhaps there is only a single one growing in an isolated place. In that case, asexual reproduction is definitely the best option. Do remember, though, that even a single plant may be able to reproduce sexually, by using self-pollination.

However, if the plant is not doing very well in its environment, or if a new disease has come along to which it is not resistant, then it could be an advantage for its offspring to be genetically different from it. There is a good chance that at least some of the offspring may be better adapted to that environment, or be resistant to that disease.

In flowering plants, sexual reproduction produces seeds, which are likely to be dispersed over a wide area. This spreads the offspring far away from the parents, so that they are less likely to compete with them. It also allows them to colonise new areas.

You will find out more about variation, and its importance for evolution, in Chapter 14.

Questions

13.52 Do you think that cross-pollination is likely to result in more or less variation amongst the offspring than self-pollination? Explain your answer.

13.53 Suggest some advantages and disadvantages of self-pollination to a species of plant.

Key ideas

◆ Asexual reproduction involves cell division by mitosis, producing a group of genetically identical individuals called a clone. Bacteria, fungi and potatoes can reproduce asexually.

◆ Sexual reproduction involves the production of genetically different gametes by meiosis. A male gamete fertilises a female gamete, producing a zygote which is genetically different from its parents.

◆ In humans, the male gametes are sperm and are made in the testes. During sexual intercourse, semen containing sperm passes out of the penis and into a woman's vagina.

◆ The female gametes are eggs and are made in the ovaries. After sexual intercourse, sperm swim through the cervix and uterus into the oviducts, where they may meet an egg. One sperm may fertilise the egg to produce a zygote.

◆ The zygote travels to the uterus and implants in the lining, growing into an embryo attached to the uterus wall via an umbilical cord and placenta. The placenta brings the growing embryo's blood very close to the mother's blood, so that nutrients and waste products can diffuse between them.

◆ The growing embryo is protected by amniotic fluid produced by the amnion.

◆ After birth, a young mammal is fed on milk from its mother. This provides it with exactly the correct balance of nutrients, as well as antibodies which protect it from infectious diseases.

◆ An egg is released from an ovary about once a month. If it is not fertilised, the thick lining of the uterus breaks down, in menstruation.

(continued ...)

(... *continued*)

◆ The menstrual cycle is controlled by the hormones oestrogen, progesterone, FSH and LH.

◆ Birth control helps a couple to avoid having unwanted children. There are natural, surgical, mechanical, and chemical methods, each of which has its own advantages and disadvantages to be weighed up when making the choice of which to use.

◆ Hormones can be used to increase fertility.

◆ Gonorrhoea and HIV/AIDS are infectious diseases that can be transmitted by sexual contact.

◆ In plants, the flowers are the reproductive organs. Male gametes are made inside pollen grains, produced by anthers. Female gametes are made inside ovules, produced by ovaries.

◆ The movement of pollen from an anther to a stigma is called pollination, and may be brought about by insects or the wind.

◆ After landing on a suitable stigma, a pollen grain germinates and the gametes travel down the style to the ovules. Here, fertilisation takes place and a zygote is produced. The zygote develops into an embryo, and the ovule develops into a seed. The ovary develops into the fruit, containing the seeds which contain the embryos.

◆ Fruits are adapted to disperse seeds, using animals or the wind.

◆ Seeds require certain conditions before they will germinate.

Revision questions

1 Match each of these words with its definition below: zygote, mitosis, meiosis, gamete, pollination, fertilisation, pericarp, fruit, seed.
   a a sex cell, containing only half the normal number of chromosomes
   b an ovary after fertilisation
   c a diploid cell, formed by the fusion of two gametes
   d a type of cell division which produces daughter cells just like the parent cell
   e a type of cell division which produces daughter cells with only half the number of chromosomes of the parent cell
   f an ovary wall after fertilisation
   g the transfer of pollen from an anther to a stigma
   h an ovule after fertilisation
   i the fusion of two gametes

2 a Where does meiosis occur in humans?
   b Describe how the egg and sperm are brought together.
   c How is the developing fetus supplied with food?
   d Explain some advantages of breast-feeding a baby rather than using formula milk.
   e What precautions do you think should be taken when using formula milk products for babies?

3 a Name a plant that naturally reproduces asexually.
   b What advantages are there to the plant in reproducing this way?
   c Many plants also reproduce sexually. What are the advantages to a plant in reproducing in this way?

S

14 Inheritance and evolution

- ◆ about chromosomes and genes, and how they are inherited
- ◆ how and why cells divide by mitosis and meiosis
- ◆ how to predict the outcomes of a genetic cross
- ◆ what causes variation amongst living things
- ◆ how natural selection can bring about evolution
- ◆ how we can use genetic engineering to change the features of living things.

Chromosomes

14.1 Nuclei contain chromosomes carrying genes.

In the nucleus of every cell there are a number of long threads called **chromosomes**.

Most of the time, the chromosomes are too thin to be seen except with an electron microscope. But when a cell is dividing, they get shorter and fatter so they can be seen with a light microscope. Figure **14.1** shows human chromosomes seen with a powerful electron microscope.

Each chromosome contains one very long molecule of DNA. The DNA molecule carries a code that instructs the cell about which kinds of proteins it should make. Each chromosome carries instructions for making many different proteins. A part of a DNA molecule coding for one protein is called a **gene**.

It is the genes on your chromosomes which determine all sorts of things about you – what colour your eyes or hair are, whether you have a snub nose or a straight one, and whether you have a genetic disease such as cystic fibrosis. You inherited these genes from your parents.

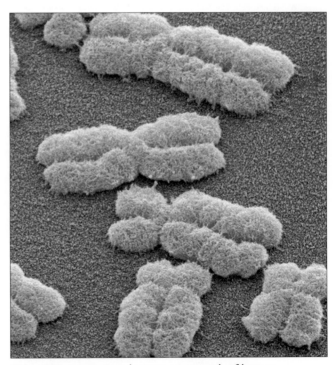

Figure 14.1 A scanning electron micrograph of human chromosomes. You can see that each one is made of two identical chromatids, linked at a point called the centromere.

Key definitions

chromosome   a thread of DNA, made up of a string of genes

gene   a length of DNA that is the unit of heredity and codes for a specific protein. A gene may be copied and passed on to the next generation

inheritance   the transmission of genetic information from generation to generation

14.2   Each species has its own set of genes.

Each species of organism has its own number and variety of genes. This is what makes their body chemistry, their appearance and their behaviour different from those of other organisms.

Humans have a large number of genes. You have 46 chromosomes inside each of your cells, all with many genes on them. Every cell in your body has an exact copy of all your genes. But, unless you are an identical twin, there is no-one else in the world with exactly the same combination of genes that you have. Your genes make you unique.

Cell division

14.3   Mitosis produces genetically identical daughter cells.

You began your life as a single cell – a zygote – formed by the fusion of an egg cell and a sperm cell. The nuclei of each of these gametes contained a single complete set of 23 chromosomes. They were **haploid** cells. The nucleus of the zygote therefore contained two sets of chromosomes. It was a **diploid** cell.

Key definitions

haploid nucleus   a nucleus containing a single set of unpaired chromosomes (e.g. sperm and egg)

diploid nucleus   a nucleus containing two sets of chromosomes (e.g. in body cells)

Figures **14.2** and **14.3** show the chromosomes in a cell of a man and of a woman. They have been arranged in order,

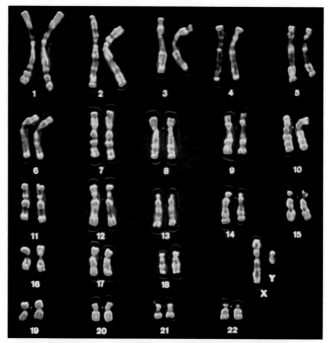

Figure 14.2 Chromosomes of a man, arranged in order.

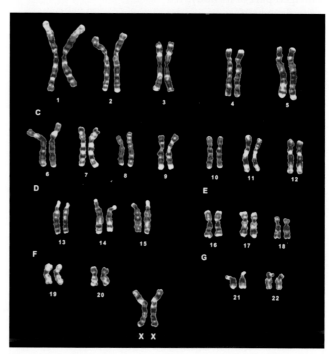

Figure 14.3 Chromosomes of a woman, arranged in order.

largest first. You can see that there are two chromosomes of each kind, because they are from diploid cells. In each pair, one is from the person's mother and the other from their father. The two chromosomes of a pair are called **homologous chromosomes**.

Soon after the zygote was formed, it began to divide over and over again, producing a ball of cells that eventually grew into you. Each time a cell divided, the two new cells produced were provided with a perfect copy of the two sets of chromosomes in the original zygote. The new cells produced were all genetically identical.

This type of cell division, which produces genetically identical cells, is called **mitosis**.

Key definition

mitosis nuclear division giving rise to genetically identical cells in which the chromosome number is maintained by the exact duplication of chromosomes

14.4 Mitosis is involved in growth and asexual reproduction.

Mitosis is the way in which any cell – plant or animal – divides when an organism is growing, or repairing a damaged part of its body.

Growth means getting bigger. An individual cell can grow a certain amount, but not indefinitely. Once a cell gets to a certain size, it becomes difficult for all parts of the cell to obtain oxygen and nutrients by diffusion. In order to grow any more, the cell divides to form two smaller cells, each of which can then grow and divide again.

Growth involves an increase in the mass of the organism. This is often recorded as **dry mass** – that is, the actual mass of biological material present, minus the water. This is done because the quantity of water in an organism's body is very variable, so finding the dry mass gives you more reliable information about the real increase in size of the organism over time.

Key definition

growth a permanent increase in size and dry mass by an increase in cell number or cell size or both

As a young organism grows, it also develops. A zygote first forms a ball of identical cells, but then as this develops into an embryo, a fetus and finally a baby, some of these cells become specialised and all the different tissues and organs of the body gradually form. The body of the organism becomes more complex. This is called **development**.

Key definition

development an increase in complexity

Mitosis is also used in asexual reproduction. You have seen, for example, how a potato plant can reproduce by growing stem tubers which eventually produce new plants (section **13.4**). All the cells in the new tubers are produced by mitosis, so they are all genetically identical.

14.5 Mitosis involves division of chromosomes.

During mitosis, the chromosomes in the parent cell are copied. Each copy remains attached to the original one, so each chromosome is made up of two identical threads joined together (Figure **14.4**). The two threads are called **chromatids**, and the point where they are held together is called the **centromere**.

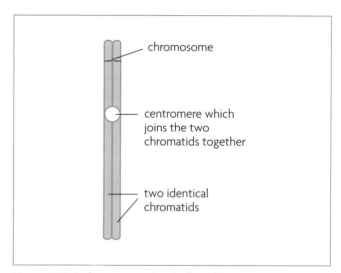

Figure 14.4 A chromosome just before division.

Figure **14.5** shows what happens when a cell with four chromosomes (two sets of two) divides by mitosis. Two new cells are formed, each with one copy of each of the four chromosomes. As the new cells grow, they make new copies of each chromosome, ready to divide again.

14.6 Gametes are produced by meiosis.

In section **13.6**, we saw that gametes have only half the number of chromosomes of a normal body cell. They have one set of chromosomes instead of two. This is so that when they fuse together, the zygote formed has two sets.

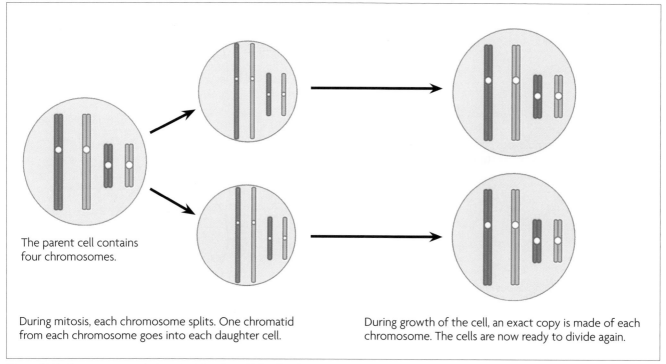

The parent cell contains four chromosomes.

During mitosis, each chromosome splits. One chromatid from each chromosome goes into each daughter cell.

During growth of the cell, an exact copy is made of each chromosome. The cells are now ready to divide again.

Figure 14.5 Chromosomes during the life of a cell dividing by mitosis.

Human gametes are formed by the division of cells in the ovaries and testes. The cells divide by a special type of cell division called **meiosis**. Meiosis shares out the chromosomes so that each new cell gets just one of each type.

Figure **14.6** (overleaf) summarises what happens during meiosis.

> **Key definition**
>
> **meiosis**   reduction division in which the chromosome number is halved from diploid to haploid

You may remember that one of each pair of homologous chromosomes came from the person's mother, and one from their father. During meiosis, the new cells get a mixture of these. So a sperm cell could contain a chromosome 1 from the man's father and a chromosome 2 from his mother, and so on. There are all sorts of different possible combinations. This is one of the reasons why gametes are genetically different from the parent cell. Meiosis produces genetic variation.

Inheritance

14.7 Each cell has two copies of each gene.

We have seen that chromosomes each contain many genes. We think there are about 20 000 human genes, carried on our two sets of 23 chromosomes.

Each cell in the body contains identical chromosomes, and therefore identical genes. But not all of these genes are switched on in every cell. Cells in different tissues use different collections of genes. For example, if you have blue eyes you must have genes for blue eyes in all of your cells. But in most of your cells, these genes are switched off. Your heart cells, for example, don't produce a blue colour. Only the cells in your iris use these 'blue eyes' genes.

Because you have two complete sets of chromosomes in each of your cells, you have two complete sets of genes. Each chromosome in a homologous pair contains genes for the same characteristic in the same positions (Figure **14.7**, overleaf). In section **14.8**, next, we look at one kind of gene to see how it behaves, and how it is inherited.

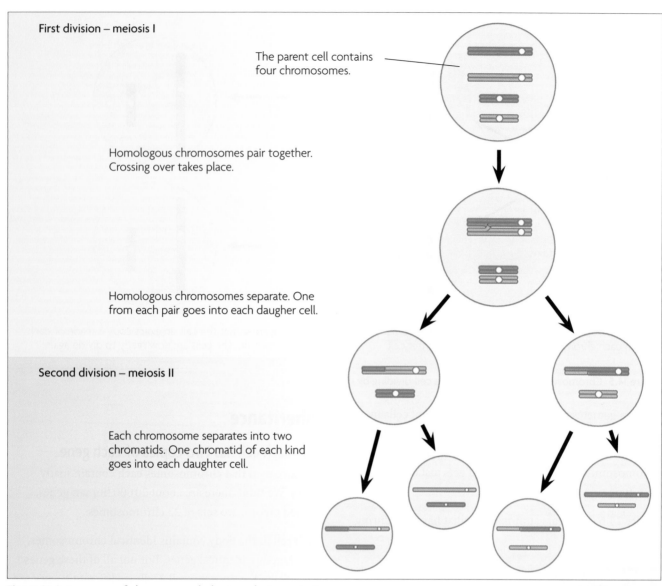

First division – meiosis I

The parent cell contains four chromosomes.

Homologous chromosomes pair together. Crossing over takes place.

Homologous chromosomes separate. One from each pair goes into each daugher cell.

Second division – meiosis II

Each chromosome separates into two chromatids. One chromatid of each kind goes into each daughter cell.

Figure 14.6 Summary of chromosome behaviour during meiosis.

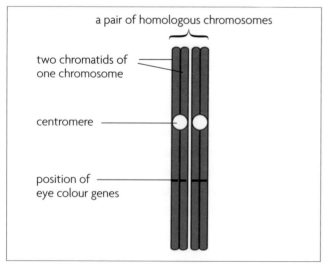

a pair of homologous chromosomes

two chromatids of one chromosome

centromere

position of eye colour genes

Figure 14.7 Homologous chromosomes have genes for the same characteristic in the same position.

14.8 Cystic fibrosis is caused by an unusual protein.

In humans, cells in the lungs make mucus. The kind of mucus that is made partly depends on a protein in the cell membrane of some of the cells in the lungs. In some people, this protein is made incorrectly. This causes the disease **cystic fibrosis**. The lungs of a person with this disease make mucus that is thicker than usual. The mucus collects in the lungs, making it difficult to get enough oxygen into the blood. The mucus makes a good breeding ground for bacteria, so infections can build up. The thick mucus is also made in the pancreas, where it blocks the pancreatic duct. This prevents pancreatic juice, containing digestive enzymes, from flowing into the duodenum, so food cannot be digested properly.

Instructions for making the protein are given by a gene. There are two varieties of the gene for this protein – one for making the normal protein, and one for making the incorrect one. Different varieties of a gene are called **alleles**. We can give letters to alleles to use as symbols. So we can call the normal allele of the gene F, and the one which codes for the abnormal protein f.

14.9 Each cell has two genes for any characteristic.

In each of your cells, there are two genes giving instructions about which of these two kinds of protein to make. This means that there are three possible combinations of alleles. You might have two F alleles, FF. You might have one of each, Ff. Or you might have two f alleles, ff (Figure **14.8**).

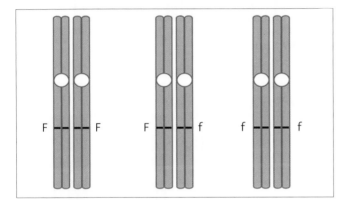

Figure 14.8 Genotypes for the cystic fibrosis gene.

If the two alleles for this gene in your cells are the same – that is, FF or ff – you are said to be **homozygous**. If the two alleles are different – that is, Ff – then you are **heterozygous**.

14.10 Genotype can determine phenotype.

The genes that you have are your **genotype**. Your genotype could be FF, Ff or ff.

The genotype determines the kind of protein you make, and therefore whether you will have cystic fibrosis or not. If your genotype is FF, then your protein is normal, and you will not have the disease. If your genotype is ff, then your protein is the incorrect one, and you will have cystic fibrosis. If your genotype is Ff, some of your protein is the incorrect one, and some of it is normal. You will have enough normal protein to ensure that the right sort of mucus is made, and you will not suffer from cystic fibrosis.

The features you have are called your **phenotype**. This can include what you look like – for example, what colour your hair is, or how tall you are – as well as things which we cannot actually see, such as what kind of protein you have in your cell membranes. In our example, your phenotype is either being normal or having cystic fibrosis.

You can see that, in this example, your phenotype depends entirely on your genotype. This is not always true. Sometimes, other things, such as what you eat, can affect your phenotype. However, for the moment, we will only consider the effect which genotype has on phenotype, and not worry about effects which the environment might have.

14.11 Alleles can be dominant or recessive.

You have seen that there are three different possible genotypes for the cell surface membrane protein involved in cystic fibrosis, but only two phenotypes. We can summarise this as follows:

genotype	phenotype
FF	normal
Ff	normal
ff	cystic fibrosis

This happens because the allele F is **dominant** to the allele f. A dominant allele has just as much effect on phenotype when there is only one of it as when there are two of it. A person who is homozygous for a dominant allele has the same phenotype as a person who is heterozygous. A heterozygous person is said to be a **carrier** of cystic fibrosis, because he or she has the allele for it but shows no symptoms.

The allele f is **recessive**. A recessive allele only affects the phenotype when there is no dominant allele present. Only people with the genotype ff – homozygous recessive – have cystic fibrosis.

> **Key definitions**
>
> **dominant**   an allele that is expressed if it is present (e.g. F)
>
> **recessive**   an allele that is only expressed when there is no dominant allele of the gene present (e.g. f)

14.12 Some alleles show codominance.

Sometimes, neither of a pair of alleles is completely dominant or completely recessive. Instead of one of them completely hiding the effect of the other in a heterozygote, they both have an effect on the phenotype. This is called **codominance** (Figure **14.9**).

Imagine a kind of flower which has two alleles for flower colour. The allele C^W produces white flowers, while the allele C^R produces red ones. If these alleles show codominance, then the genotypes and phenotypes are:

genotype	phenotype
$C^W C^W$	white flowers
$C^W C^R$	pink flowers
$C^R C^R$	red flowers

14.13 The inheritance of blood groups is another example of codominance.

The inheritance of the ABO blood group antigens in humans is another example of codominance. There are three alleles of the gene governing this instead of the usual two. Alleles I^A and I^B are codominant, but both are dominant to I^o. A person with the genotype $I^A I^B$ has the blood type AB, in which characteristics of both A and B antigens are expressed (Figure **14.10**).

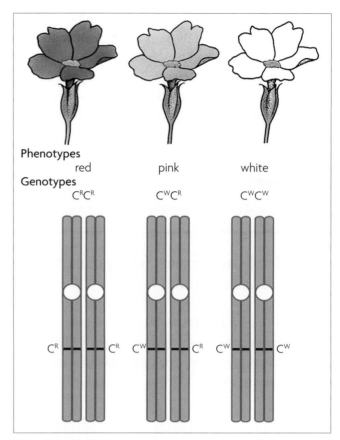

Figure 14.9 Codominance.

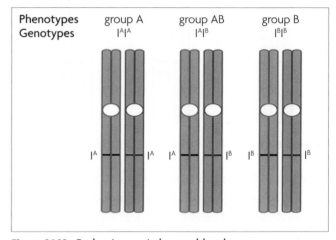

Figure 14.10 Codominance in human blood groups.

14.14 Gametes have only one allele of the gene for any characteristic.

Gametes (reproductive cells) are produced by meiosis, as described in section **14.6**. Each gamete has only one of each kind of chromosome instead of two as in the body cells. So, for example, human egg and sperm cells have 23 chromosomes, not 46 as in other cells. These cells, therefore, only carry *one* of each pair of alleles of all the genes.

Questions

14.1 What are chromosomes made of?

14.2 What are homologous chromosomes?

14.3 What are alleles?

14.4 **a** The allele for brown eyes is dominant to the allele for blue eyes. Write down suitable symbols for these alleles.
   b What is the phenotype of a person who is heterozygous for this characteristic?

14.5 What is codominance?

14.6 Alleles of the gene for the ABO blood group antigens in humans show two unusual characteristics. What are these?

14.7 Figure **14.10** shows three possible genotypes for blood group. Write down all the other possible genotypes, and the phenotype that is associated with each one.

Imagine a man who has the genotype Ff. He is a carrier for cystic fibrosis. In his testes, sperm are made by meiosis. Each sperm cell gets either an F allele or f allele. Half of his sperm cells have the genotype F and half have the genotype f.

14.15 Genes and fertilisation.

If this heterozygous man marries a woman who has cystic fibrosis (genotype ff), will their children have cystic fibrosis or not?

The eggs that are made in the woman's ovaries are also made by meiosis. She can only make one kind of egg. All of the eggs will carry an f allele.

During sexual intercourse, hundreds of thousands of sperm will begin a journey towards the egg. About half of them will carry an F allele, and half will carry an f allele. If there is an egg in the woman's oviduct, it will probably be fertilised. There is an equal chance of either kind of sperm getting there first.

If a sperm carrying an F allele wins the race, then the zygote will have an F allele from its father and an f allele from its mother. Its genotype will be Ff. After nine months, a baby will be born with the genotype Ff.

But if a sperm carrying an f allele manages to fertilise the egg, then the baby will have the genotype ff, like its mother (Figure **14.11**).

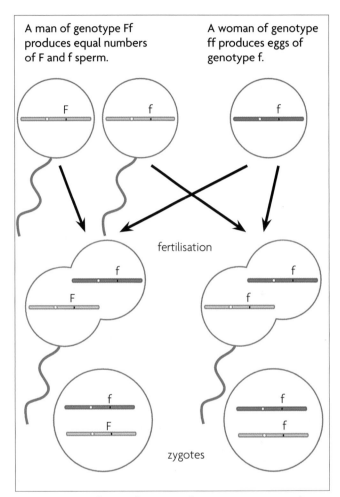

Figure 14.11 Fertilisation between a heterozygous man and a woman with cystic fibrosis.

14.16 Genetic crosses must be written clearly.

There is a standard way of writing out all this information. It is called a **genetic diagram**. First, write down the phenotypes and genotypes of the parents. Next, write down the different types of gametes they can make, like this.

Parents' phenotypes	normal	cystic fibrosis
Parents' genotypes	Ff	ff
Gametes	F or f	f

The next step is to write down what might happen during fertilisation. Either kind of sperm might fuse with an egg.

Offspring genotypes and phenotypes

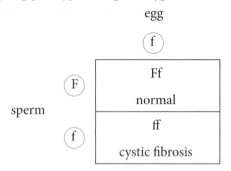

To finish your summary of the genetic cross, write out in words what you would expect the offspring from this cross to be.

So we would expect approximately half of the couple's children to be heterozygous carriers of cystic fibrosis, and half to be homozygous, with cystic fibrosis. Another way of putting this is to say that the expected ratio of heterozygous carriers to people with cystic fibrosis would be 1:1.

14.17 If both parents are heterozygous, more offspring genotypes are possible.

What happens if both parents are carriers?

Parents' phenotypes	normal (carrier)	normal (carrier)
Parents' genotypes	Ff	Ff
Gametes	(F) or (f)	(F) or (f)

Offspring genotypes and phenotypes

eggs

	(F)	(f)
(F)	FF normal	Ff normal (carrier)
(f)	Ff normal (carrier)	ff cystic fibrosis

sperm

About one quarter of the children would be expected to have cystic fibrosis, and three quarters would be normal.

This example illustrates the inheritance of one pair only of contrasting characteristics. This is known as **monohybrid inheritance**.

14.18 Offspring ratios are only probabilities.

In the last example, there were four possible offspring genotypes at the end of the cross. This does not mean that the man and woman will have four children. It simply means that each time they have a child, these are the possible genotypes that it might have.

When they have a child, there is a 1 in 4 chance that its genotype will be FF, and a 1 in 4 chance that its genotype will be ff. There is a 2 in 4, or rather 1 in 2, chance that its genotype will be Ff.

However, as you know, probabilities do not always work out. If you toss a coin up four times you might expect it to turn up heads twice and tails twice. But does it always do this? Try it and see.

With small numbers like this, probabilities do not always match reality. If you had the patience to toss your coin up a few thousand times, though, you will almost certainly find that you get much more nearly equal numbers of heads and tails.

The same thing applies in genetics. The offspring genotypes which you work out are only probabilities. With small numbers, they are unlikely to work out exactly. With very large numbers of offspring from one cross, they are more likely to be accurate.

So, if the man and woman in the last example had eight children, they might expect six of them to be normal and two to have cystic fibrosis. But they should not be too surprised if they have three children with cystic fibrosis.

14.19 Test crosses help to determine genotype.

An organism that shows a dominant characteristic could have either of two possible genotypes. It could be homozygous for the dominant allele, or it could be heterozygous.

We can find out the genotype of an individual with the dominant phenotype for a particular gene by crossing it with one known to have the homozygous recessive genotype for the same gene. This is called a **test cross**.

For example, if we know that tallness is dominant to dwarfness in a certain species of peas, then the genotype of any tall plant could be determined by crossing it with a dwarf plant. If any of the offspring are dwarf, then this must mean that the tall parent had an allele for dwarfness. It must have been heterozygous. Try this out for yourself, using a genetic diagram.

If none of the offspring are dwarf, this almost certainly means that the tall parent was homozygous for the tallness allele. However, unless there are large numbers of offspring, this could also happen if the tall parent is heterozygous but, just by chance, none of its gametes carrying the recessive allele were successful in fertilisation.

14.20 'Pure breeding' means homozygous.

Some populations of animals or plants always have offspring just like themselves. For example, a rabbit breeder might have a strain of rabbits which all have brown coats. If he or she interbreeds them with one another, all the offspring always have brown coats as well. The breeder has a **pure-breeding** strain of brown rabbits. Pure-breeding strains are always homozygous for the pure-breeding characteristics.

The offspring of two different pure-breeding (homozygous) strains are sometimes called the first filial generation, or **F1 generation**. They are always heterozygous.

14.21 Sex is determined by X and Y chromosomes.

The last pairs of chromosomes in Figures **14.2** and **14.3** are responsible for determining what sex a person will be. They are called the **sex chromosomes** (Figure **14.12**). A woman's chromosomes are both alike and are called X chromosomes. She has the genotype XX. A man, though, only has one X chromosome. The other, smaller one is a Y chromosome. He has the genotype XY.

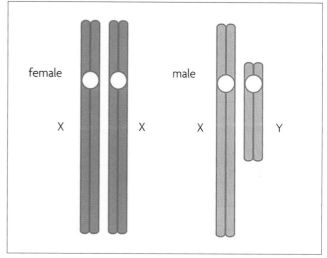

Figure 14.12 The sex chromosomes.

Questions

14.8 If a normal human cell has 46 chromosomes, how many chromosomes are there in a human sperm cell?

14.9 Using the symbols W for normal wings, and w for vestigial wings, write down the following:
   a the genotype of a fly which is heterozygous for this characteristic.
   b the possible genotypes of its gametes.

14.10 Using the method shown in sections **14.16** and **14.17**, work out what kind of offspring would be produced if the heterozygous fly in question **14.9** mated with one which was homozygous for normal wings.

14.11 In humans, the allele for red hair, b, is recessive to the allele for brown hair, B. A man and his wife both have brown hair. They have five children, three of whom have red hair, while two have brown hair. Explain how this may happen, using a genetic diagram to explain your answer.

14.12 In Dalmation dogs, the allele for black spots is dominant to the allele for liver spots. If a breeder has a black-spotted dog, how can he or she find out whether it is

homozygous or heterozygous for this characteristic? Use genetic diagrams to explain your answer.

14.13 A man of blood type A married a woman of blood type B. They had three children, of blood types O, B and AB, respectively. What are the genotypes of the parents and children? Use genetic diagrams to explain your answer.

14.14 The pedigree diagram shows the known blood groups in three generations of a family. Squares represent males and circles represent females. What are the genotypes of 1 and 3? What is the blood group of 2?

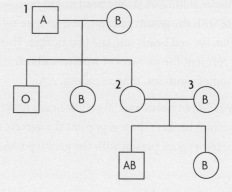

14.22 Sex is inherited.

You can work out sex inheritance in just the same way as for any other characteristic, but using the letter symbols to describe whole chromosomes, rather than individual alleles.

Parents' phenotypes	male	female
Parents' genotypes	XY	XX
Gametes	Ⓧ or Ⓨ	Ⓧ

Offspring genotypes and phenotypes

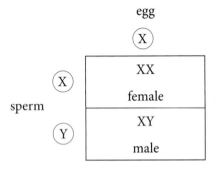

So each time a child is conceived, there is a 1:1 chance of its being either sex.

Variation

14.23 Individuals of the same species show differences.

You have only to look around a group of people to see that they are different from one another. Some of the more obvious differences are in height or hair type. We also vary in intelligence, blood groups, whether we can roll our tongues or not, and many other ways.

There are two basic kinds of variation. One kind is **discontinuous variation**. Blood groups are an example of discontinuous variation. Everyone fits into one of four definite categories – they have blood group A, B, AB or O. There are no in-between categories.

The other kind is **continuous variation**. Height is an example of continuous variation. There are no definite heights that a person must be. People vary in height, between the lowest and highest extremes.

You can try measuring and recording discontinuous and continuous variation in Activity **14.2** (page **206**). Your results for continuous variation will probably look similar

---

Activity 14.1
'Breeding' beads

skills
> C2 *Observing, measuring and recording*
> C3 *Interpreting and evaluating*

In this investigation, you will use two containers of beads. Each container represents a parent. The beads represent the gametes they make. The colour of a bead represents the genotype of the gamete. For example, a red bead might represent a gamete with genotype A, for 'tongue rolling'. A yellow bead might represent a gamete with the genotype a, for 'non-tongue rolling'.

1   Put 100 red beads into the first beaker. These represent the gametes of a person who is homozygous for 'tongue rolling', AA.

2   Put 50 red beads and 50 yellow beads into the second beaker. These represent the gametes of a heterozygous person with the genotype Aa.

3   Close your eyes, and pick out one bead from the first beaker, and one from the second. Write down the genotype of the 'offspring' they produce. Put the two beads back.

4   Repeat step 3 100 times.

5   Now try a different cross – for example, Aa crossed with Aa.

Questions

1   In the first cross, what kinds of offspring were produced, and in what ratios?

2   Is this what you would have expected? Explain your answer.

3   Why must you close your eyes when choosing the beads?

4   Why must you put the beads back into the beakers after they have 'mated'?

to Figure **14.13**. This is called a **normal distribution**. Most people come in the middle of the range, with fewer at the lower or upper ends. Human height (Figure **14.14**) shows a normal distribution.

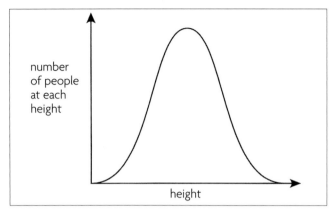

Figure 14.13 A normal distribution curve. This is a graph that shows the numbers of people of different heights.

Figure 14.14 Human height shows continuous variation. What characteristic here shows discontinuous variation?

14.24 What causes variation?

By describing variation as continuous or discontinuous, we can begin to explain *how* organisms vary. But the *cause* of the variation is another question altogether.

Genetic variation One reason for the differences between individuals is that their genotypes are different. Tongue rolling, for example, is controlled by genes. There are also genes for hair colour, eye colour, blood groups, height and many other characteristics (Figure **14.15**).

Environmental variation Another important reason for variation is the difference between the environments of individuals. Pine trees possess genes that enable them to grow to a height of about 30 m. But if a pine tree is

Figure 14.15 a The presence of horns in cattle is controlled by a dominant allele of a gene. **b** Polled (hornless) cattle have two copies of the recessive allele of this gene.

grown in a very small pot, and has its roots regularly pruned, it will be permanently stunted (Figure **14.16**, overleaf). The tree's genotype gives it the potential to grow tall, but it will not realise this potential unless its roots are given plenty of space and it is allowed to grow freely.

Characteristics caused by an organism's environment are sometimes called **acquired characteristics**. They are not caused by genes, and so they cannot be handed on to the next generation.

In general, discontinuous variation is caused by genes alone. Continuous variation is often influenced by both genes and the environment.

Questions

14.15 Decide whether each of these features shows continuous variation or discontinuous variation.
  a blood group in humans
  b foot size in humans
  c leaf length in a species of tree
  d presence of horns in cattle

14.16 For each of the examples in **(a)** to **(d)** above, suggest whether the variation is caused by genes alone, or by both genes and environment.

Variation caused by the environment is not inherited.

A cutting from a bonsai pine tree would grow into a full size tree, if given sufficient space.

bonsai

A bonsai pine tree is dwarfed by being grown in a very small pot, and continually pruned.

A dwarf pony, such as a Shetland pony, is small because of its genes. The offspring of Shetland ponies are small like their parents, no matter how well they are fed and cared for.

Figure 14.16 The inheritance of variation.

Activity 14.2
Measuring variation

skills

C2 *Observing, measuring and recording*

C3 *Interpreting and evaluating*

1   Make a survey of at least 30 people, to find out whether or not they can roll their tongue. Record your results.

2   Measure the length of the third finger of the left hand of 30 people. Take the measurement from the knuckle to the finger tip, not including the nail.

3   Divide the finger lengths into suitable categories, and record the numbers in each category, like this.

length / cm	number of measurements
8.0 – 8.4	2
8.5 – 8.9	4

and so on ...

4   Draw a histogram of your results.

Questions

1   Which characteristic shows continuous variation, and which shows discontinuous variation?

2   Your histogram may be a similar shape to the curve in Figure **14.13**. This is called a normal distribution. The category, or class, which has the largest number of individuals in it is called the modal class. What is the modal class for finger length in your results?

3   The mean finger length is the total of all the finger lengths, divided by the number of people in your sample. What is the mean finger length of the sample?

14.25 Genetic variation arises in several ways.

Meiosis   During sexual reproduction, gametes are formed by **meiosis**. In meiosis, homologous chromosomes exchange genes, and separate from one another, so the gametes which are formed are not all exactly the same.

Fertilisation   Any two gametes of opposite types can fuse together at **fertilisation**, so there are many possible combinations of genes which may be produced in the zygote. In an organism with a large number of genes the possibility of two offspring having identical genotypes is so small that it can be considered almost impossible.

Mutation   Sometimes, a gene may suddenly change. This is called **mutation**. Mutations are the only source of brand-new characteristics in the gene pool. So they are really the final source of all genetic variation.

Another type of mutation affects whole chromosomes. For example, when eggs are being made by meiosis in a woman's ovaries, the chromosome 21s sometimes do not separate from one another. One of the daughter cells therefore gets two chromosome 21s and the other one gets none. The cell with none dies. The other one may survive, and eventually be fertilised by a sperm. The zygote from this fertilisation will have three copies of chromosome 21. The child that grows from the zygote has Down's syndrome. Children with Down's have characteristic facial features and are usually very happy and friendly people. However, they often have heart problems and other physical and physiological difficulties.

> **Key definition**
>
> **mutation**   a change in a gene or chromosome

14.26   Environmental factors can increase mutation rate.

Mutations often happen for no apparent reason. However, we do know of many factors which make mutation more likely. One of the most important of these is ionising radiation. Radiation can damage the bases in DNA molecules. If this happens in the ovaries or testes, then the altered DNA may be passed on to the offspring.

Many different chemicals are known to increase the risk of a mutation happening. The heavy metals lead and mercury and their compounds can interfere with the process in which DNA is copied. If this process goes wrong, the daughter cells will get faulty DNA when the cell divides. Chemicals which can cause mutations are called **mutagens**.

14.27   Variation is the raw material for natural selection to work on.

Over the many millions of years that living things have existed, there have been gradual changes in organisms and populations. Fossils tell us that many animals and plants that once lived no longer exist.

In the 19th century, several ideas were put forward to suggest how this might have happened. By far the most important was suggested by Charles Darwin (Figure **14.17**). He put forward his theory in a book called *On the Origin of Species*, which was published in 1859.

Figure 14.17 A portrait of Charles Darwin at the age of 72.

Darwin's theory of how evolution could have happened may be summarised like this.

Variation   Most populations of organisms contain individuals which vary slightly from one to another. Some slight variations may better adapt some organisms to their environment than others.

Over-production   Most organisms produce more young than will survive to adulthood.

Struggle for existence Because populations do not generally increase rapidly in size, there must therefore be considerable competition for survival between the organisms.

Survival of the fittest Only the organisms which are really well adapted to their environment will survive (Figures **14.18** and **14.19**).

Advantageous characteristics passed on to offspring Only these well-adapted organisms will be able to reproduce successfully, and will pass on their advantageous characteristics to their offspring.

Gradual change In this way, over a period of time, the population will lose all the poorly adapted individuals. The population will gradually become better adapted to its environment.

The theory is often called the theory of **natural selection**, because it suggests that the best-adapted organisms are selected to pass on their characteristics to the next generation.

Key definition

natural selection the greater chance of passing on of genes by the best-adapted organisms

---

Darwin proposed his theory before anyone understood how characteristics were inherited. Now that we know something about genetics, his theory can be stated slightly differently. We can say that natural selection results in the genes producing advantageous phenotypes being passed on to the next generation more frequently than the genes which produce less advantageous phenotypes (Figures **14.18** and **14.19**).

14.28 Melanic moths were selected near cities.

Darwin's theory of natural selection provides a good explanation for our observations of the many types of animals and plants. For example, it can help us to understand some changes that have taken place in a species of moth in Britain and Ireland.

The peppered moth, *Biston betularia*, lives in most parts of Great Britain and Ireland. It flies by night, and spends the daytime resting on tree trunks. It has speckled wings, which camouflage it very effectively on lichen-covered tree trunks (Figure **14.20**).

People have collected moths for many years, so we know that up until 1849 all the moths in collections were speckled. But in 1849, a black or melanic form of the moth was caught near Manchester. By 1900, 98% of the moths near Manchester were black.

Figure 14.18 When large numbers of organisms, such as these wildebeest of East African plains, live together, there is competition for food, and the weaker ones are likely to be killed by predators. Individuals best adapted to their environment survive and reproduce.

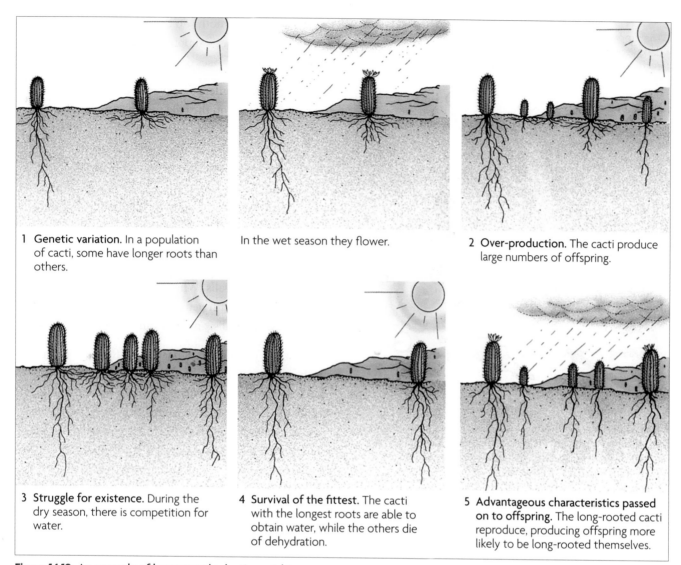

1 **Genetic variation.** In a population of cacti, some have longer roots than others.

In the wet season they flower.

2 **Over-production.** The cacti produce large numbers of offspring.

3 **Struggle for existence.** During the dry season, there is competition for water.

4 **Survival of the fittest.** The cacti with the longest roots are able to obtain water, while the others die of dehydration.

5 **Advantageous characteristics passed on to offspring.** The long-rooted cacti reproduce, producing offspring more likely to be long-rooted themselves.

Figure 14.19 An example of how natural selection might occur.

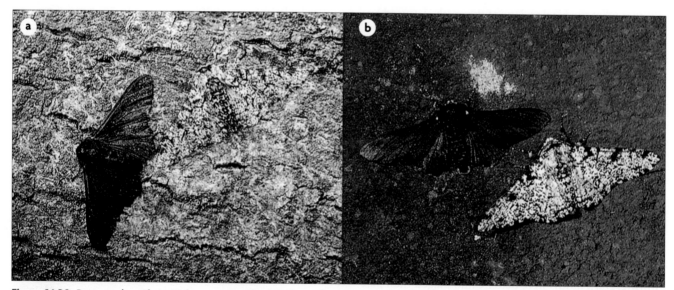

Figure 14.20 Peppered moths. **a** Lichen-covered bark hides a speckled moth perfectly. **b** Dark moths are better camouflaged on lichen-free trees.

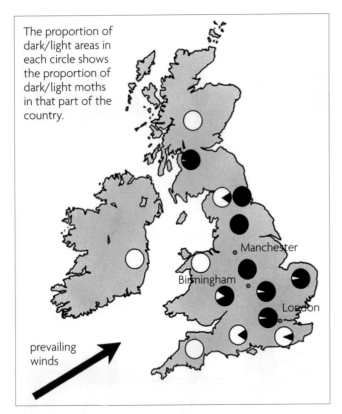

The proportion of dark/light areas in each circle shows the proportion of dark/light moths in that part of the country.

prevailing winds

Figure 14.21 The distribution of the pale and dark forms of the peppered moth, *Biston betularia*, in 1958. Since then, the number of dark moths has dramatically decreased, because now there is much less air pollution.

The distribution of the black and speckled forms in 1958 is shown in Figure **14.21**.

How can we explain the sudden rise in numbers of the dark moths, and their distribution today?

We know that the black colour of the moth is caused by a single dominant allele of a gene. The mutation from a normal to a black allele happens fairly often, so it is reasonable to assume that there have always been a few black moths around, as well as pale speckled ones.

Up until the beginning of the Industrial Revolution, the pale moths had the advantage, as they were better camouflaged on the lichen-covered tree trunks.

But in the middle of the 19th century, some areas became polluted by smoke. Because the prevailing winds in Britain blow from the west, the worst affected areas were to the east of industrial cities like Manchester and Birmingham. The polluted air prevented lichens from growing. Dark moths were better camouflaged than pale moths on trees with no lichens on them.

Proof that the dark moths do have an advantage in polluted areas has been supplied by several investigations. Figure **14.22** summarises one of them.

The factor which confers an advantage on the dark moths, and a disadvantage on the pale moths in polluted areas, is predation by birds. This is called a **selection pressure**, because it 'selects' the dark moths for survival. In unpolluted areas, the pale moths are more likely to survive.

14.29 Antibiotic resistance in bacteria is selected.

Another example of natural selection can be seen in the way that bacteria may become resistant to **antibiotics**, such as **penicillin**. Penicillin works by stopping bacteria from forming cell walls. When a person infected with bacteria is treated with penicillin, the bacteria are unable to grow new cell walls, and they burst open.

However, the population of bacteria in the person's body may be several million. The chances of any one of them mutating to a form which is not affected by penicillin is quite low, but because there are so many bacteria, it could well happen. If it does, the mutant bacterium will have a tremendous advantage. It will be able to go on reproducing while all the others cannot. Soon, its descendants may form a huge population of penicillin-resistant bacteria (Figure **14.23**, page 212).

This does, in fact, happen quite frequently. This is one reason why there are so many different antibiotics available – if some bacteria become resistant to one, they may be treated with another.

The more we use an antibiotic, the more we are exerting a selection pressure which favours the resistant forms. If antibiotics are used too often, we may end up with resistant strains of bacteria that are very difficult to control.

14.30 Natural selection does not always cause change.

Natural selection does not always produce change. Natural selection ensures that the organisms which are best adapted to their environment will survive. Change will only occur if the environment changes, or if a new mutation appears which adapts the organism better to the existing environment.

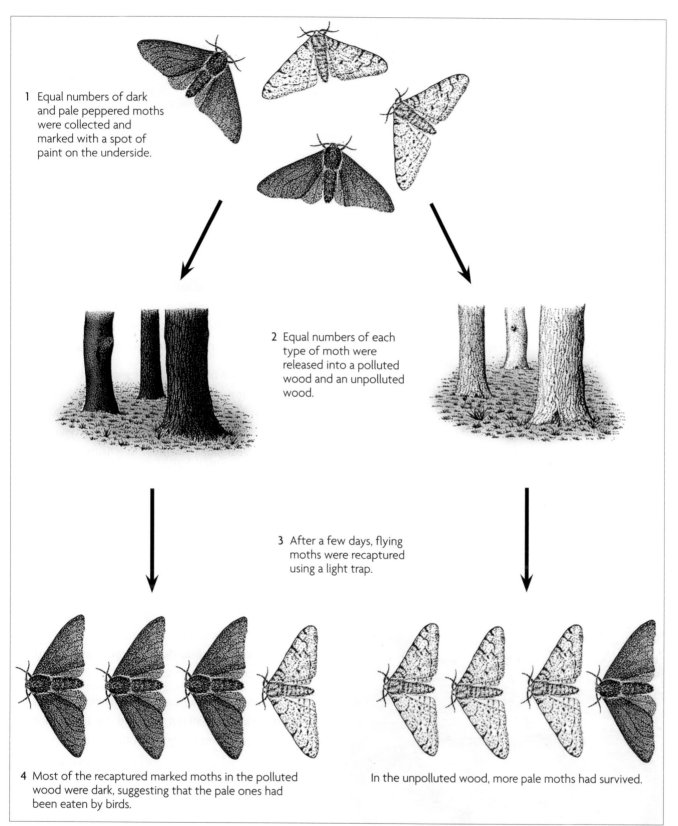

1 Equal numbers of dark and pale peppered moths were collected and marked with a spot of paint on the underside.

2 Equal numbers of each type of moth were released into a polluted wood and an unpolluted wood.

3 After a few days, flying moths were recaptured using a light trap.

4 Most of the recaptured marked moths in the polluted wood were dark, suggesting that the pale ones had been eaten by birds.

In the unpolluted wood, more pale moths had survived.

Figure 14.22 An investigation to measure the survival of dark and pale peppered moths in polluted and unpolluted environments.

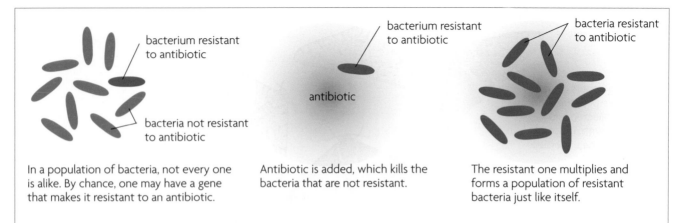

bacterium resistant
to antibiotic

bacteria not resistant
to antibiotic

bacterium resistant
to antibiotic

bacteria resistant
to antibiotic

antibiotic

In a population of bacteria, not every one is alike. By chance, one may have a gene that makes it resistant to an antibiotic.

Antibiotic is added, which kills the bacteria that are not resistant.

The resistant one multiplies and forms a population of resistant bacteria just like itself.

Figure 14.23 How resistance to antibiotics increases in a population of bacteria.

For example, in the south-west of Britain, the environment of the peppered moth has never changed very much. The air has not become polluted, so lichens have continued to grow on trees. The best camouflaged moths have always been the pale ones. So selection has always favoured the pale moths in this part of Britain. Any mutant dark moths which do appear are at a disadvantage, and are unlikely to survive.

Most of the time, natural selection tends to keep populations very much the same from generation to generation. It is sometimes called **stabilising selection**. If an organism is well adapted to its environment, and if that environment stays the same, then the organism will not evolve. Coelacanths, for example, have remained virtually unchanged for 350 million years. They live deep in the Indian Ocean, which is a very stable environment (Figure **14.24**).

Figure 14.24 Coelacanths, which live deep in the Indian Ocean, have existed almost unchanged for 350 million years. Humans have existed for only about 4 million years.

14.31 The sickle cell allele is sometimes selected for.

A genetic disease called sickle cell anaemia is a good example of how natural selection can work in humans.

Some people have a mutation in the gene that codes for the production of haemoglobin. The normal allele, Hb^A, codes for normal haemoglobin. The mutant allele, Hb^S, codes for an allele that produces a faulty type of haemoglobin. This faulty haemoglobin has a tendency to produce fibres inside red blood cells when oxygen concentration is low. The red blood cells get pulled into a 'sickle' shape and get stuck in blood capillaries. This is very painful and very dangerous. The condition is called sickle cell anaemia.

The two alleles are codominant. A person with genotype Hb^{A}Hb^A has normal haemoglobin, a person with genotype Hb^{A}Hb^S has a mix of normal and sickle cell haemoglobin, and a person with genotype Hb^{S}Hb^S has all sickle cell haemoglobin. Heterozygous people don't usually show any symptoms.

If sickle cell anaemia is such a dangerous disease, then why has natural selection not removed it from the human population? The answer lies with another disease – malaria.

Malaria is a serious disease caused by a single-celled parasite that is injected into the blood when an infected mosquito bites. Millions of people are killed by this disease each year, most of them children. A person who lives in a part of the world where malaria is present, and who has some resistance to the disease, will be at an advantage compared with others who are susceptible.

Malaria is common in many parts of the world where the sickle cell allele is present in the population (Figure 14.25). In the past, people homozygous for the sickle cell allele often died early from sickle cell disease. People homozygous for the normal allele often died early from malaria. Those, however, who were heterozygous (with one Hb^S allele and one Hb^A allele) were more resistant to malaria than those with all normal haemoglobin. In parts of the world where malaria was present, people with the heterozygous genotype were most likely to survive until they were old enough to reproduce.

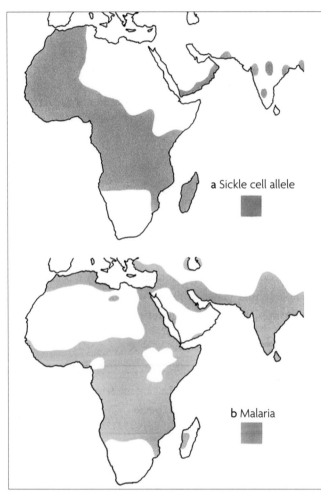

Figure 14.25 The distribution of **a** the sickle cell allele and **b** malaria.

Therefore, in each generation, the people most likely to reproduce are heterozygous people. Some of their children will also be heterozygous, but some will be homozygous dominant and some homozygous recessive. This will continue for generation after generation – until someone finds a really good cure or prevention for malaria, or a cure for sickle cell anaemia.

Questions

14.17 When was the idea of natural selection first suggested?

14.18 Using the six points listed in section **14.27**, explain why the proportion of dark peppered moths near Manchester in Britain increased at the end of the 19th century.

14.19 Why is it unwise to use antibiotics unnecessarily?

14.20 What is meant by stabilising selection? Give **one** example.

14.32 Humans have used artificial selection for thousands of years.

Humans can also bring about changes in living organisms, by selecting certain varieties for breeding. Figures **14.26** and **14.27** (overleaf) show examples of the results of this kind of selection. From the varied individuals amongst a herd of cattle, the breeder chooses the ones with the characteristics he or she wants to appear in the next generation. He or she then allows these individuals, and not the others, to breed. Over many generations, these characteristics will become the most common ones in the population.

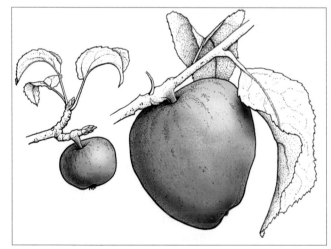

Figure 14.26 Wild and cultivated apples.

This process is called **artificial selection**. It has been going on for thousands of years, ever since humans first began to cultivate plants and to domesticate animals. It works in just the same way as natural selection. Individuals with 'advantageous' characteristics breed, while those with 'disadvantageous' ones do not.

Figure 14.27 a White Park cattle, like these in England, are a very old breed. They are thought to be quite similar to original wild cattle. **b** Friesian cattle have been bred for high milk yield.

However, what humans think are desirable characteristics would often not be at all advantageous to the plant or animal if it was living in the wild. Modern varieties of cattle, for example, selected over hundreds of years for high milk yield or fast meat production, would stand little chance of surviving for long in the wild.

Some farmers are now beginning to think differently about the characteristics they want in their animals and plants. Instead of enormous yields as their first priority, they are now looking for varieties which can grow well with less fertiliser or pesticides in the case of food plants, and with less expensive housing and feeding in the case of animals. Luckily, many of the older breeds, which had these characteristics, have been conserved, and can now be used to breed new varieties with 'easy-care' characteristics.

14.21 Imagine you are a farmer with a herd of dairy cattle. You want to build up a herd with a very high production of milk. You have access to sperm samples from bulls, for each of which there are records of the milk production of his offspring. What will you do?

14.22 Wheat is attacked by many different pests, including a fungus called yellow rust.

   a Describe how you could use artificial selection to produce a new variety of wheat which is naturally resistant to yellow rust.

   b How could the growing of resistant varieties reduce pollution?

   c When resistant varieties of wheat are produced, it is found that after a few years they are infected by yellow rust again. Explain how this might happen.

Genetic engineering

14.33 Genes can be transferred from one organism to another.

You have seen that a gene is a length of DNA that codes for the production of a particular protein by a cell. We are now able to take genes from one organism and put them into another. This is called **genetic engineering**.

> **Key definition**
>
> **genetic engineering**   taking a gene from one species and putting it into another species

To explain how this is done, we will look at the way in which genetic engineering is used to produce insulin.

14.34 Human insulin genes are inserted into bacteria.

Some people are not able to make the protein insulin. Insulin is a hormone that helps to regulate the concentration of glucose in your blood. People whose bodies cannot make insulin have the disease diabetes mellitus. They have to have injections of insulin every day.

For a long time, the only source of insulin was from animals which had been killed for food, such as pigs. Now, genetic engineering has produced bacteria which make human insulin (Figure **14.28**).

The process begins with the extraction of the gene for making insulin from human cells. This is done using enzymes which chop up DNA molecules into short lengths. The particular length of DNA which codes for making insulin is identified, and separated from all the unwanted DNA.

Now the DNA carrying the gene for insulin must be inserted into a bacterium. This is not easy – you cannot just suck up some DNA with a syringe and inject it into a bacterial cell. One way of getting DNA into a bacterium is to use a plasmid. A plasmid is a ring of DNA, found in bacteria, which is able to reproduce itself inside other living cells.

First, some of the DNA in the plasmid is cut out, using enzymes like the ones used for cutting up the human DNA. This leaves a gap in the ring. The human DNA is then mixed up with the plasmid, and a different kind of enzyme added. This enzyme sticks DNA together. It sticks the human DNA into the gap in the plasmid. (You can think of the first kind of enzymes as acting like scissors, and this second kind as acting like glue.)

Next, these genetically engineered plasmids are added to a culture of a bacterium. A few of the bacteria will take up one or more plasmids into their cells. These bacteria now contain the human insulin gene. They can be grown in large vats where they secrete insulin into the culture solution.

14.35 There are social, ethical and moral concerns about genetic engineering.

If you look at newspapers, magazines and the Internet, you will probably find many more examples of genetic engineering. For example, it is being used to produce new varieties of crop plants.

We can also insert genes into animal cells, including human cells. Although research is still in its early stages, there are hopes that we may be able to use a technique called gene therapy to help people who have genetic diseases like cystic fibrosis. There have been some fairly successful experiments with this so far.

People with cystic fibrosis are lacking a gene needed for making normal mucus in their lungs. This gene has been extracted from human cells, and inserted into a virus. The virus is one which can infect human cells, but it has been damaged in such a way that it should not make you ill. The viruses, containing the human gene, have then been inserted into the air passages of volunteers with cystic fibrosis. The idea is that the viruses inject their DNA into the cells lining the respiratory passages. This DNA, of course, will include the human gene which the volunteers are lacking. There are, however, still several problems with this technique, such as the viruses still causing mild illness, or not enough genes getting into the human cells.

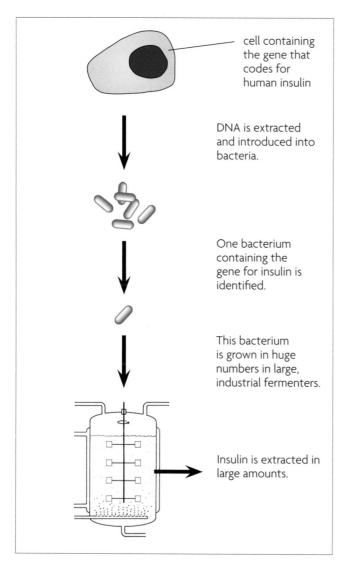

cell containing the gene that codes for human insulin

DNA is extracted and introduced into bacteria.

One bacterium containing the gene for insulin is identified.

This bacterium is grown in huge numbers in large, industrial fermenters.

Insulin is extracted in large amounts.

Figure 14.28 How genetic engineering can be used to produce bacteria that make human insulin.

So genetic engineering has opened up many possibilities of improving people's lives. It has also opened up many possible dangers.

For example, genetic engineering means producing bacteria and viruses with different genes from their usual ones. This could cause health hazards. What if some of these microorganisms were changed in such a way that they became pathogenic? They might cause a new disease, for which there is no cure. To try to make sure that this does not happen, strict regulations are enforced about the kinds of microorganisms which can be used in genetic engineering, the kinds of genes which can be put into them, and the conditions under which they can be kept. However, what if someone wanted deliberately to produce a new pathogen, and release it into the environment?

As well as health hazards, genetic engineering could produce environmental hazards. For example, imagine that a new variety of maize plant is produced, which has been genetically engineered to be resistant to a particular insect which feeds on it. This gene might be passed – perhaps in pollen – to closely related, wild, plants growing nearby. The gene might spread through the population of wild plants. This could upset the food web in the ecosystem, because insects could no longer feed on the wild plants. So far, there have been no serious instances of this happening, and it is thought to be very unlikely – but it is just possible.

Genetic engineering also raises issues about what is morally and ethically acceptable to society. For example, in theory, it will eventually be possible to check the genes in a human zygote, to make sure that there are no major genetic faults. If a fault was found, then the 'right' sort of gene could be inserted into the zygote, so that all the cells in the embryo which developed from it contained this 'right' gene. Should this be allowed? Should it be allowed just for very unpleasant diseases such as cystic fibrosis? Or should it be allowed for things like hair colour or anything else? This is all a long way in the future, but perhaps we should think about it now, before things begin to happen that we are not prepared for.

Like all major new scientific discoveries and inventions, genetic engineering has tremendous potential to provide all sorts of benefits for people, and perhaps also for other living things, as well as equally tremendous potential for harm. Scientists and everyone else must remain well aware of this. If as many people as possible, whether scientists or not, try to stay well informed about what developments are taking place, then we can do our best to ensure that 'good' uses of genetic engineering go ahead, while potential problems are stopped in their tracks.

Key ideas

- Chromosomes are long threads of DNA made up of strings of genes. In a diploid cell, each of a pair of homologous chromosomes carries the same genes in the same position. A diploid cell therefore has two copies of each gene.

- Gametes have only one set of chromosomes, and so they have ony one copy of each gene.

- Different forms of a particular gene are called alleles. They may be dominant or recessive. The genotype of an organism tells us the alleles of genes that it carries. If the two alleles of a gene are the same in the organism, then it is homozygous. If they are different, it is heterozygous.

- If two heterozygous organisms breed together, we expect a 3 : 1 ratio of offspring showing the dominant characteristic to offspring showing the recessive characteristic. If one parent is heterozygous and the other is homozygous recessive, we expect to see a 1 : 1 ratio in the offspring.

- Variation is caused by genes and environment. Continuous variation, such as human height, has no distinct categories and is usually caused by both genes and environment. Discontinuous variation, such as human blood groups, involves a small number of discrete categetories and is caused by genes alone.

- New alleles of genes, or changes in chromosomes, can be caused by mutation. Most mutations are harmful. Ionising radiation and certain chemicals increase the risk of mutation happening.

(continued ...)

(*... continued*)

◆ In a population of organisms, those with the characteristics best adapted to the environment are most likely to survive and reproduce. This is called natural selection.

◆ If the environment changes, or if a new advantageous allele appears, natural selection can lead to change over many generations. This is called evolution.

◆ Sickle cell anaemia is caused by a recessive allele of the gene for haemoglobin. People who are homozygous recessive often die before they can reproduce. People who are homozygous dominant have a greater chance of getting malaria if they live in places where this disease is present. People who are heterozygous have a selective advantage, because they are less likely to get malaria. Natural selection therefore maintains this allele in the population in parts of the world where people may be killed by malaria.

◆ Humans select plants and animals with desirable characteristics and breed from them. Over many generations, this produces new strains of plants or animals with features that we require. This is called artificial selection.

◆ Genetic engineering involves taking a gene from one species and inserting it into another. This has been done with the human insulin gene, to give bacteria that produce insulin for harvest and sale, for use by people with diabetes.

Revision questions

1 Two women gave birth in the same hospital on the same afternoon. Their babies were taken away, and then brought back to them one hour later.

One of the women was worried that she had been given the wrong baby. She asked for blood tests to be carried out.

The hospital found that she was group A and her husband was group O. The other mother was group AB and her husband was group A.

The woman with blood group A had been given the baby with blood group O. The woman with group AB was given the baby with blood group B.

Use genetic diagrams to determine whether the women had been given the right babies.

2 Match each of the following terms with its definitions.

| evolution | natural selection |
| artifical selection | genetic engineering |

a gradual changes in the characteristics of living organisms over a long period of time
b the transfer of genes from one organism to another of a different species
c the choice, by humans, of which animals or plants to allow to breed
d a process in which only those organisms best adapted to their environment survive and reproduce

3 Natural selection causes species to become well adapted to their environment. The best-adapted organisms are the most successful. Explain each of the following.
a A population of organisms that can reproduce sexually often becomes adapted to a new environment more quickly than a population of organisms that reproduce asexually.
b Evolution does not always come to a halt once a population of organisms has become adapted to its environment.

15 Living organisms in their environment

In this chapter, you will find out:

- about studying organisms in their natural environment
- how energy is passed from one organism to another through food chains and food webs
- that energy is lost as it passes along food chains
- how understanding these energy losses can help us to improve the efficiency of producing food by agriculture
- how carbon and water are cycled in ecosystems
- how nitrogen is cycled in ecosystems
- about the factors that affect the size of a population of organisms, including humans.

Ecology

15.1 Organisms interact with their environment.

One very important way of studying living things is to study them where they live. Animals and plants do not live in complete isolation. They are affected by their surroundings, or **environment**. Their environment is also affected by them. The study of the interaction between living organisms and their environment is called **ecology**.

15.2 Ecology uses special terms.

There are many words used in ecology with which it is useful to be familiar.

The area where an organism lives is called its **habitat**. The habitat of a tadpole might be a pond. There will probably be many tadpoles in the pond, forming a **population** of tadpoles. A population is a group of organisms of the same species, living in the same area at the same time.

But tadpoles will not be the only organisms living in the pond. There will be many other kinds of animals and plants making up the pond **community**. A community is all the organisms, of all the different species, living in the same habitat.

The living organisms in the pond, the water in it, the stones and the mud at the bottom, make up an **ecosystem**. An ecosystem consists of a community and its environment (Figure 15.1).

Within the ecosystem, each living organism has its own life to live and role to play. The way in which an organism lives its life in an ecosystem is called its **niche**. Tadpoles, for example, eat algae and other weeds in the pond; they disturb pebbles and mud at the bottom of shallow areas in the pond; they excrete ammonia into the water; they breathe in oxygen from the water, and breathe out carbon dioxide. All these things, and many others, help to describe the tadpoles' role, or niche, in the ecosystem.

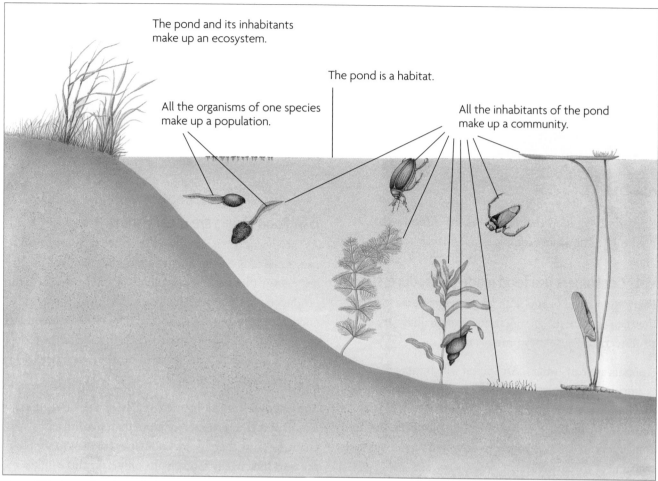

The pond and its inhabitants make up an ecosystem.

The pond is a habitat.

All the organisms of one species make up a population.

All the inhabitants of the pond make up a community.

Figure 15.1 A pond and its inhabitants – an example of an ecosystem.

Key definitions

ecosystem a unit containing all of the organisms and their environment, interacting together, in a given area e.g. decomposing log or a lake

population a group of organisms of one species, living in the same area at the same time

Questions

15.1 What is ecology?

15.2 What is a population?

15.3 Give **two** examples of an ecosystem, other than a pond.

Food and energy in an ecosystem

15.3 Energy passes along food chains.

All living organisms need energy. They get energy from food, by respiration. All the energy in an ecosystem comes from the Sun. Some of the energy in sunlight is captured by plants, and used to make food – glucose, starch and other organic substances such as fats and proteins. These contain some of the energy from the sunlight. When the plant needs energy, it breaks down some of this food by respiration.

Animals get their food, and therefore their energy, by eating plants, or by eating animals which have eaten plants.

The sequence by which energy, in the form of food, passes from a plant to an animal and then to other animals, is called a **food chain**. Figure **15.2** (overleaf) shows one example of a food chain.

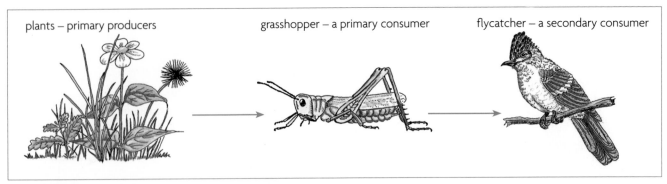

plants – primary producers   grasshopper – a primary consumer   flycatcher – a secondary consumer

Figure 15.2 A food chain.

Many different food chains link to form a **food web**. Figure **15.3** shows an example of a food web.

15.4 Consumers use food made by producers.

Every food chain begins with green plants because only they can capture the energy from sunlight. They are called **producers**, because they produce food.

Animals are **consumers**. An animal which eats plants is a **primary consumer**, because it is the first consumer in a food chain. An animal which eats that animal is a **secondary consumer**, and so on along the chain. Primary consumers are also called **herbivores**, and higher level consumers are **carnivores**.

Key definitions

food chain a chart showing the flow of energy (food) from one organism to the next (e.g. mahogany tree → caterpillar → song bird → hawk) beginning with a producer

food web a network of interconnected food chains showing the energy flow through part of an ecosystem

producer an organism that makes its own organic nutrients, usually using energy from sunlight, through photosynthesis

consumer an organism that gets its energy by feeding on other organisms

herbivore an animal that gets its energy by eating plants

carnivore an animal that gets its energy by eating other animals

---

15.5 Food chains are usually short.

As the energy is passed along the chain, each organism uses some of it. So the further along the chain you go, the less energy there is. There is plenty of energy available for producers, so there are usually a lot of them. There is less energy for primary consumers, and less still for secondary consumers. This means that towards the end of the food chain, the organisms get fewer in number, or smaller in total size (Figure **15.4**).

The loss of energy along the food chain also limits the length of it. There are rarely more than five links in a chain, because there is not enough energy left to supply the next link. Many food chains only have three links.

15.6 Consumers feed at different trophic levels.

In Figure **15.5**, the number of organisms in the food chain is shown as a pyramid. The area of each block in the pyramid represents the number of organisms. It is called a **pyramid of numbers**. Each level in the pyramid is called a **trophic level** ('trophic' means feeding).

The pyramid is this shape because there is less energy available as you go up the trophic levels, so there are fewer organisms at each level.

Many organisms feed at more than one trophic level. You, for example, are a primary consumer when you eat vegetables, a secondary consumer when you eat meat or drink milk, and a tertiary consumer when you eat a predatory fish such as a salmon.

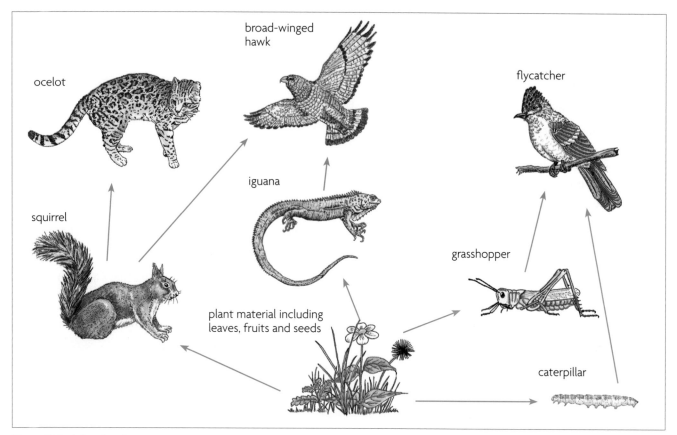

Figure 15.3 A food web.

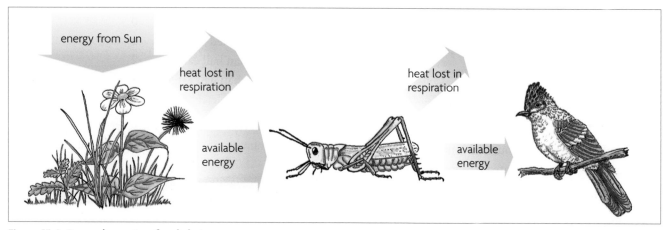

Figure 15.4 Energy losses in a food chain.

trophic level the position of an organism in a food chain, food web or pyramid of biomass, numbers or energy

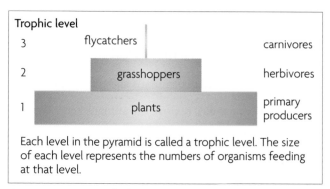

Trophic level

3	flycatchers	carnivores
2	grasshoppers	herbivores
1	plants	primary producers

Each level in the pyramid is called a trophic level. The size of each level represents the numbers of organisms feeding at that level.

Figure 15.5 A pyramid of numbers.

Activity 15.1
Studying an ecosystem

C2 *Observing, measuring and recording*

C3 *Interpreting and evaluating*

In this activity, you will try to work out some food chains in an ecosystem. Remember that you must disturb the ecosystem as little as possible. Do not take plants or animals away from the ecosystem unless your teacher tells you that you can do this. If you have a digital camera, take photographs of the organisms rather than collecting them.

1 Search the area thoroughly and try to identify all the types of plants in the area. If you cannot identify a plant, and there appears to be a lot of it, then collect samples of leaves and flowers to take back to your laboratory, where you can spend longer trying to find out what it is. Better still, take photographs of the plant so that you do not need to take samples from it.

2 Try to identify any small animals you see. Where possible, take photographs of each kind of animal.

3 Make notes about the large animals in the area, such as the types of birds present and what they are feeding on.

4 In the laboratory, with your teacher's assistance, try to identify all the organisms you found.

5 Use books or the Internet to find out what some of the animals feed on.

6 Construct a food web for this ecosystem.

Activity 15.2
Investigating the food preferences of slugs

C2 *Observing, measuring and recording*

C3 *Interpreting and evaluating*

1 Collect 12 slugs or other small herbivores of the same species.

2 Collect leaves from four different kinds of plant growing in the same area.

3 Identify the plants and call them A, B, C and D.

4 Place pairs of undamaged leaves into six jars as follows:

A and B   B and C   A and C

B and D   A and D   C and D

Make sure that you label the jars.

5 To each jar add two animals and put the lid on.

6 On the next day, remove the animals and examine the leaves.

7 Draw up a suitable results table. For each leaf record the amount eaten as follows:

No damage	0
Leaf nibbled	1
Less than half eaten	2
More than half eaten	3
Leaf completely eaten	4

8 For each type of leaf, add the scores. Construct a histogram or pie chart to show the results.

Questions

1 Which kind of plant did your animals prefer?

2 Can you suggest why your animals preferred this kind of plant?

3 Why must undamaged leaves be used in the experiment?

4 Why were the leaves used in pairs rather than one at a time?

5 Do you think that it would have been better to give the animals all the leaves at one time?

15.7 Pyramids of numbers may be 'upside-down'.

Figure **15.6** shows a different shaped pyramid of numbers. The pyramid is this shape because of the sizes (biomass) of the organisms in the food chain. Although there is only a single tree, it is huge compared with the caterpillars which feed on it.

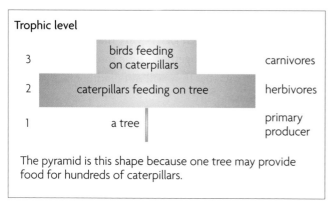

The pyramid is this shape because one tree may provide food for hundreds of caterpillars.

Figure 15.6 An inverted pyramid of numbers.

If you make the areas of the blocks represent the mass of the organisms, instead of their numbers, then the pyramid becomes the right shape again. It is called a **pyramid of biomass** (Figure **15.7**), and gives a much better idea of the actual quantity of animal or plant material at each trophic level.

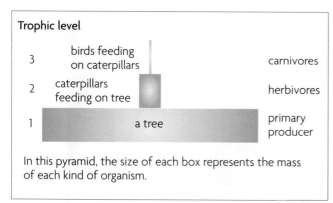

In this pyramid, the size of each box represents the mass of each kind of organism.

Figure 15.7 A pyramid of biomass.

15.8 Understanding energy flow helps agriculture.

Understanding how energy is passed along a food chain can be useful in agriculture. We can eat a wide variety of foods, and can feed at several different trophic levels. Which is the most efficient sort of food for a farmer to grow, and for us to eat?

The nearer to the beginning of the food chain we feed, the more energy there is available for us. This is why our staple foods, such as wheat, rice and potatoes, are plants.

When we eat meat, eggs or cheese, or drink milk, we are feeding further along the food chain. There is less energy available for us from the original energy provided by the Sun. It would be more efficient in principle to eat the grass in a field, rather than to let cattle eat it, and then eat them.

In fact, however, although there is far more energy in the grass than in the cattle, it is not available to us. We simply cannot digest the cellulose in grass, so we cannot release the energy from it. The cattle can; they turn the energy in cellulose into energy in protein and fat, which we can digest.

However, there are many plant products which we can eat. Soya beans, for example, yield a large amount of protein, much more efficiently and cheaply than cattle or other animals. A change towards vegetarianism would enable more food to be produced on the Earth, if the right crops were chosen.

Questions

15.4 Where does all the energy in living organisms originate from?

15.5 Write down a food chain **(a)** which ends with humans, **(b)** in the sea, and **(c)** with five links in it.

15.6 Why are green plants called producers?

15.7 Why are there rarely more than five links in a food chain?

Nutrient cycles

15.9 Decomposers release minerals from dead organisms.

One very important group of organisms which it is easy to overlook when you are studying an ecosystem, is the **decomposers**. They feed on waste material from animals and plants, and on their dead bodies. Many fungi and bacteria are decomposers.

Decomposers are extremely important, because they help to release substances from dead organisms, so that they can be used again by living ones. Two of these substances are carbon and nitrogen.

Key definition

decomposer   an organism that gets its energy from dead or waste organic matter

15.10 Carbon is recycled.

Carbon is a very important component of living things, because it is an essential part of carbohydrates, fats and proteins.

Figure **15.8** shows how carbon circulates through an ecosystem. The air contains about 0.04% carbon dioxide. When plants photosynthesise, carbon atoms from carbon dioxide become part of glucose or starch molecules in the plant.

Some of the glucose is then broken down by the plant in respiration. The carbon in the glucose becomes part of a carbon dioxide molecule again, and is released back into the air.

Some of the carbon in the plant will be eaten by animals. The animals respire, releasing some of it back into the air as carbon dioxide.

When the plant or animal dies, decomposers will feed on them. The carbon becomes part of the decomposers' bodies. When they respire, they release carbon dioxide into the air again.

15.11 Few organisms can use nitrogen gas.

Living things need nitrogen to make proteins. There is plenty of nitrogen around. The air is about 78% nitrogen gas. Molecules of nitrogen gas, N_2, are made of two nitrogen atoms joined together. These molecules are very inert, which means that they will not readily react with other substances.

So, although the air is full of nitrogen, it is in such an unreactive form that plants and animals cannot use it at all. It must first be changed into a more reactive form, such as ammonia (NH_3) or nitrates (NO_3^-).

Changing nitrogen gas into a more reactive form is called **nitrogen fixation** (Figure **15.9**). There are several ways that it can happen.

Lightning   Lightning makes some of the nitrogen gas in the air combine with oxygen, forming nitrogen oxides. They dissolve in rain, and are washed into the soil, where they form nitrates.

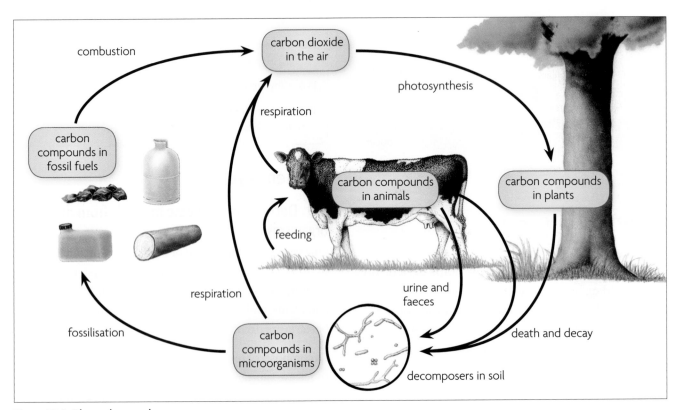

Figure 15.8 The carbon cycle.

Artificial fertilisers　Nitrogen and hydrogen can be made to react in an industrial chemical process, forming ammonia. The ammonia is used to make ammonium compounds and nitrates, which are sold as fertilisers.

Nitrogen-fixing bacteria　These bacteria live in the soil, or in root nodules (small swellings) on plants like peas, beans and clover. One kind is called *Rhizobium* ('rhizo' means root, 'bium' means living). They use nitrogen gas from the air spaces in the soil, and combine it with other substances to make ammonium ions and other compounds.

15.12 Fixed nitrogen moves round the nitrogen cycle.

Once the nitrogen has been fixed, it can be absorbed by the roots of plants, and used to make proteins. Animals eat the plants, so animals get their nitrogen in the form of proteins.

When an animal or plant dies, bacteria and fungi decompose the body. The protein, containing nitrogen, is broken down to ammonia and this is released. Another group of bacteria, called **nitrifying bacteria**, turn the ammonia into nitrates, which plants can use again.

Nitrogen is also returned to the soil when animals excrete nitrogenous waste material. It may be in the form of ammonia or urea. Again, nitrifying bacteria will convert it to nitrates.

15.13 Denitrifying bacteria make nitrogen gas.

A third group of bacteria complete the nitrogen cycle. They are called **denitrifying bacteria**, because they undo the work done by nitrifying bacteria. They turn nitrates and ammonia in the soil into nitrogen gas, which goes into the atmosphere.

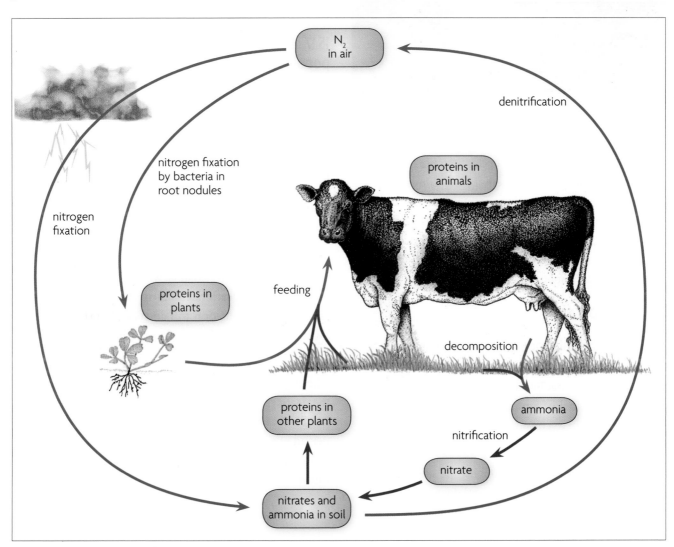

Figure 15.9　The nitrogen cycle.

15.14 Carnivorous plants get nitrogen from insects.

If the soil is waterlogged, nitrogen-fixing bacteria cannot live there, but denitrifying ones can. So boggy soil is usually very short of nitrates. Plants living in these places either have to manage with very little nitrogen, or get it from somewhere else. Some of them have become carnivorous (Figure 15.10). Plants like the Venus fly trap, or the sundews, supplement their diet with insects. They digest them with enzymes, and get extra nitrogen from the protein in the insects' bodies.

Figure 15.10 An insectivorous pitcher plant.

15.15 The water cycle.

Figure 15.11 shows how water cycles between living organisms and their environment. Living things, especially trees, play a very important role in this cycle. When rain falls, the tree roots absorb water from the soil. The water travels up through xylem vessels into the leaves, where some of it evaporates and diffuses out of the stomata as water vapour (Figure 8.31, page 103).

Population size

15.16 Most populations stay about the same size.

We have seen that a population is all the individuals of a particular species that live together in a habitat. In this section, we will look at how and why population sizes change, and begin to consider the implication of Earth's rapidly increasing human population.

Most populations tend to stay roughly the same size over a period of time. They may go up and down (fluctuate) but the average population will probably stay the same over a number of years. The population of greenfly in

a garden, for example, might be much greater one year than the next. But their numbers will almost certainly be back to normal in a year or so. Over many years, the sizes of most populations tend to remain at around the same level.

Yet if all the offspring of one female greenfly survived and reproduced, she could be the ancestor of 600 000 000 000 greenfly in just one year! Why doesn't the greenfly population shoot upwards like this? Why isn't the world overrun with greenfly?

The answers to those questions are of great importance to human beings, because our own population is doing just that; it is shooting upwards at an alarming rate. Every hour, there are 9000 extra people in the world. We need to understand why this is happening, and what is likely to happen next. Can we slow down the increase? What happens if we don't?

15.17 Birth rate and death rate determine population size.

The size of a population depends on how many individuals leave the population, and how many enter it.

Individuals leave a population when they die, or when they migrate to another population. Individuals enter a population when they are born, or when they migrate into the population from elsewhere. Usually, births and deaths are more important in determining population sizes than immigration and emigration.

A population increases if new individuals are born faster than the old ones die – that is, when the birth rate is greater than the death rate. If birth rate is less than death rate, then the population will decrease. If birth rate and death rate are equal, the population will stay the same size.

This explains why we are not knee-deep in greenfly. Although the greenfly population's birth rate is enormous, the death rate is also enormous. Greenfly are eaten by ladybirds and birds, and sprayed by gardeners. Over a period of time, the greenfly's birth and death rates stay about the same, so the population doesn't change very much.

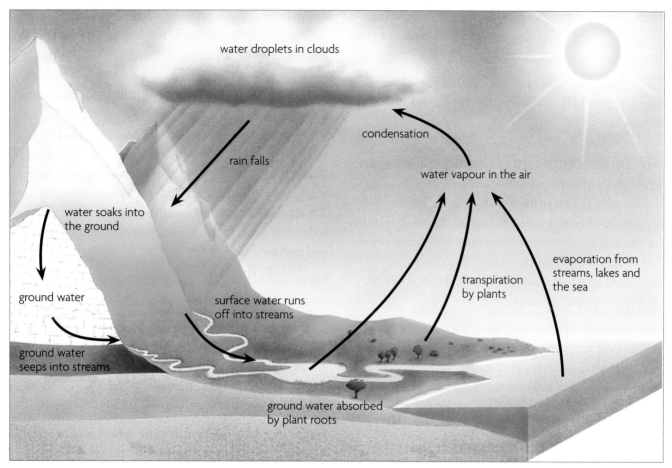

Figure 15.11 The water cycle.

15.18 Yeast experiments give some clues about population growth.

By looking at changes in population sizes in other organisms, we can learn quite a lot about our own. Many experiments on population sizes have been done on organisms like bacteria and yeast, because they reproduce quickly and are easy to grow. Figure **15.12** shows the results of an experiment in which a few yeast cells are put into a container of nutrient broth. The cells feed on the broth, grow and reproduce. The numbers of yeast cells are counted every few hours.

At the beginning of the experiment, the population only grows quite slowly, because there are not many cells there to reproduce. This is called the **lag phase**.

But once they get going, growth is very rapid. Each cell divides to form 2, then 4, then 8, then 16. There is nothing to hold them back except the time it takes to grow and divide. This is called the **log phase**, or **exponential phase**.

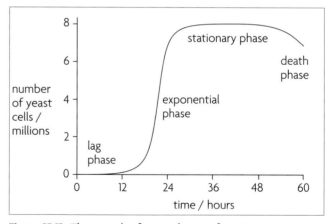

Figure 15.12 The growth of a population of yeast.

As the population gets larger, the individual cells can no longer reproduce as fast, and begin to die off more rapidly. This may be because there is not enough food left for them all, or it might be that they have made so much alcohol that they are poisoning themselves. The cells are now dying off as fast as new ones are being produced, so the population stops growing and levels off. This is called the **stationary phase**.

Eventually, the death rate exceeds the birth rate, so the number of living yeast cells in the population starts to fall. This is called the **death phase**.

This curve is sometimes called a **sigmoid** growth curve. 'Sigma' is the old Greek letter s, so 'sigmoid' means S-shaped.

15.19 Environmental factors control population size.

Although the experiment with the yeast is done in artificial conditions, a similar pattern is found in the growth of populations of many species in the wild. If a few individuals get into a new environment, then their population curve may be very like the one for yeast cells in broth. The population increases quickly at first, and then levels off.

The levelling off is always caused by some kind of environmental factor. In the case of the yeast, the factor may be food supply. Other populations may be limited by disease, or the number of nest sites, or the number of predators, for example. The factor that stops the population from getting any larger is called a **limiting factor**.

15.20 Population sizes often oscillate.

It is usually very difficult to find out which environmental factors are controlling the size of a population. Almost always, many different factors will interact. A population of rabbits, for example, might be affected by the number of foxes and other predators, the amount of food available, the amount of space for burrows, and the amount of infection by the virus which causes myxomatosis.

Figure **15.13** shows an example of how the size of population of a predator may be affected by its prey. This information comes from the number of skins which were sold by fur traders in Northern Canada to the Hudson Bay Company, between 1845 and 1925. Snowshoe hares and northern lynxes were both trapped for their fur, and the numbers caught probably give a very good idea of their population sizes.

Snowshoe hare populations tend to vary from year to year. No-one is quite sure why this happens, but it may be related to their food supply. Whenever the snowshoe hare population rises, the lynx population also rises shortly afterwards, as the lynxes now have more food. A drop in the snowshoe hare population is rapidly followed by a drop in the lynx population.

The numbers tend to go up and down, or oscillate, but the average population sizes stay roughly the same over many years.

15.21 Age pyramids show whether a population is increasing or decreasing.

When scientists begin to study a population, they want to know whether the population is growing or shrinking. This can be done by counting the population over many years, or by measuring its birth rate and death rate. But often it is much easier just to count the numbers of individuals in various age groups at one point in time, and to draw an age pyramid.

Figure **15.14** shows two examples of age pyramids. The area of each box represents the numbers of individuals in that age group.

Figure **15.14a** is a bottom-heavy pyramid, because there are far more young individuals than old ones. This indicates that birth rate is greater than death rate, so this population is increasing.

Figure **15.14b** shows a much more even spread of ages. Birth rate and death rate are probably about the same. This population will remain about the same size.

If an age pyramid is drawn for the human population on Earth, it is bottom-heavy, like Figure **15.14a**. Age pyramids for many of the world's developing countries are also this shape, showing that their populations are increasing. But an age pyramid for a European country such as France looks more like Figure **15.14b**. The human population in France is staying about the same.

15.22 The human death rate has decreased dramatically.

Figure **15.15** shows how the human population of the world has changed since about 3000 BC. For most of that time, human populations have been kept in check by a combination of disease, famine and war. Nevertheless, there has still been a steady increase.

Twice there have been definite 'spurts' in this growth. The first was around 8000 BC, not shown on the graph, when people in the Middle East began to farm, instead of just hunting and finding food. The second began around 300 years ago, and is still happening now.

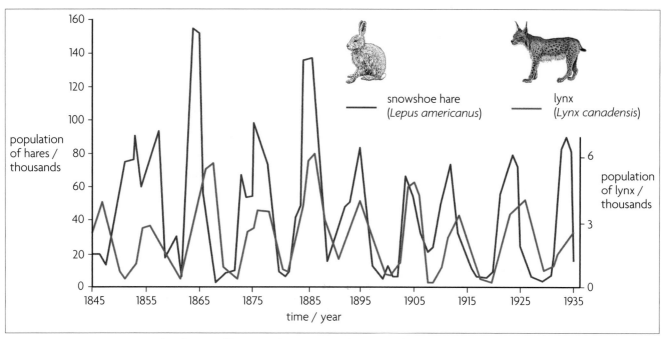

Figure 15.13 Variations in snowshoe hare and lynx populations in northern Canada.

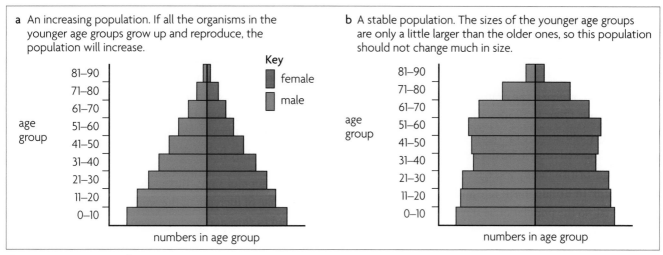

a An increasing population. If all the organisms in the younger age groups grow up and reproduce, the population will increase.

b A stable population. The sizes of the younger age groups are only a little larger than the older ones, so this population should not change much in size.

Figure 15.14 Age pyramids.

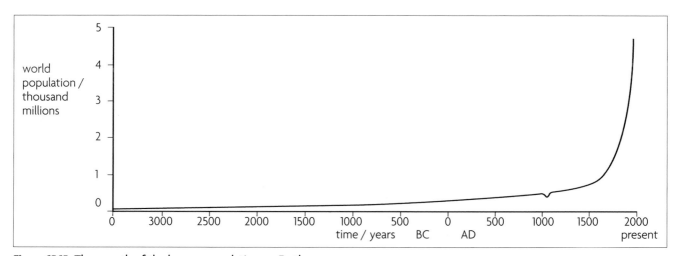

Figure 15.15 The growth of the human population on Earth.

There are two main reasons for this recent growth spurt. The first is the reduction of disease. Improvements in water supply, sewage treatment, hygienic food handling and general standards of cleanliness have virtually wiped out many diseases in countries such as the USA and most European countries – for example typhoid and dysentery. Immunisation against diseases such as polio has made these very rare indeed. Smallpox has been totally eradicated. And the discovery of antibiotics has now made it possible to treat most diseases caused by bacteria.

Secondly, there has been an increase in food supply. More and more land has been brought under cultivation. Moreover, agriculture has become more efficient, so that in many parts of the world each hectare of land is now producing more food than ever before.

15.23 Birth rate now exceeds death rate.

The human population has increased dramatically because the death rate has been brought down. More and more people are now living long enough to reproduce. If the birth rate doesn't drop by the same amount as the death rate, then the world population will continue to increase.

In developed countries, the dramatic fall in the death rate began in about 1700. To begin with, the birth rate stayed high, so the population grew rapidly. But since 1800, there has been a marked drop in birth rate. In 1870, for example, the 'average' British family was 6.6 children, but by 1977 it was only 1.8. In Britain, birth rate and death rate are now about equal, so the population is staying the same.

However, in many of the developing countries, the fall in the death rate only began about 50 years ago. As yet, the birth rates have not dropped, and so the populations are rising rapidly.

15.24 Birth rate must be reduced to slow population growth.

The human population could be brought back under control in two ways – increasing the death rate or decreasing the birth rate. There is no question as to which of these is the best.

In the developed countries, the single largest factor which brought down the birth rate was the introduction of contraceptive techniques. Considerable efforts are being made to introduce these to people in the developing countries, with some success. But many people are suspicious of contraceptive methods, or barred from using them by their religion, or simply want to have large families. It looks as though the population will go on rising for some time.

If we do not control the overall human birth rate, then it may happen that famine, war or disease will increase the death rate. This cannot be the best thing for the human race. We must do our best to stabilise the world population at a level at which everyone has a fair chance of a long, healthy life.

There are hopeful signs. Birth rates are steadily falling, and the rate of increase in the world population is predicted to slow and perhaps even start to fall later in this century.

Key ideas

◆ Energy enters ecosystems in sunlight. Producers (photosynthetic plants) capture some of this energy and transfer it to organic substances such as carbohydrates. Consumers (animals and fungi) get their energy by eating producers or other consumers.

◆ Food chains and food webs show how energy flows through an ecosystem. The level at which an organism feeds in a food chain is its trophic level.

◆ Energy is lost as it passes along food chains.

◆ The energy losses in food chains limit the length of the chain, so few food chains have more than five trophic levels.

◆ It would be more energy-efficient for humans to harvest and eat plant crops, rather than feeding the crops to animals and then eating those.

◆ Pyramids of numbers and pyramids of biomass are ways of showing the relative numbers or biomass at different trophic levels in a food chain.

(continued ...)

(... continued)

◆ The carbon cycle shows how carbon dioxide from the air is used in photosynthesis to make organic compounds in plants, which are then eaten by animals. Decomposers obtain their carbon by feeding on dead plants or animals, or on their waste materials. Respiration by all organisms returns carbon dioxide to the air.

◆ Nitrogen gas is very inert, and must be fixed (to produce ammonium ions or nitrate ions) before it can be used by living organisms. Some plants have nitrogen-fixing bacteria in their roots, and other nitrogen-fixing bacteria live freely in the soil. Plants absorb ammonium or nitrate ions and use them to make proteins, which can then be eaten by animals. Decomposers and nitrifying bacteria convert proteins to ammonia and nitrates, which can be re-used by plants. Denitrifying bacteria convert nitrates to nitrogen gas which is returned to the air.

◆ The size of a population of organisms is affected by environmental factors such as food supply, predation and disease.

◆ When a resource is in limited supply, the growth of a population often shows a lag phase, exponential phase, stationary phase and death phase.

◆ Age pyramids show the structure of a population at one moment in time, and can be used to predict how the population is likely to change in the future. The global human population is increasing, but there is hope that by the end of this century the growth will have slowed significantly.

Revision questions

1 **a** Why do living organisms need carbon?
  b Explain how carbon atoms become part of a plant.
  c What happens to some of these carbon atoms when a plant respires?
  d Explain the role of decomposers in the carbon cycle.

2 Explain the difference between each of the following pairs, giving examples where you can: **(a)** producer, consumer, **(b)** primary consumer, secondary consumer, **(c)** community, population, **(d)** food chain, food web, **(e)** pyramid of biomass, pyramid of numbers.

3 **a** Why do living organisms need nitrogen?
  b Explain why plants and animals cannot use the nitrogen in the air
  c What is nitrogen fixation?
  d Where do nitrogen-fixing bacteria live?
  e Explain how animals obtain nitrogen.
  f What do nitrifying bacteria do?
  g Which type of bacteria return nitrogen to the air?

4 **a** In what form do each of the following obtain their nitrogen? **(i)** a green plant
    (ii) nitrogen-fixing bacteria **(iii)** a mammal
  b In the sea, the main nitrogen-fixing organisms are blue-green algae, which float near the top of the water in the plankton. Construct a diagram or chart similar to Figure **15.9**, showing how nitrogen is circulated amongst marine organisms.

5 The graph below shows population changes over one summer, for two insects. One is a type of greenfly, and the other is a ladybird which feeds on it.

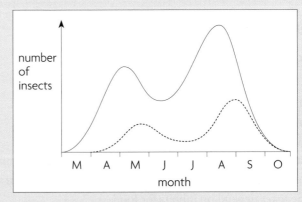

  a Which curve represents the ladybird population, and which the greenfly population?
  b Give a reason for your answer to part **(a)**.
  c Explain why the two curves are similar.
  d Why do the two curves rise and fall at slightly different times?

16 Humans and the environment

In this chapter, you will find out:

- what is meant by the greenhouse effect, and how adding extra carbon dioxide and methane to the atmosphere is causing global warming
- S about the causes and effects of acid rain
- about pollution from nuclear fall-out
- about the damaging effects of deforestation
- S how water pollution by sewage or fertilisers can lead to eutrophication
- why pesticides and herbicides are used, and how they can cause pollution
- about the importance of conservation of species and their habitats
- S why we should try to recycle limited and non-renewable resources such as paper and water.

Environmental impacts

16.1 All organisms affect their environment.

All living things affect the living and non-living things around them. For example, earthworms make burrows and wormcasts, which affect the soil, and therefore the plants growing in it. Rabbit fleas carry the virus which causes myxomatosis, so they can affect the size of a rabbit population, and perhaps the size of the fox population if the foxes depend on rabbits for food.

Perhaps the biggest ever effect of living organisms on the environment happened about 1500 million years ago. At this time, the first living cells that could photosynthesise evolved. Until then, there had been no oxygen in the atmosphere. These organisms began to produce oxygen, which gradually accumulated in the atmosphere. The oxygen in the air we now breathe has been produced

by photosynthesis. The appearance of oxygen in the air meant that many anaerobic organisms could now only live in particular parts of the Earth which were oxygen-free, such as in deep layers of mud. It meant that many other kinds of organism could evolve, which used the oxygen for respiration. All this oxygen excreted by photosynthetic organisms could be considered to be the biggest case of pollution ever!

16.2 Humans affect the environment.

Within the past 10 000 years or so, another organism has had an enormous impact on the environment. Ever since humans learnt to hunt with weapons, to domesticate animals and to farm crops, we have been changing the environment around us in a very significant way. We kill wild animals for food, decreasing their populations and making some species extinct. We cut down forests. We build cities, roads and dams. We release harmful substances into the water, air and soil.

As the human population continues to increase, and expectations of living standards become higher, the effects we have on the environment become ever greater. It is very important that we do our best to understand what we are doing, and what the effects might be. The more we understand, the more we can do to prevent too much damage before it happens, and keep the Earth a pleasant place for humans, plants and other animals to live.

Table **16.1** summarises some of the damaging effects we have had on our environment. The rest of this chapter describes some of these in more detail, and explains what we can do to limit further damage.

Global warming

16.3 The greenhouse effect is essential to life.

The Earth's atmosphere contains several different gases that act like a blanket, keeping the Earth warm. They are sometimes called **greenhouse gases**. The most important of these gases is carbon dioxide. Methane is also a significant greenhouse gas.

Carbon dioxide is transparent to shortwave radiation from the Sun. The sunlight passes freely through the atmosphere (Figure **16.1**, overleaf), and reaches the ground. The ground is warmed by the radiation, and emits longer wavelength, infrared radiation. Carbon dioxide does not let all of this infrared radiation pass through. Much of it is kept in the atmosphere, making the atmosphere warmer.

This is called the **greenhouse effect**, because it is similar

Table 16.1 A summary of the harmful effects of humans on the environment

damage	example	main causes	possible solutions
air pollution	damage to the ozone layer	CFCs	stop using CFCs; find harmless alternatives
	global warming	enhanced greenhouse effect, caused by release of carbon dioxide, methane, CFCs and nitrogen oxides	reduce use of fossil fuels; stop using CFCs; produce less organic waste and/or collect and use methane produced from landfill sites
	acid rain	sulfur dioxide and nitrogen oxides from the burning of fossil fuels	burn less fossil fuel; use catalytic converters on cars
habitat destruction	deforestation	destruction of forests, especially rainforests, for wood and for land for farming, roads and houses	provide alternative sources of income for people living near rainforests
	loss of wetlands	draining wetlands for housing and land for farming	protect areas of wetlands
water pollution	toxic chemicals	untreated effluent from industry; run-off from mining operations	impose tighter controls on industry and mining
	eutrophication	sewage and fertilisers running into streams	treat all sewage before discharge into streams; use fewer fertilisers
	oil spills	shipwrecks; leakages from undersea oil wells	impose tighter controls on shipping and the oil industry
species destruction	loss of habitat	see deforestation and wetlands above	see above
	damage from pesticides	careless use of insecticides and herbicides	development of more specific and less persistent pesticides; more use of alternative control methods, such as biological control
	damage from fishing	overfishing, greatly reducing populations of species caught for food; accidental damage to other animals such as dolphins	impose controls on methods and amount of fishing

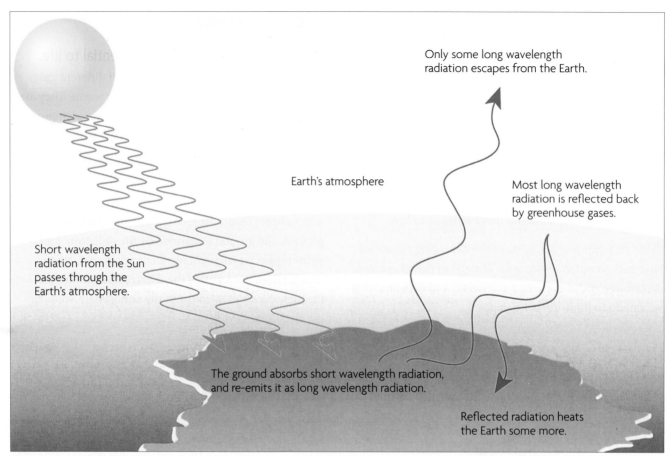

Only some long wavelength radiation escapes from the Earth.

Earth's atmosphere

Most long wavelength radiation is reflected back by greenhouse gases.

Short wavelength radiation from the Sun passes through the Earth's atmosphere.

The ground absorbs short wavelength radiation, and re-emits it as long wavelength radiation.

Reflected radiation heats the Earth some more.

Figure 16.1 The greenhouse effect. Short wavelength radiation from the Sun passes through the atmosphere and reaches the ground. Some of it is absorbed by the ground, and is re-emitted as long wavelength radiation. Much of this cannot pass through the blanket of greenhouse gases in the atmosphere. It is reflected back towards the Earth, warming the atmosphere.

to the effect which keeps an unheated greenhouse warmer than the air outside. The glass around the greenhouse behaves like the carbon dioxide in the atmosphere. It lets shortwave radiation in, but does not let out the longwave radiation. The longwave radiation is trapped inside the greenhouse, making the air inside it warmer.

We need the greenhouse effect. If it did not happen, then the Earth would be frozen and lifeless. The average temperature on Earth would be about 33 °C lower than it is now.

16.4 The enhanced greenhouse effect may cause global warming.

Although we need the greenhouse effect, people are worried that it may be increasing. The amount of carbon dioxide and other greenhouse gases in the atmosphere is getting greater. This may trap more infrared radiation, and make the atmosphere warmer. This is called the **enhanced greenhouse effect**, and its possible effect on the Earth's temperature is called **global warming**.

Over recent years, the amount of fossil fuels which have been burnt by industry, and in engines of vehicles such as cars, trains and aeroplanes, has increased greatly. This releases carbon dioxide into the atmosphere. Figure **16.2** shows what has happened to the amount of carbon dioxide in the atmosphere since 1750.

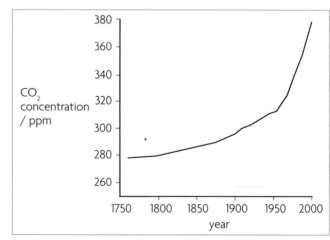

Figure 16.2 How carbon dioxide concentrations in the atmosphere have changed since 1750.

Other gases that contribute to the greenhouse effect have also been released by human activities. These include methane, nitrogen oxides and CFCs. Table **16.2** shows where these gases come from. The concentrations of all of these gases in the atmosphere are steadily increasing.

Table 16.2 Gases contributing to the greenhouse effect

gas	% estimated contribution	main sources
carbon dioxide	55	burning fossil fuels
methane	15	decay of organic matter, e.g. in waste tips and paddy fields; waste gases from digestive processes in cattle and insects; natural gas leaks
CFCs	24	refrigerators and air conditioning systems; plastic foams
nitrogen oxides	6	fertilisers; burning fossil fuels, especially in vehicles

As the concentration of these gases increases, the temperature on Earth will also increase. At the moment, we are not able to predict just how large this effect will be. There are all sorts of other processes, many of them natural, which can cause quite large changes in the average temperature of the Earth, and these are not fully understood. For example, every now and then the Earth has been plunged into an Ice Age. Perhaps we are due for another Ice Age soon. Perhaps the enhanced greenhouse effect might help to delay this.

But most people think that we should be very worried about the enhanced greenhouse effect and global warming. If the Earth's temperature does rise significantly, there will be big changes in the world as we know it. For example, the ice caps might melt. This would release a lot more water into the oceans, so that sea levels would rise. Many low-lying areas of land might be flooded. This could include large parts of countries like Bangladesh, almost the whole of the Maldive islands, and major cities such as London.

A rise in temperature would also affect the climate in many parts of the world. No-one is sure just what would happen where – there are too many variables for scientists to be able to predict the consequences. It would probably mean that some countries which already have low rainfall might become very dry deserts. Others might have more violent storms than they do now. This would mean that animals and plants living in some areas of the world might become extinct. People in some places might not be able to grow crops.

There might be some beneficial effects, too. For example, extra carbon dioxide in the atmosphere and higher temperatures might increase the rate of photosynthesis in some parts of the world. This could mean that higher yields could be gained from crops.

16.5 Can we avoid global warming?

It is important that we cut down the emission of greenhouse gases. One obvious way to do this is to reduce the amount of fossil fuels that are burnt. This would reduce the amount of carbon dioxide we pour into the air. Agreements have been made between countries to try to do this, but they are proving very difficult to implement.

Deforestation has also been blamed for increasing the amount of carbon dioxide in the air. (You can read more about deforestation in sections **16.11** to **16.13**.) It has been argued that cutting down rainforests leaves fewer trees to photosynthesise and remove carbon dioxide from the air. Moreover, if the tree is burnt or left to rot when it is chopped down, then carbon dioxide will be released from it.

Methane is produced by farming activities (Figure **16.3**).

Figure 16.3 Methane is produced by anaerobic microorganisms growing in the mud in paddy fields.

It is released by bacteria which live on organic matter, such as in paddy fields (flooded fields which are used for growing rice), by animals which chew the cud, such as cattle, and by some insects, such as termites. There is probably not much that we can do about this. Methane is also produced by decaying rubbish in landfill sites. We can reduce this problem by decreasing the amount of rubbish we throw away, and by collecting the methane from these sites. It can be used as fuel (Figure **16.4**). Although burning it for fuel does release carbon dioxide, this carbon dioxide does not trap so much infrared radiation as the methane would have done.

Figure 16.4 Bacteria feeding on rubbish in landfill sites produce methane, which can be piped off and used as a fuel.

Questions

16.1 Explain the difference between the greenhouse effect, the enhanced greenhouse effect and global warming.

16.2 Each of the following has been suggested as a way of reducing global warming. For each suggestion, explain why it would work, and discuss the problems which would probably occur in trying to implement it.

 a reducing the top speed limit for cars and trucks
 b improving traffic flow in urban areas
 c insulating houses in countries with cold climates
 d increasing the number of nuclear power stations
 e encouraging people to recycle more of their rubbish

Acid rain

16.6 Burning fossil fuels releases sulfur and nitrogen oxides.

Fossil fuels, such as coal, oil and natural gas, were formed from living organisms. They all contain sulfur; coal contains the most. When they are burnt, the sulfur combines with oxygen in the air and forms sulfur dioxide. Nitrogen oxides are also formed.

Sulfur dioxide is a very unpleasant gas. If people breathe it in, it can irritate the linings of the breathing system. If you are prone to asthma or bronchitis, sulfur dioxide can make it worse. Sulfur dioxide is also poisonous to many kinds of plants, sometimes damaging their leaves so badly that the whole plant dies.

16.7 Sulfur and nitrogen oxides produce acid rain.

Rain is usually slightly acidic, with a pH a little below 7. This is because carbon dioxide dissolves in it to form carbonic acid.

Sulfur dioxide and nitrogen oxides also dissolve in rain. They form an acidic solution, called **acid rain**. The pH of acid rain can be as low as 4.

Acid rain damages plants. Although the rain usually does not hurt the leaves directly when it falls onto them, it does affect the way in which plants grow. This is because it affects the soil in which the plants are growing. The acid rainwater seeps into the soil, and washes out ions such as calcium, magnesium and aluminium. The soil becomes short of these ions, so the plant becomes short of nutrients. It also makes it more difficult for the plant to absorb other nutrients from the soil. So acid rain can kill trees and other plants.

The ions which are washed out of the soil by the acid rain often end up in rivers and lakes. Aluminium, in particular, is very poisonous to fish, because it affects their gills. Young fish are often killed if the amount of aluminium in the water is too great. Other freshwater organisms are often killed, too. At the same time, the water itself becomes more acidic, which means that many kinds of plants and animals cannot live in it (Figure **16.5**).

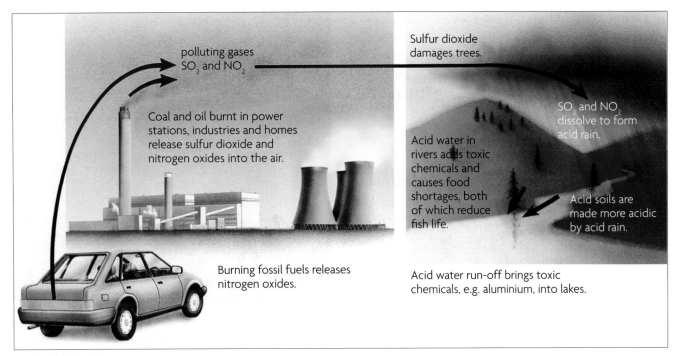

Figure 16.5 The causes and effects of acid rain.

16.8 Sulfur dioxide is carried long distances.

One of the biggest problems in trying to do anything about the problems of acid rain is that it does not usually fall anywhere near the place which is causing it. A coal-burning power station might release a lot of sulfur dioxide, which is then carried high in the air for hundreds of miles before falling as acid rain. Sulfur dioxide produced in England might fall as acid rain in Norway.

16.9 Sulfur dioxide emissions are being reduced.

Acid rain is, in many ways, a much easier problem to solve than the enhanced greenhouse effect. The answer is simple – we must cut down emissions of sulfur dioxide and nitrogen oxides.

Coal-burning power stations have been the worst culprits. The number of coal-burning power stations in some European countries has been going down – more because of the cost of coal than to try to stop pollution – and more of them are burning oil which produces less sulfur dioxide. However, this has meant a reduction in the demand for coal, which has been part of the reason for the closure of many coal mines, and the loss of jobs for many miners in countries such as Britain.

Whatever kind of fossil fuel is burnt in power stations or other industries, the waste gases can be 'scrubbed' to remove sulfur dioxide. This often involves passing the gases through a fine spray of lime.

The burning of petrol (gasoline) in car engines also produces nitrogen oxides. These can be removed by catalytic converters fitted to the exhaust system. In most countries, all new petrol-burning cars now have to have catalytic converters.

Questions

16.3 What causes acid rain?

16.4 How does acid rain damage trees?

16.5 How does acid rain damage fish?

16.6 Summarise what is being done to try to reduce the production of acid rain.

Nuclear fall-out

16.10 Ionising radiation may cause cancer.

Accidents at nuclear power stations may release radioactive substances into the atmosphere. Exposure to large amounts of radiation from these substances can cause radiation sickness and burns. This type of radiation can also increase mutation rates in DNA in our cells, which may lead to cancer.

This happens because ionising radiation – such as alpha, beta and gamma radiation – damages the DNA molecules in living cells. Alpha is the most ionising and so causes the most damage, but only if it gets inside the body. This is because it is not able to penetrate the skin. Gamma is the least ionising but the most penetrating.

Deforestation

16.11 People cut down trees for fuel and farmland.

Humans have always cut down trees. Wood is an excellent fuel and building material. The land on which trees grow can be used for growing crops for food, or to sell. One thousand years ago, most of Europe was covered by forests. Now, most of them have been cut down. The cutting down of large numbers of trees is called **deforestation** (Figure **16.6**).

Figure 16.7 This rainforest is growing in a part of Chile where the climate is temperate (with cold winters and warm summers) and there is very high rainfall. It has an enormous species diversity.

Figure 16.6 When rainforest is cut down and burnt, as here in Brazil, large amounts of carbon dioxide are released and soil nutrients are lost.

16.12 Rainforests are special places.

Rainforests occur in temperate and tropical regions of the world (Figure **16.7**). Recently, most concern about deforestation has been about the loss of tropical rainforests. In the tropics, the relatively high and constant temperatures, and high rainfall, provide perfect conditions for the growth of plants (Figure **16.8**). A rainforest is a very special place, full of many different species of plants and animals. More different species live in a small area of rainforest than in an equivalent area of

Figure 16.8 Unspoiled tropical rainforest in Sarawak, Malaysia.

any other habitat in the world. We say that rainforest has a high **species diversity**.

When an area of rainforest is cut down, the soil under the trees is exposed to the rain. The soil of a rainforest is very thin. The soil is quickly washed away once it loses its cover of plants. This soil erosion may make it very difficult for the rainforest to grow back again, even if the land is left alone. The soil can also be washed into rivers, silting it and causing flooding (Figure **16.9**).

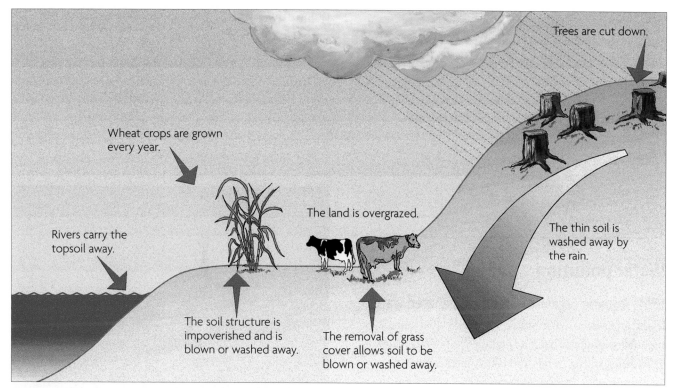

Wheat crops are grown every year.

Trees are cut down.

The land is overgrazed.

The thin soil is washed away by the rain.

Rivers carry the topsoil away.

The soil structure is impoverished and is blown or washed away.

The removal of grass cover allows soil to be blown or washed away.

Figure 16.9 How human activities can increase soil erosion.

The loss of part of a rainforest means a loss of a habitat for many different species of animals. Even if small 'islands' of forest are left as reserves, these may not be large enough to support a breeding population of the animals. Deforestation threatens many species of animals and plants with extinction.

The loss of so many trees can also affect the water cycle (Figure **15.11**, page **227**). While trees are present, when rain falls a lot of it is taken up by the trees, and transported into their leaves. It then evaporates, and goes back into the atmosphere in the process of transpiration. If the trees have gone, then the rain simply runs off the soil and into rivers. Much less goes back into the air as water vapour. The air becomes drier, and less rain falls. This can make it much more difficult for people to grow crops and keep livestock.

16.13 Developing countries need help to conserve rainforests.

When people in industrialised countries get concerned about the rate at which some countries are cutting down their forests, it is very important they should remember that they have already cut down most of theirs.

Most tropical rainforests grow in developing countries, and in some countries many of the people are very

poor. The people may cut down the forests to clear land on which they can grow food. It is difficult to expect someone who is desperately trying to produce food, to keep their family alive, not to do this, unless you can offer some alternative. International conservation groups such as the World Wide Fund for Nature, and governments of the richer, developed, countries such as the USA, can help by providing funds to the people or governments of developing countries to try to help them to provide alternative sources of income for people. Many of the most successful projects involve helping local people to make use of the rainforest in a sustainable way.

The greatest pressure on the rainforest may come from the country's government in the big cities, rather than the people living in or near the rainforest. The government may be able to obtain large amounts of money by allowing logging companies to cut down forests and extract the timber. A way of getting round this could be to allow countries to sell 'carbon credits' to other, richer countries. In 2009, Indonesia did this. The idea is that other countries give money to Indonesia to use in conserving their forests, and that these countries are then allowed to produce more carbon dioxide from their industrial activities.

Questions

16.7 Why are tropical rainforests often considered to be 'very special places'?

16.8 Explain why people cut down large areas of forest. (You can probably think of other reasons as well as those which are described above.)

16.9 Outline the damage which can be caused when rainforests are cut down.

16.10 How can developed countries help to prevent the loss of rainforest?

Water pollution

16.14 Aquatic organisms need oxygenated water.

Many organisms live in water. They are called aquatic organisms. Aquatic habitats include fresh water, such as streams, rivers, ponds and lakes; and also marine environments – the sea and oceans.

Most organisms which live in water respire aerobically, and so need oxygen. They obtain their oxygen from oxygen gas which has dissolved in the water. Anything which reduces the amount of oxygen available in the water can make it impossible for fish or other aquatic organisms to live there.

There are two main sources of pollution which can reduce oxygen levels in fresh water. They are fertilisers and untreated sewage.

Farmers and horticulturists use fertilisers to increase the yield of their crops. The fertilisers usually contain nitrates and phosphates. Nitrates are very soluble in water. If nitrate fertiliser is put onto soil, it may be washed out in solution when it rains. This is called **leaching**. The leached nitrates may run into streams and rivers.

Algae and green plants in the river grow faster when they are supplied with these extra nitrates. They may grow so much that they completely cover the water. They block out the light for plants growing beneath them, which die. Even the plants on the top of the water eventually die. When they do, their remains are a good source of food for bacteria.

The bacteria breed rapidly. The large population of bacteria respires aerobically, using up oxygen from the water. Soon, there is very little oxygen left for other living things. Those which need a lot of oxygen, such as fish, have to move to other areas, or die.

This whole process is called **eutrophication** (Figures **16.10** and **16.11**). It can happen whenever food for plants or bacteria is added to water. As well as fertilisers, other pollutants from farms, such as slurry from buildings where cattle or pigs are kept, or from pits where grass is rotted down to make silage, can cause eutrophication.

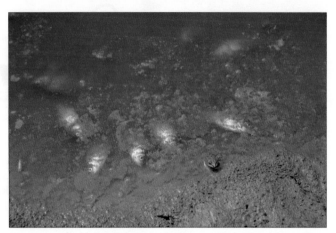

Figure 16.10 The huge growth of algae in this polluted pond has provided food for aerobic bacteria. These have used up most of the oxygen in the water, so the fish have died.

Untreated sewage can also cause eutrophication (Figure **16.12**). Sewage does not usually increase the growth of algae, but it does provide a good food source for many kinds of bacteria. Once again, their population grows, depleting the oxygen levels.

16.15 Nitrate use must be carefully controlled.

Eutrophication is not the only problem caused by the leaching of nitrate fertilisers. Some of the nitrates are carried deeper into the soil, where they find their way into water in rocks deep underground, called **aquifers**. Water in aquifers may be extracted to use as drinking water. There is some concern that, if people drink water containing a lot of nitrate, they may become ill.

Could we stop using nitrate fertilisers? It is not really sensible at the moment to suggest that we could. People expect to have plentiful supplies of relatively cheap food. Although fertilisers are expensive, by using them farmers get so much higher yields that they make more profit. If they did not use fertilisers at all, their yields would be much lower and they would have to sell their crops for a higher price, in order to make any profit at all.

Some farmers are doing just this. They do not use inorganic fertilisers, such as ammonium nitrate, at all. Instead, they use organic fertilisers, such as manure.

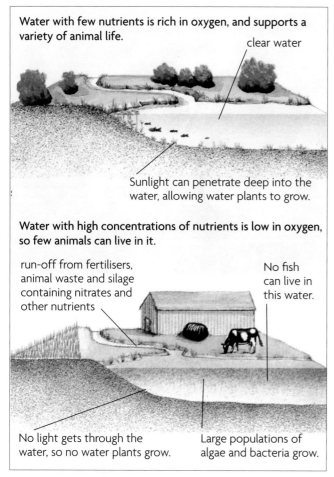

Water with few nutrients is rich in oxygen, and supports a variety of animal life.

clear water

Sunlight can penetrate deep into the water, allowing water plants to grow.

Water with high concentrations of nutrients is low in oxygen, so few animals can live in it.

run-off from fertilisers, animal waste and silage containing nitrates and other nutrients

No fish can live in this water.

No light gets through the water, so no water plants grow.

Large populations of algae and bacteria grow.

Figure 16.11 Eutrophication. Nutrients flowing into the water increase algal and bacterial growth. This reduces oxygen concentration, killing fish.

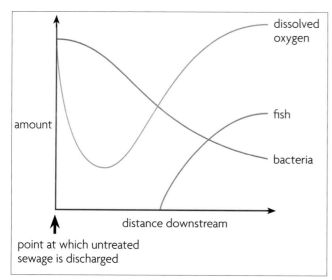

dissolved oxygen

amount

fish

bacteria

distance downstream

point at which untreated sewage is discharged

Figure 16.12 The effect of raw sewage on a stream.

Organic fertilisers are better than inorganic ones in that they do not contain many nitrates which can easily be leached out of the soil. Instead, they release their nutrients gradually, over a long period of time, giving crops time to absorb them efficiently. Nevertheless, manures can still cause pollution, if a lot is put onto a field at once, at a time of year when there is a lot of rain or when crops are not growing and cannot absorb the nutrients from them.

The yields obtained when using organic fertilisers are not usually as great as when using inorganic ones, so the crops are usually sold for a higher price. Many people are now prepared to pay this extra money for food from crops grown in this way, but many cannot afford to.

If nitrate fertilisers are used, there is much which can be done to limit the harm they do. Care must be taken not to use too much, but only to apply an amount which the plants can take up straight away. Fertilisers should not be applied to empty fields, but only when plants are growing. They should not be applied just before rain is forecast. They should not be sprayed near to streams and rivers.

16.16 Sewage can be treated before it is released into streams.

As well as eutrophication, untreated sewage can cause other problems. It contains urine and faeces, both of which may contain harmful pathogens. If a person swims in, or drinks, water contaminated with untreated sewage, they run the risk of catching a wide variety of diseases, some of which – such as poliomyelitis – are very serious.

In many countries, most sewage is now treated to remove these harmful pathogens, and most of the nutrients which could cause eutrophication, before it is released into rivers. Sewage treatment is described in sections **16.23** and **16.24**.

16.17 Chemical waste may contain toxins.

A very different kind of water pollution may result from the discharge of chemical waste into waterways. Chemical waste may contain heavy metals, such as lead, cadmium or mercury. These substances are very poisonous (toxic) to living organisms, because heavy metals stop enzymes from working. If they get into streams, rivers or the sea, they may kill almost every living thing in that area of water.

Questions

16.11 Explain what is meant by 'eutrophication'.

16.12 Discuss how each of the following can reduce water pollution:
a treating sewage
b using organic fertilisers rather than inorganic ones.

Pesticides

16.18 Pesticides help to increase crop yields.

A **pesticide** is a substance that kills organisms which damage crops. Insects that eat crops can be killed with insecticides. Fungi that grow on crops are controlled with fungicides. Weeds that compete with crop plants for water, light and minerals can be controlled with herbicides. Pesticides may also be used to control organisms which transmit disease, such as mosquitoes.

In a natural ecosystem, a wide variety of plant species will probably grow in a particular habitat. A wide variety of animals will live in that habitat too, feeding on different plants and on each other. A natural ecosystem often has a large species diversity. Factors such as predation or food supply will prevent the population of any one species from growing too large (sections **15.19** to **15.20**).

On a farm, only a few plant species are allowed to grow in many of the fields. One field might contain nothing but wheat, for example. This is called a **monoculture**. Insects or fungi which can feed on the wheat have an almost inexhaustible food supply. The usual limits on their population growth do not apply. The populations of the insects or fungi may grow very rapidly, until they are so big that they cause extensive damage to the crop (Figure **16.13**).

If nothing is done about this, then crop yields can be very badly reduced. It has been estimated that, in developing countries, at least one-third of potential crops are destroyed by pests. If they did not use pesticides, then this would be even worse.

16.19 Pesticides can harm the environment.

By definition, a pesticide is a harmful substance. If they are not used with care, some pesticides can do a lot of damage to the environment. For example, DDT is a pesticide that kills insects. It is a **persistent** insecticide,

Figure 16.13 These caterpillars, called African army worms, can cause huge amounts of damage to a maize crop.

which means that it does not break down, but remains in the bodies of the insects or in the soil. When a bird or other organism eats the insects, they eat the DDT too. The DDT stays in their bodies; each time they eat an insect, more DDT accumulates in their tissues. If a bird of prey eats the insect-eating bird, it too begins to accumulate DDT. Birds and other animals near the ends of food chains can build up very large concentrations of DDT in their bodies (Figure **16.14**).

Unfortunately, as well as being persistent, DDT is also **nonspecific**. This means that it not only harms the insects it is meant to kill, but is also harmful to other living things. In high concentrations it is very harmful to birds, for example. In Britain, it affected the breeding success of peregrine falcons, by making their egg shells very weak, so that they very rarely hatched. The peregrine falcon population dropped very rapidly.

Once it was realised that DDT was doing so much harm, its use in Britain was stopped. Now DDT is banned in many parts of the world. However, it is still used in many developing countries, because without it, insects would be such a problem that more people would starve or die of diseases like malaria. Other insecticides need to be developed which are as cheap and effective as DDT, but that do not harm other living organisms.

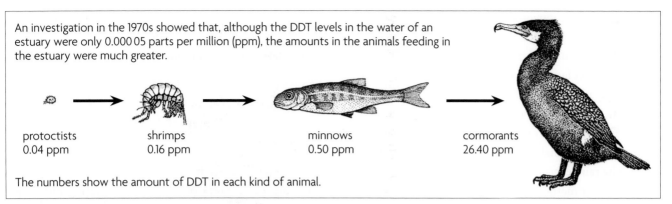

An investigation in the 1970s showed that, although the DDT levels in the water of an estuary were only 0.000 05 parts per million (ppm), the amounts in the animals feeding in the estuary were much greater.

protoctists
0.04 ppm

shrimps
0.16 ppm

minnows
0.50 ppm

cormorants
26.40 ppm

The numbers show the amount of DDT in each kind of animal.

Figure 16.14 DDT accumulation along a food chain.

Questions

16.13 What is a pesticide?

16.14 Why do we use pesticides?

16.15 Explain how pesticides can damage the environment.

Conservation

16.20 Biodiversity should be conserved.

Conservation is the process of looking after the natural environment. Conservation attempts to maintain or increase the range of different species living in an area, known as **biodiversity**.

One of the greatest threats to biodiversity is the loss of habitats. Each species of living organism is adapted to live in a particular habitat. If this habitat is destroyed, then the species may have nowhere else to live, and will become extinct.

Tropical rainforests have a very high biodiversity compared with almost anywhere else in the world. This is one of the main reasons why people think that conserving them is so important. When tropical rainforests are cut down or burnt, the habitats of thousands of different species are destroyed.

We have already seen how and why deforestation occurs. Another kind of habitat that is under threat is wetland, such as swamps. People drain wetland so that it can be more easily farmed. We build roads and houses, destroying whatever used to grow on that land. We farm animals in large numbers on land that cannot really produce enough vegetation to support them, so that the land becomes a semi-desert.

Many governments and also world-wide organisations such as the World Wide Fund for Nature are aware of these problems and are attempting to make sure that especially important habitats are not damaged. Most countries have special areas where people's activities are carefully controlled, ensuring that wildlife can continue to live there. Often, the loss of money from agriculture in these areas can be regained by allowing tourists to visit them. The most successful projects actively involve local people, who are usually delighted to see their environment being cared for, so long as they can still make a living from it.

16.21 Timber production can be sustainable.

A major threat to all kinds of forest, not only tropical rainforests, comes from logging companies who cut down huge areas of trees. The wood has many uses; for example, for building, making furniture and making paper. Tropical hardwoods fetch especially high prices.

If logging of forests cannot be entirely stopped, then at least we should try to limit its damaging effects.

For example, in many parts of Europe a system of woodland management called **coppicing** has been used for hundreds of years. Trees are cut down to just above ground level, and then left to regrow. This can be done every 10 or 15 years, repeatedly harvesting wood from the same trees, which just keep on regrowing. Often, a wood will be divided up into several different areas, and each year the trees in just one area are coppiced. This means that, at any one time, different parts of the wood contain trees of different sizes. This provides different habitats that suit many different species, so that biodiversity may be even higher than it would be if the wood was not harvested at all.

The trouble with coppicing is that the wood you harvest is of quite small diameter. It is not much use for building, though it can be used to make furniture or to make charcoal. To get larger pieces of timber, you need to cut down large, mature trees. If all the trees in one area are cut down, this is known as **clear-felling**. Clear-felling is harmful to the environment, because it completely destroys the forest habitat. It also leaves the soil open to erosion by rain, especially if the area is on a slope (Figure **16.9**). A better method is selective felling, in which particular trees are cut down but others growing around them are left to grow for a few more years. However, even selective felling can cause considerable loss of biodiversity, because it usually involves heavy machinery and the construction of roads into the forest, both of which can cause considerable damage to the habitat.

16.22   Natural resources should be conserved.

As well as trees and other living organisms, we need to be aware of other natural resources that we use, and attempt to conserve them. For example, we use very large amounts of fossil fuels such as coal and oil, that are extracted from the ground. They are burnt to release energy that is used to generate electricity. Oil is also used to make other substances, such as plastics.

These fossil fuels formed in the Earth many millions of years ago. If we keep on using them, they will eventually run out. Moreover, we have seen how burning fossil fuels can produce acid rain (sections **16.6** to **16.9**) and can contribute to global warming (section **16.4**).

For all of these reasons, it is important that we should try to reduce our use of these limited resources. We can do this by reducing the amount of energy that we use, and also looking for other sources of renewable energy, such as wind, water or solar power.

16.23   Sewage treatment can conserve water.

Water is a scarce resource in many parts of the world. Water that we have used can be recycled.

Sewage is waste liquid which has come from houses, industry and other parts of villages, towns and cities. Some of it has just run off streets into drains when it rains. Some of it has come from bathrooms and kitchens in people's houses and offices. Some of it has come from factories. Sewage is mostly water, but also contains many other substances. These include urine and faeces, toilet paper, detergents, oil and many other chemicals.

Sewage should not be allowed to run into rivers or the sea before it has been treated. This is because it can harm people and the environment. Untreated sewage is called **raw sewage**.

Raw sewage contains many bacteria and other microorganisms, some of which may be pathogenic. People who come into contact with raw sewage, especially if it gets into their mouths, may get ill. Raw sewage also contains many substances which provide nutrients for plants and microorganisms. We have seen how this can cause eutrophication if it gets into waterways (section **16.14**).

It is therefore very important that sewage is treated to remove any pathogenic organisms, and most of the nutrients, before it is released as effluent. Microorganisms play an important part in all the most commonly used methods of sewage treatment.

When sewage has been treated, the water in it can be used again, so sewage treatment enables water to be recycled. It may not be a nice thought to know that the water you drink was once inside someone else's body, but if we did not recycle water in this way then significant water shortages would occur in many parts of the world.

16.24   Liquids from sewage can be treated by two different methods.

Figure **16.15** shows how sewage is treated to make it safe. First, the raw sewage is passed through screens. These trap large objects such as grit which may have been washed off roads. The screened liquid is then left for a while in settlement tanks, where any other insoluble particles drift to the bottom and form a sediment.

There are two different ways in which the resulting liquid can now be treated.

Trickling filters   The liquid from the settlement tanks is sprinkled over a trickling filter bed. This is made of small stones and clinker. Many different aerobic microorganisms live on the surface of the stones. Some of them are aerobic bacteria, which feed on various nutrients in the sewage. Protoctists (single-celled animal-like organisms, such as *Amoeba*) feed on the bacteria.

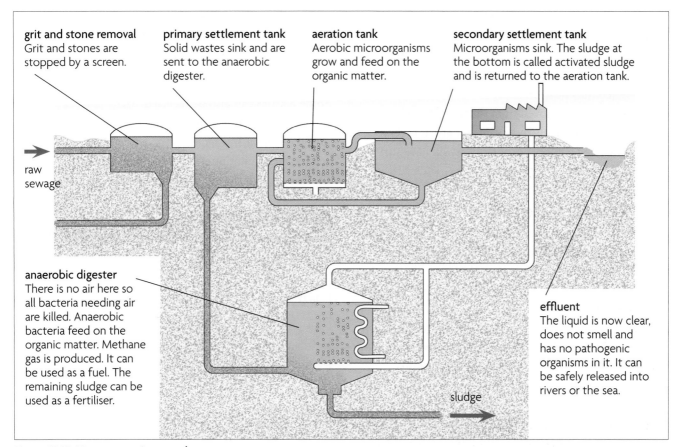

grit and stone removal
Grit and stones are stopped by a screen.

primary settlement tank
Solid wastes sink and are sent to the anaerobic digester.

aeration tank
Aerobic microorganisms grow and feed on the organic matter.

secondary settlement tank
Microorganisms sink. The sludge at the bottom is called activated sludge and is returned to the aeration tank.

raw sewage

anaerobic digester
There is no air here so all bacteria needing air are killed. Anaerobic bacteria feed on the organic matter. Methane gas is produced. It can be used as a fuel. The remaining sludge can be used as a fertiliser.

effluent
The liquid is now clear, does not smell and has no pathogenic organisms in it. It can be safely released into rivers or the sea.

sludge

Figure 16.15 How sewage is treated.

Fungi feed on soluble nutrients. These microorganisms make up a complex ecosystem in the trickling filter bed.

The liquid is trickled onto the surface of the stones through holes in a rotating pipe. This makes sure that air gets mixed in with the liquid. The liquid trickles quite slowly through the stones, giving the microorganisms plenty of time to work on it. By the time the water drains out of the bottom of the bed, it looks clear, smells clean, contains virtually no pathogenic organisms, and can safely be allowed to run into a river or the sea.

Activated sludge In this method (Figure 16.15), the liquid from the settlement tanks runs into a tank called an aeration tank. Like the trickling filter bed, this contains aerobic microorganisms, mostly bacteria and protoctists. Oxygen is provided by bubbling air through the tank. As in the trickling filter bed, these aerobic microorganisms make the sewage harmless.

Why is this method called 'activated sludge'? 'Activated' means that microorganisms are present. Some of the liquid from the tank, containing these microorganisms, is kept to add to the next lot of sewage coming in.

'Sludge' means just what it sounds like! It is a word which describes the semi-solid waste materials in sewage.

Both the trickling filter and the activated sludge methods can run into problems if the sewage contains substances which harm the microorganisms. These include heavy metals such as mercury, disinfectants, or large quantities of detergents. Heavy metals and disinfectants are toxic to many of the microorganisms. Detergents may cause foaming, which stops oxygen getting into the liquid. To solve these problems, the contaminated sewage can be diluted before being allowed to enter the trickling filter bed or the activated sludge tank.

16.25 Sludge can be digested anaerobically.

So far, we have described how the liquid part of the sewage is treated. What about the solid part?

Solids – sludge – first drop out of the sewage in the settlement tank. The activated sludge method also produces sludge. This material contains lots of living and dead microorganisms. It contains valuable organic material. It is a pity to waste it.

The sludge can be acted on by anaerobic bacteria. The sludge is put into large, closed tanks. Inside the tanks, several different kinds of bacteria act on the sludge. Some of them produce methane, which can be used as a fuel. When they have finished, the remaining solid material has to be removed from the tank. It is often used as fertiliser – it is usually quite safe, because it is very unlikely that any pathogenic organisms will have survived all these processes.

16.26 Paper can be recycled.

Another substance that is often recycled is paper. Paper is made from trees, so if used paper can be pulped and used to make paper again, this should reduce the number of trees that are cut down. However, things are not quite so straightforward. The trees that are used to make paper are often especially planted, and when they are cut down new ones are planted to take their place. So paper-making need not actually threaten natural forests. Another

problem is that collecting and transporting waste paper so that it can be recycled uses a lot of energy, so the paper that is made may end up costing more than 'first time around' paper.

Some plastics can also be recycled. Many plastics are **non-biodegradable** – they do not rot and so remain as unsightly rubbish when discarded. Recycling plastics reduces this pollution problem, and also helps to conserve the fossil fuels from which many plastics are made.

Questions

16.16 Explain the meaning of (a) biodiversity, and (b) conservation.

16.17   Explain the advantages of (a) coppicing, and (b) selective felling in the conservation of forests.

16.18 For **one** named resource, explain the advantages and problems of recycling it.

Key ideas

- Carbon dioxide and methane are greenhouse gases, trapping outgoing long wavelength radiation in the atmosphere and warming the Earth. Increased concentrations of these gases are causing global warming.

- Sulfur dioxide is produced when coal and other fossil fuels burn.

- Sulfur dioxide and nitrogen oxides dissolve and react in water droplets in the atmosphere, and fall to the ground as acid rain. This leaches aluminium ions from soils, and kills plants and aquatic organisms.

- Fall-out from accidents at nuclear reactors emits ionising radiation, which damages DNA and can cause mutations, cancer and radiation burns.

- Deforestation reduces the amount of carbon dioxide that is taken out of the atmosphere for photosynthesis, and so may increase global warming. Combustion of the felled trees releases carbon dioxide into the atmosphere. Deforestation also

destroys habitats for animals, possibly leading to their extinction. It increases soil erosion and flooding.

- Water pollution by fertilisers or raw sewage can cause eutrophication, in which large populations of aerobic bacteria form, reducing the amount of dissolved oxygen in the water and making it impossible for most animals to live there. Other chemical wastes, such as heavy metals, can also cause water pollution.

- Pesticides and herbicides are used to increase crop yields. They may, however, also harm organisms that we do not want to kill. Some pesticides are non-biodegradable and can build up along food chains.

- Conservation involves trying to maintain biodiversity, looking after species and their habitats.

- Recycling water by sewage treatment can help to reduce the quantity of water we take from the environment. Paper can also be recycled, possibly reducing the damage done to forests when harvesting trees for paper-making.

Revision questions

1 Explain the difference between each of the following pairs of terms.
   a the greenhouse effect, global warming
   b fertilisers, pesticides

2 The graph shows the amount of dissolved oxygen in the water of a river in a city.

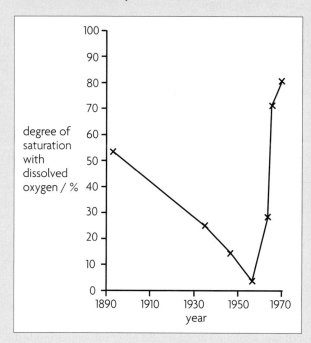

In the 19th century, sewage from the city drained directly into the river. At the begining of the 20th century sewage treatment plants were installed, which removed some of the organic material from the sewage before it entered the river. These plants have gradually become more efficient.

   a Give **two** ways in which water obtains dissolved oxygen.
   b Explain how pollution by sewage causes dissolved oxygen levels to decrease.
   c Suggest why dissolved oxygen levels in the river:
      i   decreased until 1948
      ii have increased since the 1950s.
   d What effect would you expect a decrease in dissolved oxygen to have on the fish population in the river?
   e Apart from affecting the levels of dissolved oxygen, what other harmful effects can the discharge of untreated sewage into rivers have?

Glossary

absorption the movement of digested food molecules through the wall of the intestine into the blood or lymph

accommodation the change of shape of the lens, in order to focus on objects at different distances

active site the part of an enzyme molecule into which its substrate fits

active transport the movement of ions in or out of a cell through the cell membrane, from a region of their lower concentration to a region of their higher concentration against a concentration gradient, using energy released during respiration

adrenaline a hormone secreted by the adrenal glands, which prepares the body for 'flight or fight'

aerobic respiration the release of a relatively large amount of energy in cells by the breakdown of food substances in the presence of oxygen

allele any of two or more alternative forms of a gene

alveolus (plural: alveoli) an air sac in the lungs, where gas exchange occurs

amino acids molecules that can link together in long chains to form proteins; they contain carbon, hydrogen, oxygen and nitrogen, and sometimes sulfur

amniotic fluid fluid secreted by the amnion, which supports and protects a developing fetus

amylase an enzyme which breaks down starch to maltose

anaemia an illness caused by a lack of haemoglobin

anaerobic respiration the release of a relatively small amount of energy by the breakdown of food substances in the absence of oxygen

antagonistic muscles muscles that work as a pair – for example, one may cause extension and the other flexing of a joint

anther the part of a stamen in which pollen is produced

antibiotic a drug that kills bacteria in the human body, without damaging human cells

antibodies chemicals secreted by lymphocytes, which attach to antigens and help to destroy them

antigens chemicals on the surfaces of pathogens, which are recognised as foreign by the body

artificial selection the choice by a farmer or grower of only the 'best' parents to breed, generation after generation

asexual reproduction the process resulting in the production of genetically identical offspring from one parent

assimilation the movement of digested food molecules into the cells of the body where they are used, becoming part of the cells

auxin a plant hormone which causes cells to elongate

axon a nerve fibre that conducts impulses away from the cell body

balanced diet a diet containing some of each of the different types of nutrients, in a suitable quantity and proportions

bile a liquid made in the liver, stored in the gall bladder and emptied into the small intestine, where it helps to emulsify fats

bile salts substances in bile that help to emulsify fats

biodegradable able to be broken down (digested) by microorganisms

breathing muscular movements which cause air to move into and out of the lungs

bronchioles the small tubes into which the bronchi branch

bronchus (plural: bronchi) one of the two tubes into which the trachea branches, carrying air into each lung

cancer a disease in which cells divide uncontrollably, producing lumps (tumours)

carbohydrase an enzyme that catalyses the breakdown of carbohydrates

carbohydrates starches and sugars

carcinogen a substance which increases the risk of a person's body developing cancer

cardiac muscle the muscle of which the heart is made

carnivore an animal that gets its energy by eating other animals

carpel the female part of a flower

catalase an enzyme found in almost all living tissues, which catalyses the breakdown of hydrogen peroxide to water and oxygen

catalyst a substance that speeds up a chemical reaction and is not changed by the reaction

cell sap a solution of sugars and other substances inside the vacuole of a plant cell

cell surface membrane a very thin layer of fat and protein that surrounds every living cell

cellulose a polysaccharide carbohydrate which forms fibres and is found in the cell walls of plant cells

central nervous system the brain and spinal cord

chemical digestion the breakdown of large molecules of food into smaller ones, done by enzymes

chlorophyll a green, light-absorbing pigment found inside chloroplasts in plant cells

chloroplast an organelle found in some plant cells, which contains chlorophyll and where photosynthesis takes place

chromosome a thread of DNA, made up of a string of genes

cilia tiny extensions on the surface of a cell, which can wave in unison and cause fluids to move

ciliary muscle a ring of muscle around the lens, which can change its shape

cirrhosis a disease of the liver in which the cells are permanently damaged

clone a group of genetically identical organisms

codominance a situation in which both alleles in a heterozygote have an effect on the phenotype

colon the first part of the large intestine, in which water and ions are absorbed

community all the organisms, of all the different species, living in an area at the same time

competition an interaction between organisms that occurs when both need the same resource which is in short supply

consumer an organism that gets its energy by feeding on other organisms

continuous variation differences in the features of a group of organisms in which there are no definite categories; each individual's features can lie anywhere between two extremes

corpus luteum the structure that forms in an ovary after an egg has been released; it secretes progesterone

cortex in a kidney, the outer layer; in a plant stem or root, a tissue made of typical plant cells (usually, however, without chloroplasts)

cotyledons food storage structures in a seed, which sometimes come above ground during germination and begin to photosynthesise

cross-pollination the transfer of pollen from the anther of one plant to the stigma of another plant of the same species

cuticle a layer of wax on a leaf

deamination a metabolic reaction that takes place in the liver, in which the nitrogen-containing part of amino acids is removed to form urea, followed by the release of energy from the remainder of the amino acid

decomposer an organism that gest its energy from dead or waste organic matter

denatured an enzyme is said to be denatured when its molecule has changed shape so much that the substrate can no longer fit into it

denitrifying bacteria bacteria that obtain their energy by converting nitrate ions into nitrogen gas

deoxygenated blood blood containing only a little oxygen

depressant a drug that inhibits the nervous system and slows it down

development an increase in complexity

dialysis exchange of substances between two solutions through a partially permeable membrane; dialysis machines are used in the treatment of people with kidney failure

diastole the stage of a heart beat in which the muscles in the heart relax

diffusion the net movement of molecules from a region of their higher concentration to a region of their lower concentration down a concentration gradient, as a result of their random movement

digestion the break-down of large, insoluble food molecules into small, water-soluble molecules using mechanical and chemical processes

diploid nucleus a nucleus containing two sets of chromosomes (e.g. in body cells)

disaccharide a complex sugar; a carbohydrate whose molecules are made of two sugar units

discontinuous variation differences in the features of a group of organisms where each fits into one of a few clearly defined categories

DNA the chemical from which genes and chromosomes are made

dominant an allele that is expressed if it is present (e.g. T or G)

dormant a condition in which an organism shuts its metabolism down, so that it can survive in adverse conditions

double circulatory system a system in which blood passes twice through the heart on one complete circuit of the body

drug a substance taken into the body that modifies or affects chemical reactions in the body

dry mass the mass of an organism after all water has been removed

ductless glands glands of the endocrine system, which secrete hormones directly into the blood

ecosystem a unit containing all of the organisms and their environment, interacting together, in a given area e.g. decomposing log or a lake

ectothermic poikilothermic; unable to regulate body temperature physiologically; the organism's temperature varies with that of its environment

effector a part of the body that responds to a stimulus, e.g. a muscle or a gland

egestion the passing out of food that has not been digested, as faeces, through the anus

egg a female gamete

embryo a young organism before birth, and before all the body organs have formed

emphysema a disease in which the walls of the alveoli in the lungs break down, reducing the surface area for gas exchange

emulsification breaking large globules of fat into tiny droplets, so that they mix easily with water

endocrine system the endocrine glands, which secrete hormones

endothermic homeothermic; able to regulate body temperature; the body temperature is independent of the temperature of the environment

environment all the living (biotic) and non-living (abiotic) factors an organism encounters during its life

enzymes proteins that function as biological catalysts

epidermis (mammal) the outer layer of the skin

epidermis (plant) a tissue made up of a single layer of cells which covers the top and bottom of a leaf, and the outside of the stem and root

epithelium a layer of cells covering a surface in an animal, e.g. the cells lining the trachea

euphoria a condition in which a person forgets all their worries and feels completely happy

excretion removal from organisms of toxic materials, the waste products of metabolism (chemical reactions in cells including respiration) and substances in excess of requirements

extensor muscle a muscle that causes a limb to straighten when it contracts

F1 generation the offspring from a parent homozygous for a dominant allele and a parent homozygous for the recessive allele

fermentation the breakdown of glucose by yeast, using anaerobic respiration; it produces carbon dioxide and alcohol

fertilisation the fusion of the nuclei of two gametes

fetus a young organism before birth, once all the body organs have formed

filament the stalk of a stamen

flaccid a term used to describe a cell that has lost a lot of water, becoming soft

flexor muscle a muscle that causes a limb to bend when it contracts

follicle a space inside an ovary in which an egg develops

food chain a chart showing the flow of energy (food) from one organism to the next beginning with a producer (e.g. mahogany tree → caterpillar → song bird → hawk)

food web a network of interconnected food chains showing the energy flow through part of an ecosystem

fossil fuel a substance that can be combusted to release energy, formed millions of years ago from the partially decomposed and compressed bodies of organisms

fruit an ovary of a plant after fertilisation; it contains seeds

FSH follicle stimulating hormone; a hormone secreted by the pituitary gland which causes the development of eggs in the ovaries

fully permeable able to let most substances pass through

gametes sex cells, e.g. eggs and sperm

gas exchange the entry of oxygen into an organism's body, and the loss of carbon dioxide

gene a length of DNA that is the unit of heredity and codes for a specific protein. A gene may be copied and passed on to the next generation

genetic diagram the conventional way to set out a genetic cross

genetic engineering taking a gene from one species and putting it into another species

genotype the genetic makeup of an organism in terms of the alleles present (e.g. Tt or GG)

geotropism a response in which a plant grows towards or away from gravity

glomerulus a tangle of blood capillaries in a Bowman's capsule in the kidney

glucagon a hormone secreted by the pancreas, which increases blood glucose level

glycogen the polysaccharide that is used as an energy store in animal cells and fungi

goblet cells cells which secrete mucus

greenhouse effect the warming effect of carbon dioxide, methane and other greenhouse gases, on the Earth

growth a permanent increase in size and dry mass by an increase in cell number or cell size or both

guard cell one of two sausage-shaped cells in the epidermis in plants, between which there is a hole called a stoma; the guard cells can change shape to open and close the stoma

habitat the place where an organism lives

haploid nucleus a nucleus containing a single set of unpaired chromosomes (e.g. sperm and egg)

hepatic relating to the liver

herbivore an animal that gets its energy by eating plants

heterozygous having two different alleles of a gene (e.g. Tt or Gg), not pure-breeding

hilum the scar where a seed was attached to a fruit

HIV/AIDS HIV is the human immunodeficiency virus, which causes AIDS

homeostasis the maintenance of a constant internal environment

homeothermic endothermic; able to regulate body temperature; the body temperature is independent of the temperature of the environment

homologous chromosomes the two chromosomes of a pair in a diploid cell; they have genes for the same features at the same positions

homozygous having two identical alleles of a particular gene (e.g. TT or gg). Two identical homozygous individuals that breed together will be pure-breeding

hormone a chemical substance produced by a gland, carried by the blood, which alters the activity of one or more specific target organs and is then destroyed by the liver

hypha (plural: hyphae) one of the long, thin threads of which the body of a fungus is made; each hypha is just one cell thick

immune able to fight off a particular type of pathogen before it causes any symptoms in the body

implantation the movement of a young embryo into the lining of the uterus, and its attachment there

infection the entry of a pathogen to the body

infectious disease a disease caused by a pathogen, which can be passed from one person to another

ingestion taking substances (e.g. food, drink) into the body through the mouth

inheritance the transmission of genetic information from generation to generation

inorganic a term used to describe substances that are not made by living organisms

insulin a hormone secreted by the pancreas, which reduces blood glucose level

intercostal muscles muscles between the ribs, which help to produce breathing movements

iris the coloured part of the eye, which controls the amount of light allowed through to the lens and retina

islets of Langerhans groups of cells in the pancreas which secrete insulin and glucagon

lactase an enzyme that breaks down the disaccharide lactose into glucose and galactose

lactation production of milk by mammary glands

LH luteinising hormone; a hormone secreted by the pituitary gland which causes an egg to be released from an ovary

ligament a strong, stretchy cord that joins two bones together at a synovial joint

lignin a tough, waterproof material that makes up the walls of xylem vessels; wood is mostly lignin

limiting factor something present in the environment in such short supply that it restricts life processes

lipase an enzyme that digests fats (lipids) to fatty acids and glycerol

lumen the space in the centre of a tube

lymph the fluid found inside lymph vessels, formed from tissue fluid

lymph nodes organs in which large numbers of white blood cells (which can destroy bacteria or toxins) collect

lymphocytes white blood cells that secrete antibodies

maltose a disaccharide produced by the digestion of starch

mechanical digestion the breakdown of large pieces of food to smaller ones, increasing their surface area; it is done by teeth in the mouth and by the contraction of muscles in the stomach wall

meiosis reduction division in which the chromosome number is halved from diploid to haploid

menstruation the loss of the uterus lining through the vagina

mesophyll the tissues in the centre of a leaf, where photosynthesis takes place

metabolic reactions the chemical reactions that take place inside a living organism

micropyle a tiny hole in the testa of a seed

mitosis nuclear division giving rise to genetically identical cells in which the chromosome number is maintained by the exact duplication of chromosomes

monosaccharide a simple sugar; a carbohydrate whose molecules are made of one sugar unit

movement an action by an organism or part of an organism causing a change of position or place

mucus a viscous, sticky substance which is secreted in many parts of the body for lubrication or the removal of dust or bacteria

mutagen a substance that causes mutations

mutation a change in a gene or a chromosome

mycelium the mass or network of hyphae that makes up the body of a fungus

myelin a fatty substance surrounding the axons of many neurones, enabling the nerve impulse to travel faster

natural selection the greater chance of passing on of genes by the best-adapted organisms

nectary a gland producing a sugary fluid, found in many insect- or bird-pollinated flowers

negative feedback a mechanism used in homeostasis, in which a change in a parameter brings about actions that push it back towards normal

nephron one of the thousands of tiny tubules in a kidney, in which urine is produced

nerve a bundle of axons or dendrons belonging to many different neurones

neurone a nerve cell; a cell specialised for the rapid transfer of electrical impulses

niche the role of an organism in an ecosystem

nitrifying bacteria bacteria that obtain their energy by converting ammonia or nitrite ions to nitrate ions

nitrogen-fixing able to change unreactive nitrogen gas into a more reactive nitrogen compound such as nitrates or ammonia

nitrogenous waste excretory products containing nitrogen – for example, ammonia, urea, uric acid

non-biodegradable not able to be broken down by microorganisms

normal distribution a curve in which the largest number occurs near the midpoint, with approximately equal quantities on either side of this point and a gradual decrease towards the extremes

nutrition the taking in of nutrients which are organic substances and mineral ions, containing raw materials or energy for growth and tissue repair, absorbing and assimilating them

oestrogen a hormone secreted by the ovaries that helps to control the menstrual cycle

omnivore an animal that eats food of both animal and plant origin

optimum temperature the temperature at which something happens most rapidly

organ a structure made up of a group of tissues, working together to perform specific functions

organ system a group of organs with related functions, working together to perform body functions

organelle a structure within a cell

organic a term used to describe substances that have been made by living organisms, or whose molecules contain carbon, hydrogen and oxygen

organism a living thing

osmosis the diffusion of water molecules from a region of their higher concentration (dilute solution) to a region of their lower concentration (concentrated solution), through a partially permeable membrane

ovary an organ in which female gametes are made

oviduct the tube leading from an ovary to the uterus

ovulation the release of an egg from an ovary

ovule a structure in the ovary of a flower which contains a female gamete

oxygen debt the extra oxygen that must be taken in by the body following strenuous exercise, when anaerobic respiration took place; the oxygen is needed to break down the lactic acid that accumulated as a result of anaerobic respiration

oxygenated blood blood containing a lot of oxygen; in humans, blood becomes oxygenated in the lungs

palisade layer the upper mesophyll layer in a leaf, made up of rectangular cells containing many chloroplasts

pancreas an organ lying close to the stomach, which is both an endocrine gland (producing insulin and glucagon) and an exocrine gland (producing pancreatic juice)

pancreatic juice the liquid secreted into the pancreatic duct by the pancreas; it flows into the duodenum where its enzymes help with digestion of fats, proteins and carbohydrates

particulates tiny pieces of carbon and other substances found in smoke, which can irritate the lungs

pathogen a microorganism that causes disease

penicillin an antibiotic which destroys bacteria by damaging their cell walls

pepsin a protease enzyme found in the stomach

peristalsis rhythmic contractions of muscles that ripple along a tube – for example, peristalsis pushes food through the alimentary canal

petiole a leaf stalk

phagocytes white blood cells that surround, engulf and digest pathogens

phenotype the physical or other features of an organism due to both its genotype and its environment (e.g. tall plant or green seed)

phloem tubes long tubes made up of living cells with perforated end walls, which transport sucrose and other substances in plants

photosynthesis the fundamental process by which plants manufacture carbohydrates from raw materials using energy from light

phototropism a response in which a plant grows towards or away from the direction from which light is coming

pigment a coloured substance – for example, chlorophyll, haemoglobin

placenta in mammals, an organ made up of tissues of both the mother and embryo, through which the mother's and embryo's bodies exchange nutrients and waste materials

plasma the liquid part of blood, in which the cells float

plasmolysed the condition of a plant cell that has lost so much water that its cytoplasm shrinks and pulls the cell membrane away from the cell wall

platelets tiny fragments of cells found in blood, which help with clotting

pleural membranes two strong, slippery membranes which surround the lungs

plumule the young shoot in an embryo plant

poikilothermic ectothermic; unable to regulate body temperature physiologically; the organism's temperature varies with that of its environment

pollen grains tough, resistant structures containing the male gametes of a flower

pollination the transfer of pollen from the male part of the flower (anther of stamen) to the female part of the plant (stigma)

polysaccharide a carbohydrate whose molecules are made of hundreds of sugar units linked in long chains – for example, starch, glycogen and cellulose

population a group of organisms of one species, living in the same area at the same time

predator an animal that kills and eats other animals

primary consumers herbivores

producer an organism that makes its own organic nutrients, usually using energy from sunlight, through photosynthesis

progesterone the pregnancy hormone; a hormone secreted by the ovaries and placenta which maintains the lining of the uterus

prostate gland a gland close to a male's bladder, that secretes fluid in which sperm can swim

protease an enzyme that catalyses the breakdown of proteins

protein a substance whose molecules are made of long chains of amino acids; proteins contain carbon, hydrogen, oxygen and nitrogen, and sometimes sulfur

puberty the stage of development during which sexual maturity is reached

pulmonary relating to the lungs

pure-breeding homozygous

pyramid of biomass a sideways-on graph, in which the size of the boxes represents the dry mass of organisms in each trophic level of a food chain

pyramid of numbers a sideways-on graph, in which the size of the boxes represents the number of organisms in each trophic level of a food chain

radicle the young root in an embryo plant

receptor a cell that is able to detect changes in the environment; often part of a sense organ

recessive an allele that is only expressed when there is no dominant allele of the gene present (e.g. t or g)

reflex action a fast, automatic response to a stimulus

reflex arc the arrangement of neurones along which an impulse passes during a reflex action

relay neurone a neurone in the central nervous system which passes an impulse between a sensory neurone and a motor neurone

renal relating to the kidneys

renal capsule the cup-shaped structure at the start of a nephron, where filtration occurs

reproduction the processes that make more of the same kind of organism

respiration the chemical reactions that break down nutrient molecules in living cells to release energy

retina the part of the eye that contains receptor cells

rickets a disease caused by a lack of vitamin D or calcium, in which bones are not as hard as they should be and can grow in a bent shape

root cap a tough, protective covering over the tip of a root

sebaceous gland an oil-producing gland in the skin

secondary consumers carnivores that eat herbivores

secondary sexual characteristics features of the body that develop at puberty, as a result of the increased secretion of sex hormones

seed an ovule after fertilisation; it contains an embryo plant

selection pressure an environmental factor that causes organisms with certain characteristics to have a better chance of survival than others

self-pollination the transfer of pollen from the anther to the stigma on the same plant (but not necessarily the same flower)

semen a mixture of sperm and fluids from the prostate gland and seminal vesicles

seminal vesicles glands that secrete fluid in which sperm can swim

sense organs groups of receptor cells responding to specific stimuli: light, sound, touch, temperature and chemicals

sensitivity the ability to detect or sense changes in the environment (stimuli) and to make responses

sexual reproduction the process involving the fusion of haploid nuclei to form a diploid zygote and the production of genetically dissimilar offspring

sickle cell anaemia a condition caused by a codominant allele of the gene that codes for haemoglobin, in which a person has two copies of the gene and suffers serious health problems

simple sugar a monosaccharide; a carbohydrate whose molecules are made of one sugar unit

species a group of organisms with similar characteristics, which can interbreed with each other to produce fertile offspring

sperm a male gamete

sphincter muscle a muscle surrounding a tube, which can contract to close the tube

spongy layer the tissue beneath the palisade layer in a leaf; it is made up of cells that contain chloroplasts and can photosynthesise, with many air spaces between them

stamen the male parts of a flower

starch the polysaccharide that is used as an energy store in plant cells

stem tuber a swollen part of a stem, which stores food

stigma the part of a flower that receives pollen

stimulant a drug that makes the nervous system work faster

stimulus a change in an organism's surroundings that can be detected by its sense organs

stoma (plural: stomata) a gap between two guard cells, usually in the epidermis on the lower surface of a leaf

stroke damage caused to the brain by an interruption in blood supply, caused either by a blood vessel bursting or a blood vessel becoming blocked by a blood clot

style the connection between the stigma and ovary of a flower

substrate the substance on which an enzyme acts

succulent a plant with swollen stems or leaves, in which water is stored

sucrase a carbohydrase found in the small intestine, which breaks down sucrose to glucose and fructose

sucrose a disaccharide, non-reducing sugar, made of a glucose molecule and a fructose molecule linked together; the form in which carbohydrates are transported in the phloem of plants

suspensory ligaments a ring of ligaments linking the ciliary muscles to the lens

synovial joint a joint at which the two bones can move freely

systole the stage of a heart beat in which the muscles in the walls of the heart chambers contract

target organ an organ that is affected by a hormone

tendons strong, inelastic cords of tissue, which attach muscles to bones; they are also found in the heart, where they attach the atrioventricular valves to the wall of the ventricle

tertiary consumers organisms that feed at the fourth stage in a food chain; they eat carnivores

test cross breeding an offspring with the dominant phenotype with an organism with the recessive phenotype; the offspring of the cross can help to determine the genotype of the parent with the dominant phenotype

testa the tough waterproof covering of a seed

testis (plural: testes) an organ in which sperm are made

testosterone a hormone secreted by the testes, which causes male characteristics

tissue a group of cells with similar structures, working together to perform specific functions

tissue fluid the fluid that surrounds all the cells in the body, formed from blood plasma that leaks out of capillaries

trachea the tube that carries air from the nose and mouth down to the lungs

translocation the movement of sucrose and amino acids in phloem, from regions of production to regions of storage, or to regions of utilisation in respiration or growth

transpiration evaporation of water at the surfaces of the mesophyll cells followed by loss of water vapour from plant leaves, through the stomata

transpiration stream the pathway of water from the root hairs of a plant, up the root and stem and out of the leaves into the atmosphere

triceps muscle a muscle in the upper arm which causes the arm to straighten when it contracts

trophic level the position of an organism in a food chain, food web or pyramid of biomass, numbers or energy

tropism a plant growth response to a stimulus, in which the direction of growth is related to the direction of the stimulus

trypsin a protease enzyme found in pancreatic juice

turgid cell a plant cell that has absorbed water and has cytoplasm that is pressing outwards on the cell wall

umbilical cord an organ linking an embryo to the placenta, containing blood vessels

urea the main nitrogenous excretory product of mammals, produced in the liver from excess amino acids

ureter a tube that leads from a kidney to the bladder

urethra a tube that leads from the bladder to the outside

urine a solution of urea and other excretory products in water, produced by the kidneys

uterus the organ in a mammal in which the embryo develops

vaccination the introduction to the body of dead or weakened pathogens, to make a person immune to an infectious disease

vascular bundle a vein in a plant, containing xylem vessels and phloem tubes

vasoconstriction narrowing of blood vessels

vasodilation widening of blood vessels

villus (plural: villi) a tiny, finger-like process on the inner wall of the small intestine; villi increase the surface area for digestion and absorption

water potential gradient a difference in the concentration of water molecules; a dilute solution has a high water potential, and water tends to move from this, down a water potential gradient, into a concentrated solution

xerophyte a plant adapted to live in dry conditions

xylem vessels long hollow tubes made up of dead, empty cells with lignified walls, which transport water in plants and help to support them

zygote the diploid cell produced when two gametes fuse

Index

coordination and response
  in animals 126–39
  in plants 139–43
coppicing 243–4
cornea 134
coronary arteries 86
coronary heart disease 66, 86–7
corpus luteum 177
cotyledons 188, 189
crops
  destruction by pests 242
  use of fertilisers 240–1
crustaceans 6
cuticle, leaves 51, 51, 108
cystic fibrosis 198–202, 215
cytoplasm 15, 25, 26

Darwin, Charles 207–8
DDT 242–3
deamination 81, 153
death rates 226–30
decomposers 223–4
deforestation 235, 238–9
denatured enzymes 41
denitrifying bacteria 225
dentine 73, 75
deoxygenated blood 84–5
depressant drugs 160–1
dermis, human skin 147
desert plants 107–8
development 196
dialysis, kidneys 156
diastole 87
dichotomous key 11
diet 64–7, 86, 174
diffusion 20–3, 96, 124
digestion 71–3
  alimentary canal 76–82
  role of teeth 73–6
diploid cells/nuclei 168, 195
disaccharides 32
discontinuous variation 204–5
DNA 16, 194–5, 214–16
dominant alleles 199–200
dormant, plants/seeds 110, 189
double circulatory system 84
drugs, use and misuse 159–63
dry mass 196
duodenum 78, 79, 82

ecology 218–19
ecosystem 218–23
ectothermic animals 146
effectors 126, 135–7
egestion 81, 152
elbow joint 136–7

embryo
  human 172–3, 176–7
  plant 39, 187, 188
emphysema 163
emulsion test for fats 35
endocrine glands/system 137–9
endothermic animals 146
energy
  in an ecosystem 219–23
  from a balanced diet 64–5
  from foods 33, 35, 114–15
  released by respiration 112–13
enhanced greenhouse effect
    234–5
environment
  harmful effect of humans 233
  organisms in natural 218–30
enzymes 39–41
  effects of pH on 41, 42–3
  and temperature 41, 44
  uses of 45–7
epidermis
  flowering plants 102
  human skin 146–7
  of a leaf 51, 52
epiglottis 77, 118
etiolated plants 142
eutrophication 240, 241
evolution, Darwin's theory 207–8
excretion 2, 152–7
exercise
  effect on heart rate 88
  and oxygen debt 121–3
expiration 120, 121, 122
extensor muscle 136
eyes
  activities 133
  focusing by 134–5
  iris 132–3
  protection of 131
  retina 131–2

famines, causes of 68
fats 34–5, 37
  danger of saturated 66–7, 86
  digestion & absorption 72,
    79–80
fatty acids 34
fermentation 115
fermenters producing penicillin
  47
fertilisation 168
  in flowering plants 184–6
  and genetic variation 206–7
  in humans 171–2
  and inheritance 201–2

fertilisers, leaching into soil
    240–1
fertility drugs 180
fetus, placenta supplying 172–3
fibre, dietary 65
fibrin and fibrinogen 95
filtration in kidneys 155
fish 5, 66–7
flaccidity of plant cells 26
flexor muscle 136
flowering plants 9
  attracting pollinators 21
  fertilisation in 184–6
  gas exchange in 123–4
  sexual reproduction in 182–91
  transport systems in 100–10
fluoride 75, 76
focusing by the eye 134–5
follicle, ovarian 172, 176, 177
food
  additives in 70–1
  containing carbohydrates, fats
    & proteins 33–4, 35, 36
  digestion by enzymes 39
  energy content of 65, 114
  transport by the blood 97
food chains 219–20, 221
  harm caused by DDT 241–2
food production 68–71
  use of enzymes 45–6
  use of hormones 138–9, 143
food webs 220, 221
forests
  conservation of 243–4
  deforestation 235, 238–9
fossil fuels 233–7, 244
fruit 188–9
fruit growing, hormone use 143
fungi 10, 159, 166–7

gametes 168–9, 182–6, 200
  produced by meiosis 196–7,
    206
gas exchange
  in flowering plants 123–4
  in humans 116–19, 121
genes 87, 165, 194–5
genetic diagrams 201–2
genetic engineering 214–16
genotypes 199–204
geotropism 139
germination of seeds 189–91
glasshouses, crop growing 61–2
global warming 233–6
glomerulus 155
glucagon 151

glucose 29, 32, 33
   liver controlling level of 151
   and photosynthesis 50, 123–4
   in respiration 112–13, 115, 123
   uses of by plants 55–6
glycerol 34, 72
glycogen 33, 81, 151
goblet cells 78, 118
gonorrhoea 180–1
gravity 139, 140, 141, 142
greenhouse effect 233–5
growth 2, 196
   responses by plants 139–42
guard cells, leaves 51, 52
gum disease 74–6

habitats 218, 233, 239, 243
haemoglobin 45, 93, 212
haploid cells/nuclei 168, 195
heart 85–8
heart disease 66–7, 86
heat
   blood transporting 97
   from germinating peas 113
   see also energy; temperature
hepatic portal vein 81, 91
herbivores 220
hermaphrodite 169, 182
heroin 160
heterozygous 199, 200, 201–3
hilum 188, 189
HIV, transmission of 181–2
homeostasis 145–51
homeothermic animals 146
homologous chromosomes 195, 197–8
homozygous 199, 203
hormones 137, 176
   for increasing fertility 180
   and the menstrual cycle 177
   and milk yield in cows 138–9
   transport of 97
humidity & transpiration rate 104
hydrogen peroxide 39–40, 42–3
hyphae 167
hypothalamus 148, 149

ileum 78, 79, 82
immune system 95, 156–7, 181
implantation 172, 173
infections
   and antibiotics 159–60
   role of white blood cells 93–5
   sexually transmitted 180–2

ingestion 73, 76
inheritance 195, 197–204
insecticides 242–3
insects 6, 109, 182, 226
   damage to crops 242
   pollination by 185–6
inspiration 120, 121, 122
insulin
   controlling blood sugar levels 151
   genetically engineered 214–15
intercostal muscles 120, 121
iodine test for starch 34, 56–7
ionising radiation 237–8
iris of the eye 132–3
islets of Langerhans 151

joints, human 136–7

kidneys 154–7
kingdoms of living organisms 1, 3–4, 10
kwashiorkor 67

lactation 175
lactic acid 115, 123
lactose 69, 79
leaves
   adaptation for photosynthesis 54
   oxygen supply 124
   structure of 50–1
   surface loss of water 106
lens of the eye 134
ligaments in joints 136
light
   cells in retina receptive to 131–2
   controlling in glasshouses 61
   cornea and lens focusing 134–5
   effect of darkness on plants 142
   how shoots respond to 139, 141
   intensity and transpiration rate 105
   part of shoot sensitive to 140
   photosynthesis investigations 58–9, 60
   response of iris to 132–3
   sunlight for photosynthesis 50, 53–4
lignin in xylem vessels 100
limiting factors
   photosynthesis 61–2
   population size 228
lipase 40, 45, 78, 79

lipids (fats) 34–5, 37
liver 79, 91, 93, 97, 153
   control of blood sugar level 151
   damage by alcohol 161
   role in digestion 81
lumen 89, 90
lungs 84–5, 96, 117
   and breathing 119–21
   and cystic fibrosis 198
   damage from smoking 162, 163
   and gas exchange 116–19
lymph fluid and nodes 98–9
lymphocytes 92, 94–5

magnesium, plant nutrition 55
magnification 8
malaria 212–13
malnutrition 67
maltase 40, 79
maltose 32, 39, 40, 46, 72, 115
mammals, classifying 1, 3, 5
mechanical birth control methods 178
mechanical digestion 73
medicinal drugs 159–60
meiosis 168–9, 196–7, 206
membrane of cells 15
   partially permeable 22, 23, 25, 26
menstrual cycle/menstruation 176, 177
mesophyll cells/layer 51, 53–4
metabolic reactions/metabolism 15, 31
   enzymes 39
   and excretion 2, 152
   and temperature 146, 148
methane 233, 235–6
microorganisms
   enzymes obtained from 47
   methane produced from 235–6
   mycoprotein made from 70
   and sewage treatment 244–5
   yoghurt made from 69
micropyle 186, 188
microscopes 13–14
milk, breast versus formula 175
milk teeth 74
minerals 55, 65, 66, 109
mitosis 195–6, 197
molluscs 7
monosaccharides 32, 72
motor neurones 126–7, 129
mouth 76, 82
movement 2

seeds 35, 39, 139–42, 187–91
selective felling 244
selective weedkillers 143
semen 171
sense organs 130
sensitivity 2
sensory neurones 128, 129
septum, heart 85
sewage treatment 244–6
sex chromosomes 203
sexual reproduction 168–9
    advantages & disadvantages
        191–2
    in flowering plants 182–91
    in humans 169–72
sexually transmitted diseases
    180–1
shoots of plants
    effect of auxin on 143
    response to gravity 140
    response to light 141–2
sickle cell anaemia 212–13
sieve plates, phloem tubes 100–1
sigmoid growth curve 227–8
simple sugars 32
single cell protein (SCP) 70
sinks & sources, translocation
    109–10
skin 146–8
small intestine, digestion in 78,
    79–80
smoking, effects on health 86,
    162–3
smooth muscle 135, 136
soil erosion, activities causing
    238–9
sources and sinks, translocation
    109–10
species 1–4, 195, 204
    conservation of 243
    destruction by humans 233
    diversity in rainforests 238–9
    ecology 218–19
sperm 170–2
sphincter muscles 76, 77–8
stabilising selection 212
stamen 182
starch 32, 33
    amylase digesting 39–40, 46,
        76
    stored in plants 55, 110, 167
    testing foods for 34
    testing a leaf for 56–7
starvation 67
statins 87
stem tubers 110, 167

stigma 183–5, 186, 188
stimulant drugs 162
stimuli, responses to 126–43
stomach 18, 41, 77–8, 82
stomata 51, 52, 53, 61, 103
    and transpiration rate 105, 108
    on water hyacinth leaves 108–9
striated muscle 135, 136
style, flowers 183, 186
substrates 40–1, 82
sucrose 32
    role in plant nutrition 55–6
    translocation of 109
sugars 32, 33, 46
sulfur dioxide 236–7
sunlight and photosynthesis 15,
    49–50, 53–4, 61, 123–4
surgery
    for birth control 178–9
    organ transplants 95, 156–7
suspensory ligaments, eye 134
sweat glands/sweating 147, 148–9
synovial joints 136
systole 87

tar in cigarettes 162
teeth 73–6
temperature
    of body, regulation of 146–50
    effect on enzymes 41, 44
    effect on transpiration rate 104
    and rate of photosynthesis 61
tendons 88, 136
test crosses 202–3
testa 187, 188, 189
testes 170
testing foods 33, 34, 35, 37
thorax 118, 120, 121
timber production 243–4
tissue 18
    adipose 35, 147
    fluid 98
    rejection after transplants 95,
        156
tobacco, damage from 162–3
tooth decay 74–6
trachea 77, 117, 118
translocation, plants 109
transpiration 103
transpiration rates 104–8
transport systems
    in flowering plants 100–10
    in humans 84–99
tree felling 238, 244
triceps muscle 136–7
tricuspid valve 87

trophic levels 220, 221, 223
tropisms 139
tubers (potato) 110, 167
turgid cells 26, 108

umbilical cord 172
urea 81, 97, 152–3, 155, 156
ureter 154, 155
urethra 155, 169, 170, 180
urine 155
uterus 169, 172–3, 174, 176–7

valves
    in the heart 87–8
    in lymph vessels 99
    in veins 90
variation 204–7
    artificial selection 213–14
    and natural selection 207–13
vascular bundles 50, 101
vasoconstriction 148
vasodilation 148–9
veins, blood vessels 88–92
veins, leaves 50, 52
ventricles of the heart 85–6
vertebrates 5
villi 79–80
viruses 9
Visking tubing 22, 23
vitamins 65, 66
vocal cords 118

water 31
    absorption by plants 102–3
    photosynthesis equation 50
    uptake by leaves 53
    uptake by seeds 189–91
water cycle 226, 227, 239
water plants, adaptation of 108–9
water pollution 240–1
water potential gradient 23,
    103–4
weedkillers 143
white blood cells 92
    in lymph nodes 99
    producing specific antibodies
        94–5
    role in killing pathogens 93–4

xylem vessels 100–4, 109

yeast for baking and brewing 115
yeast experiments 227–8
yogurt, making 69

zygote 168, 172, 187